Power System Simulation Using Semi-Analytical Methods

IEEE Press
445 Hoes Lane
Piscataway, NJ 08854

Power System Simulation Using Semi-Analytical Methods

Edited by Kai Sun
University of Tennessee, Knoxville, TN, USA

Published by John Wiley & Sons, Inc., Hoboken, New Jersey.
Published simultaneously in Canada.

Library of Congress Cataloging-in-Publication Data Applied for:

Hardback ISBN: 9781119988014

Cover Design: Wiley
Cover Image: © thinkhubstudio/Shutterstock

Set in 9.5/12.5pt STIXTwoText by Straive, Chennai, India

Contents

About the Editor

Kai Sun is a professor at the Department of Electrical Engineering and Computer Science in the University of Tennessee, Knoxville. He received his bachelor's degree in Automation in 1999 and his PhD degree in Control Science and Engineering in 2004, both from Tsinghua University in Beijing. Before joining the university in 2012, Dr. Sun was a project manager with the Electric Power Research Institute in Palo Alto, California, from 2007 to 2012 for R&D programs in the areas of grid operations, planning, and renewable integration. Earlier, he worked as a research associate at Arizona State University in Tempe and a postdoctoral fellow at the University of Western Ontario in Canada. Dr. Sun has served in the editorial boards of *IEEE Transactions on Power Systems*, *IEEE Transactions on Smart Grid*, *IEEE Open Access Journal of Power and Energy*, and *IEEE Access*. He received the EPRI Technology Innovation Excellence Award in 2008, EPRI Chauncey Award in 2009, NSF CAREER Award in 2016, NASPI CRSTT Most Valuable Players Award in 2016, and several Best Conference Paper Awards by IEEE PES General Meetings. His research areas include power system dynamics, stability, and control.

List of Contributors

Nan Duan
Transmission Planning
Midcontinent Independent System Operator, Inc.
Carmel
IN
USA

Gurunath Gurrala
Department of Electrical Engineering
Indian Institute of Science
Bangalore
Karnataka
India

Francis C. Joseph
Department of Electrical Engineering
Indian Institute of Science
Bangalore
Karnataka
India

Chengxi Liu
School of Electrical Engineering and Automation
Wuhan University
Wuhan
China

Yang Liu
Division of Energy Systems and Infrastructure Analysis
Argonne National Laboratory
Lemont
IL
USA

Byungkwon Park
School of Electrical Engineering
Soongsil University
Dongjak-Gu
Seoul
South Korea

Feng Qiu
Division of Energy Systems
Argonne National Laboratory
Lemont
IL
USA

Kai Sun
Department of Electrical
Engineering & Computer Science
University of Tennessee
Knoxville
TN
USA

Bin Wang
Department of Electrical and
Computer Engineering
University of Texas at San Antonio
San Antonio
TX
USA

Rui Yao
Division of Energy Systems
Argonne National Laboratory
Lemont
IL
USA

Preface

Power system simulation is an essential tool for investigating electrical power systems, used by researchers, engineers, and other stakeholders. It is based on a mathematical model and uses time-domain simulation to predict the behavior of a power system under anticipated operating conditions, whether it is currently operational or undergoing a new design. Simulation results provide insights into system stability, reliability, efficiency, and the performance of new equipment and functionalities. However, traditional simulation programs are facing significant challenges in accurately and efficiently simulating increasingly complex and uncertain bulk power systems with the growing penetration of renewable energy resources, particularly when fast, real-time power system simulation is required.

The purpose of this book is to enhance the speed of time-domain simulations of bulk power systems by introducing a semi-analytical methodology that integrates various simulation techniques. This new approach differs from traditional simulations that use low-order numerical integrators. Instead, it employs a high-order analytical form that approximates the dynamic response of the system with improved accuracy. By embedding symbolic variables and parameters, the resulting semi-analytical solution allows for the decomposition of its computation into sequential or concurrent processes, reducing the time required for simulations. Additionally, a significant amount of computation can be performed offline before a simulation run is needed or be parallelized, making the simulations even faster. These semi-analytical methods can be integrated or interfaced with existing numerical integrators to create more robust power system simulators that can be used with high-performance parallel computers.

The book is structured to provide comprehensive and systematic coverage of such emerging methods for finding semi-analytical solutions and accelerating power system simulation, as documented in the literature. Chapters 2–8 describe these methods in detail, including their accuracy and time performance, and demonstrate their applications to realistic bulk power system models. These methods are compared to traditional numerical methods, providing a thorough evaluation of their effectiveness.

In the rest of the book, Chapter 1 provides an overview of power system simulation, including timescales, models, and numerical methods of simulation. It also introduces the basic idea of a semi-analytical solution for power systems and the fundamental approaches for semi-analytical simulation. Furthermore, the chapter briefs readers on approaches for parallel power system simulation and discusses where a semi-analytical method can fit.

Chapter 2 begins by presenting semi-analytical solutions in a power series form. These solutions are derived from Taylor's series expansion of true solutions for power systems modeled by nonlinear differential equations. The chapter then extends the power series-form solutions to differential-algebraic equation models to accelerate simulations of large-scale power systems.

Chapter 3 presents a systematic approach for deriving and evaluating a high-order power series-form semi-analytical solution for fast power system simulation. The approach applies the Differential Transformation Method to nonlinear differential equations and algebraic equations of power system models, which are transformed into formally linear equations about power series coefficients. Solving these coefficients from low orders to high orders in a recursive manner allows growing a power series-form semi-analytical solution to a high order for desired accuracy.

Chapter 4 describes applications of analytic continuation to prolong the time interval of accuracy with a power series-form semi-analytical solution. Two methods, Padé Approximation and Continued Fractions, are applied to transform a power series-form semi-analytical solution into analytical, fractional forms with improved accuracy and convergence.

Chapter 5 introduces the Adomian Decomposition Method, which decomposes nonlinear functions of power system models into Adomian polynomials. This method uses a truncated sum as the semi-analytical solution in either a power series form or a non-power series form. The chapter also discusses applications of a semi-analytical solution to parallel simulation, taking advantage of intrinsic parallelism in its analytical form, and to efficient stochastic simulation thanks to embedded symbolic variables.

Chapter 6 presents the Homotopy Analysis Method for semi-analytical power system simulation. This method offers more freedom in form and algorithm for a semi-analytical solution. By designing the homotopy characterized by an embedded variable, the method deforms a solved, linear model into the nonlinear power system model to derive the semi-analytical solution. Additionally, the method introduces auxiliary parameters to adjust the convergence characteristics of the solution. The chapter compares this method with numerical integrators and other similar semi-analytical methods, including the Adomian Decomposition Method.

Chapter 7 describes a Parareal algorithm enhanced by semi-analytical methods for parallel-in-time power system simulation. The chapter first introduces the prediction-correction mechanism using the coarse and fine operators of the Parareal algorithm. Then, it compares the algorithm, respectively adopting three semi-analytical methods as the coarse operator, which are the Differential Transformation Method, Adomian Decomposition Method, and Homotopy Analysis Method described in Chapters 3, 5, and 6, respectively.

Finally, Chapter 8 describes the extension of the Holomorphic Embedding Method for steady-state and quasi-steady-state analyses to a semi-analytical method for extended-term time-domain simulations. The method requires complex analyticity in the approximation of a power system solution by considering a complex embedding variable. The chapter also discusses the partitioning and parallelization of the Holomorphic Embedding Method for high-performance simulations.

As emerging tools, the semi-analytical methods described in this book suggest a new paradigm for accelerating simulations of large-scale, bulk power systems. It is hoped that the book will inspire researchers, developers, and engineers in the power system field to develop powerful power system simulators utilizing advanced high-performance computing technology.

I am deeply grateful to the talented contributors who have authored the chapters of this timely book. Their expertise and dedication have made this work possible. Specifically, I would like to thank my former students and postdoctoral researchers, Bin Wang, Nan Duan, Yang Liu, Nan Duan, Chengxi Liu, and Rui Yao, as well as my collaborators, Byungkwon Park and Feng Qiu, Gurunath Gurrala, and Francis C. Joseph.

My research on power system simulation has been supported by the US National Science Foundation under NSF Award Numbers EEC-1041877 and ECCS-1610025, as well as the US Department of Energy through Oakridge National Laboratory and Argonne National Laboratory.

This monograph is dedicated to the memory of my PhD advisor, Professor Da-Zhong Zheng, whose knowledgeable mentoring, thoughtful guidance, and insightful encouragement proved invaluable throughout my academic journey.

Finally, I want to express my deepest gratitude to my loving family – my wife Fang and my sons Yi and Rei – who have always been there to support me, spark my curiosity with insightful inquiries, and encourage me to explore new ideas and perspectives.

Knoxville, TN, USA
August 2023

Kai Sun

1

Power System Simulation: From Numerical to Semi-Analytical

Kai Sun

Department of Electrical Engineering & Computer Science, University of Tennessee, Knoxville, TN, USA

1.1 Timescales of Simulation

A power system is composed of a vast network of generators, loads, and control devices, each with their own intricate and diverse dynamics. Therefore, power system simulators usually focus on specific timescales or types of dynamic behaviors to simplify the modeling and computation of the simulated system. In general, there are three types of power system simulations based on their timescales, models, and primary objectives, which are electromagnetic transient simulation, transient stability simulation, and quasi-steady-state simulation, as illustrated in Figure 1.1.

Electromagnetic transient (EMT) simulation provides a high-resolution solution to the three-phase alternating current and voltage of each circuit element of a power system, typically at a microsecond scale. EMT simulations are indispensable for investigating protective actions against short-circuit faults and electromagnetic dynamics of existing or new equipment and controllers.

Transient stability simulation, on the other hand, is concerned with the slower electromechanical dynamics of generators, motors, and their controllers. It computes the time variation of phasors of voltages and currents, which are approximate representations of periodical quantities based on a common synchronous frequency. Although this phasor approximation can introduce errors in the actual frequency deviations, it is acceptable when the frequencies are close to the synchronous frequency. Under normal or nonextreme abnormal conditions, the frequencies at buses of a power transmission system must not deviate by more than 1 Hz before triggering

Power System Simulation Using Semi-Analytical Methods, First Edition. Edited by Kai Sun.

Published 2024 by John Wiley & Sons, Inc.

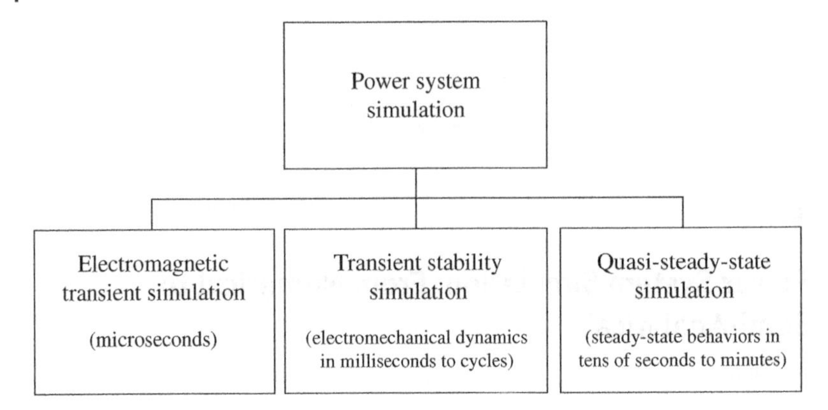

Figure 1.1 Categories of power system simulation by timescale.

an under- or over-frequency protection action. In each transient stability simulation, phasors and the system state are computed typically at a time step of milliseconds to cycles, which is sufficient to capture authentic electromechanical dynamics and much larger than the microsecond-level time step of EMT simulation. Transient stability simulation is a crucial tool in power system studies that support grid operations and planning. It is often an essential functionality of the Energy Management System (EMS) used by electric utilities and regional transmission organizations. It is a critical component of dynamic security assessment (DSA) programs that evaluate the system's angular stability, frequency regulation, and postfault voltage recovery, all of which are essential for ensuring the reliable and secure operation of the power system.

Quasi-steady-state (QSS) simulation studies how a power system behaves under slowly changing operating conditions, such as load variations and generation redispatches, at a low resolution from tens of seconds to minutes. It is important for analyzing steady-state generation, load and power-flow controls, long-term voltage stability, and the early stages of cascading outages. In such simulations, the algebraic power-flow equations form the core of the simulation model, and the dynamics of the system are either ignored or significantly simplified with differential equations.

This book focuses on accelerating transient stability simulation of electromechanical dynamics in power systems. While fast EMT dynamics are not covered, the simulation of QSS behaviors in a power system is also studied in Chapter 8, which combines transient stability simulation and QSS simulation into extended-term simulations. While many methods introduced in later chapters are primarily geared toward transient stability simulation, they can also be applied to EMT and QSS simulations. This is because all three types of simulation use nonlinear differential and algebraic equations for their models, and the semi-analytical methods introduced in the book are applicable to these equations.

The remainder of this chapter provides an introduction to power system modeling for transient stability simulation, including mathematical models of basic power system components. Power system simulation is formulated as an initial value problem based on the simulation model. Two approaches for resolving the problem are described: the conventional numerical solution approach and the emerging semi-analytical solution approach. Additionally, the chapter covers parallel power system simulation.

It should be noted that this chapter does not aim to provide a comprehensive introduction to power system modeling for transient stability simulation. Rather, it provides the minimum background information for readers to follow the formulations and methods for power system simulations in the rest of the book. Readers who are interested in more details on power system modeling and simulations may refer to books by Kundur [1], Padiyar [2], and Anderson and Fouad [3].

1.2 Power System Models

1.2.1 Overview

1.2.1.1 Simplifying a Power System Model

If the EMT dynamics, electromechanical dynamics, and QSS behaviors of a power system are to be considered, the resulting mathematical model will comprise a set of highly stiff ordinary differential equations (ODEs).

For example, let us consider a power system model in the form of a two-timescale nonlinear system:

$$\begin{cases} \dot{\mathbf{x}} = \mathbf{f}(\mathbf{x}, \mathbf{y}, \mathbf{p}, t) \\ \varepsilon\dot{\mathbf{y}} = \mathbf{g}(\mathbf{x}, \mathbf{y}, \mathbf{p}, t) \end{cases} \tag{1.1}$$

where $\mathbf{x}$ and $\mathbf{y}$ are state vectors that represent electromechanical and faster EMT dynamics, respectively. The matrix ε, which is nonsingular, has all eigenvalues close to zero and corresponds to time constants associated with EMT dynamics. The vector $\mathbf{p}$ includes parameters whose values specify the operating conditions and can vary slowly in an explicit function of time to simulate QSS behaviors of the system.

If the EMT dynamics are neglected, resulting in a value of ε equal to 0, the mathematical model takes the form of differential-algebraic equations (DAEs) shown in (1.2).

$$\begin{cases} \dot{\mathbf{x}} = \mathbf{f}(\mathbf{x}, \mathbf{y}, \mathbf{p}, t) \\ 0 = \mathbf{g}(\mathbf{x}, \mathbf{y}, \mathbf{p}, t) \end{cases} \tag{1.2}$$

This model includes ODEs about state vector $\mathbf{x}$ for the dynamical devices, such as generators, motors, and associated controllers, and their corresponding $\mathbf{f}$ functions. Additionally, it contains algebraic equations for the control laws, such as power-flow equations, and their corresponding $\mathbf{g}$ functions. The state vector $\mathbf{y}$ in (1.1), which changes rapidly with EMT dynamics, is now considered a vector of nonstate variables such as bus voltages or line currents. Their changes in the electromechanical timescale are instantaneous and dependent on state variables in $\mathbf{x}$. As a result, Eq. (1.2) is a simplification of the full ODE model (1.1) that reduces EMT dynamics.

While this book primarily focuses on simulating electromechanical dynamics and transient stability of power systems using the DAE model (1.2), it should be noted that most of the semi-analytical methods to be introduced in the book can be extended to simulate (1.1) as well. This is because there is no fundamental difference in mathematics between the ODEs on $\mathbf{x}$ and the ODEs on $\mathbf{y}$ in (1.1). In fact, semi-analytical methods are more easily applicable to a purely ODE model like (1.1) than a DAE model like (1.2).

Furthermore, to simplify the transient stability simulation process over an extended simulation period, a reduced version of the DAE model in (1.2) or (1.3) can be used, where fast elements in $\dot{\mathbf{x}}$ are ignored and assumed to be zero, resulting in a simpler DAE model with fewer state variables. Moreover, assuming $\dot{\mathbf{x}} = 0$ with all state variables for QSS simulation leads to a purely algebraic equation model, which can be expressed as:

$$\begin{cases} 0 = \mathbf{f}(\mathbf{x}, \mathbf{y}, \mathbf{p}, t) \\ 0 = \mathbf{g}(\mathbf{x}, \mathbf{y}, \mathbf{p}, t) \end{cases} \tag{1.3}$$

In QSS simulation, a slowly time-variant $\mathbf{p}(t)$ models the desired sequence of conditions of interest, such as load and generation variations. Explicitly assuming $\mathbf{p}(t)$ to change with time can reflect changes in conditions over time.

1.2.1.2 A Practical Power System Model

In practice, transient stability simulation typically involves the consideration of a time-invariant model that operates at a constant condition represented by $\mathbf{p} = \mathbf{p}_0$. A widely used model for this purpose is the bus injection representation model (1.4).

$$\begin{cases} \dot{\mathbf{x}} = \mathbf{f}(\mathbf{x}, \mathbf{V_{bus}}, \mathbf{p}_0) \\ 0 = \mathbf{I_{bus}}(\mathbf{x}, \mathbf{V_{bus}}, \mathbf{p}_0) - \mathbf{Y_{bus}} \mathbf{V_{bus}} \end{cases} \tag{1.4}$$

Here, $\mathbf{Y_{bus}}$ is the bus admittance matrix of the network and $\mathbf{I_{bus}}$, which is a complex vector-valued function of state vector $\mathbf{x}$ and bus voltage phasors in $\mathbf{V_{bus}}$, determines the current injections into the network. In contrast to the general DAE model in (1.2), model (1.4) equates $\mathbf{y}$ to the complex vector $\mathbf{V_{bus}}$ on all bus voltages, and its $\mathbf{g}$ function constrains the balance between the current from the source or load at each bus and the current injected to the bus.

In DAE models (1.2) and (1.4), the algebraic equations represent the power network that connects all buses by power lines and transformers, including AC power-flow equations. The ODEs model the dynamic devices of a power system, such as:

- Synchronous machine models for synchronous generators and condensers
- Auxiliary component/controller models of generators such as turbines, speed governors, exciters, and power system stabilizers (PSSs)
- Network controller/compensator models such as FACTS (flexible AC transmission system) devices
- Induction motor models for three-phase and one-phase motor loads
- Other dynamic or controllable loads
- Inverter-based resources such as renewable generators and battery-based energy storage systems.

In power grid simulations, power engineers often use a bulk power system model that includes the main power plants and the transmission network, while the distribution system and loads under each transmission substation are often aggregated into an equivalent composite load model. As illustrated in Figure 1.2, a bulk power system model needs to include the following elementary models:

- Synchronous generator models
- Excitation system models, which optionally include a PSS
- Turbine and speed governor models
- Transmission network model
- Bus load models (including passive shunt devices)
- Other energy resources (such as renewable generators and energy storage systems).

1.2.2 Generator Models

This section introduces the sixth-order, fourth-order, and second-order synchronous generator models that are widely used in studies on power system stability and control.

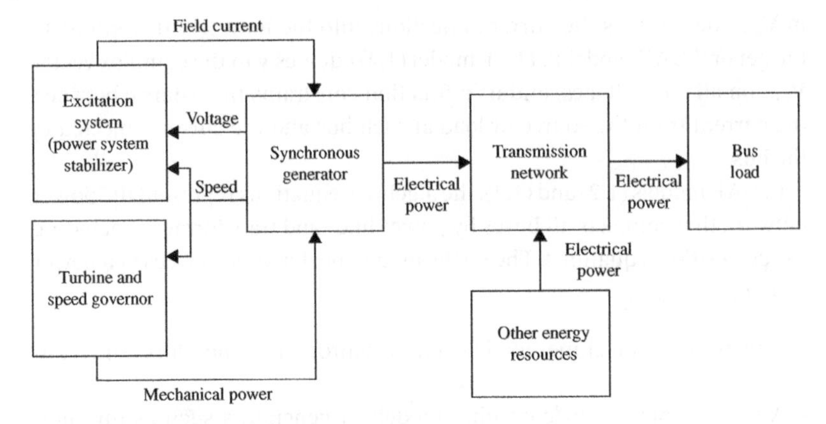

Figure 1.2 Components of a bulk power system model.

1.2.2.1 Sixth-Order Model

A detailed model for a synchronous machine is provided by equations in (1.5), which can be used to represent conventional synchronous generators or synchronous motors, including synchronous condensers used for reactive power support. This sixth-order model is based on a synchronous frequency of the system, denoted as f_0, which typically equals either 50 or 60 Hz. The model consists of six ODEs (1.5a)–(1.5f) and four algebraic equations (1.5g)–(1.5j) that describe the machine's interface with the power network.

$$\dot{\delta} = \Delta\omega = \omega - \omega_0 \tag{1.5a}$$

$$\frac{2H}{\omega_0}\Delta\dot{\omega} = P_m - P_e - D \cdot \frac{\Delta\omega}{\omega_0} \tag{1.5b}$$

$$T'_{d0}\dot{e}'_q = -\frac{x_d - x''_d}{x'_d - x''_d}e'_q + \frac{x_d - x'_d}{x'_d - x''_d} \cdot e''_q + e_{fd} \tag{1.5c}$$

$$T'_{q0}\dot{e}'_d = -\frac{x_q - x''_q}{x'_q - x''_q}e'_d + \frac{x_q - x'_q}{x'_q - x''_q} \cdot e'' \tag{1.5d}$$

$$T''_{d0}\dot{e}''_q = e'_q - e''_q - (x'_d - x''_d) \cdot i_d \tag{1.5e}$$

$$T''_{q0}\dot{e}''_d = e'_d - e''_d + (x'_q - x''_q) \cdot i_q \tag{1.5f}$$

$$P_e = v_d i_d + v_q i_q \tag{1.5g}$$

$$\begin{bmatrix} i_d \\ i_q \end{bmatrix} = \begin{bmatrix} r_a & -x''_q \\ x''_d & r_a \end{bmatrix}^{-1} \left(\begin{bmatrix} e''_d \\ e''_q \end{bmatrix} - \begin{bmatrix} v_d \\ v_q \end{bmatrix} \right) \tag{1.5h}$$

$$\begin{bmatrix} v_x \\ v_y \end{bmatrix} = \begin{bmatrix} -\sin\delta & -\cos\delta \\ \cos\delta & -\sin\delta \end{bmatrix} \begin{bmatrix} v_d \\ v_q \end{bmatrix} \quad \Leftrightarrow \quad V \overset{\text{def}}{=} (v_x + jv_y) = e^{j\left(\frac{\pi}{2}-\delta\right)}(v_d + jv_q) \tag{1.5i}$$

$$\begin{bmatrix} i_x \\ i_y \end{bmatrix} = \begin{bmatrix} -\sin\delta & -\cos\delta \\ \cos\delta & -\sin\delta \end{bmatrix} \begin{bmatrix} i_d \\ i_q \end{bmatrix} \quad \Leftrightarrow \quad I \overset{\text{def}}{=} (i_x + ji_y) = e^{j\left(\frac{\pi}{2}-\delta\right)}(i_d + ji_q) \tag{1.5j}$$

The first two ODEs, (1.5a) and (1.5b), are known as swing equations, which describe the rotor angle δ (in radians) with respect to a revolving system reference angle and the deviation $\Delta\omega = \omega - \omega_0$ of the electrical speed ω (in radians per second) from synchronous angular frequency $\omega_0 = 2\pi f_0$.

This model is also referred to as a "2.2 model" because the remaining ODEs comprise two ODEs in the d (direct)-axis plus two ODEs in the q (quadrature)-axis of the rotor, which is 90° ahead of the d-axis, as defined in ANSI/IEEE standard 100-1977. ODEs (1.5c) and (1.5d) describe internal transient voltage components e'_q and e'_d in q-axis and d-axis, respectively, with transient time constants T'_{d0} and T'_{q0} (in seconds), while ODEs (1.5e) and (1.5f) describe internal subtransient voltage components e''_q and e''_d in two axes, respectively, with faster time constants T''_{q0} and T''_{d0} (in seconds). These four time constants can be estimated by open-circuit frequency response tests on the machine. Table 1.1 lists the other constants and variables with their units, where "p.u." stands for "per unit."

In (1.5e) through (1.5j), we can eliminate v_d, v_q, i_d, and i_q in order to represent terminal voltages, and currents by v_x, v_y, i_x, and i_y, which are all based on the system-wide common reference frame. The terminal current phasor $I = i_x + ji_y$ can then be expressed as a function of variables that include the terminal voltage phasor $V = v_x + jv_y$ and the state variables in (1.4), and then a complete set of equations in (1.4) for the entire grid can be constructed after integrating the models of all sources, loads, and the network.

It is important to note that while model (1.5) accurately represents a round-rotor generator, such as a thermal generating unit, it may not be accurate for salient-pole rotor generators, such as hydraulic generating units. To model a salient-pole rotor generator, one ODE in the q-axis, i.e. (1.5d) or (1.5f), is often removed. The resulting model is referred to as a "2.1 model," with two ODEs in the d-axis and one ODE in the q-axis.

1.2.2.2 Fourth-Order Model

In academic studies, a simplified fourth-order model given by the following equations in (1.6), along with (1.5i) and (1.5j), is often used to represent a synchronous machine. This model is often referred to as a "1.1 model" and is

Table 1.1 Constants and variables of generator models.

Notation	Meaning
H (second)	Inertial time constant, which is equal to its rated kinetic energy divided by the VA base when operated at the synchronous frequency.
D (p.u.)	Damping coefficient representing integrated damping effects such as the effect of its damper windings, the control by the power system stabilizer, and nearby frequency-dependent load.
x'_d, x'_q, x''_d, x''_q (p.u.)	d- and q-axes, transient and subtransient reactances or inductances.
r_a (p.u.)	Armature resistance.
P_m (p.u.)	Mechanical power or torque controlled by its governor and turbine.
P_e (p.u.)	Electrical power injected into the grid and coupled with other sources and loads through the grid.
e_{fd} (p.u.)	Field voltage controlled by an automatic voltage regulator (AVR) with the excitor.
v_d, v_q, i_d, i_q (p.u.)	d- and q-axes components of terminal voltages and currents in the machine-based local reference frame.
v_x, v_y, i_x, i_y (p.u.)	Real (x-axis) and imaginary (y-axis) components of terminal voltages and currents in a system-wide common reference frame that lags the q-axis by rotor angle δ.

obtained from (1.5) by assuming that T''_{d0} and T''_{q0} in (1.5e) and (1.5f) are zero and by eliminating all subtransient voltages and reactances with a double prime.

$$\dot{\delta} = \Delta\omega = \omega - \omega_0 \tag{1.6a}$$

$$\frac{2H}{\omega_0}\Delta\dot{\omega} = P_m - P_e - D \cdot \frac{\Delta\omega}{\omega_0} \tag{1.6b}$$

$$T'_{d0}\dot{e}'_q = -e'_q + \left(x_d - x'_d\right) \cdot i_d + e_{fd} \tag{1.6c}$$

$$T'_{q0}\dot{e}'_d = -e'_d + \left(x_q - x'_q\right) \cdot i_q \tag{1.6d}$$

$$P_e = v_d i_d + v_q i_q \tag{1.6e}$$

$$\begin{bmatrix} i_d \\ i_q \end{bmatrix} = \begin{bmatrix} r_a & -x'_q \\ x'_d & r_a \end{bmatrix}^{-1} \left(\begin{bmatrix} e'_d \\ e'_q \end{bmatrix} - \begin{bmatrix} v_d \\ v_q \end{bmatrix} \right) \tag{1.6f}$$

Similarly, by removing (1.6d), a "1.0 model" can be obtained for a salient-pole rotor generator.

1.2.2.3 Second-Order Model

If both (1.6c) and (1.6d) are removed and x'_q is set equal to x'_d, the simplified "0.0 model" can be derived, which is commonly referred to as the second-order classical model as given in (1.7).

$$\dot{\delta} = \Delta\omega = \omega - \omega_0 \tag{1.7a}$$

$$\frac{2H}{\omega_0}\Delta\dot{\omega} = P_m - P_e - D \cdot \frac{\Delta\omega}{\omega_0} \tag{1.7b}$$

$$P_e = real(VI^*) \tag{1.7c}$$

$$E' = |E'|\angle\delta = V + \left(r_a + jx'_d\right) I \tag{1.7d}$$

This model assumes a constant internal electromotive force $E' = e'_d + je'_q$ with its angle equal to the rotor angle δ. This simplified model is frequently used for fast, approximate analysis of transient stability in power systems.

1.2.3 Controller Models

To ensure accurate power system simulation, it is crucial to model the controllers associated with each synchronous generator. These controllers include the turbine and governor for frequency and active power control, and the excitation system for automatic voltage regulation and reactive power control. In addition, for generators that contribute to undesired power system oscillations, PSSs are often installed with a nonvoltage auxiliary input to improve oscillation damping. In this section, we will introduce some commonly used examples of controller models that can be incorporated into the generator model for more accurate power system simulations.

1.2.3.1 Governor and Turbine Models

The block diagram in Figure 1.3 depicts a linear frequency control system for each synchronous generator.

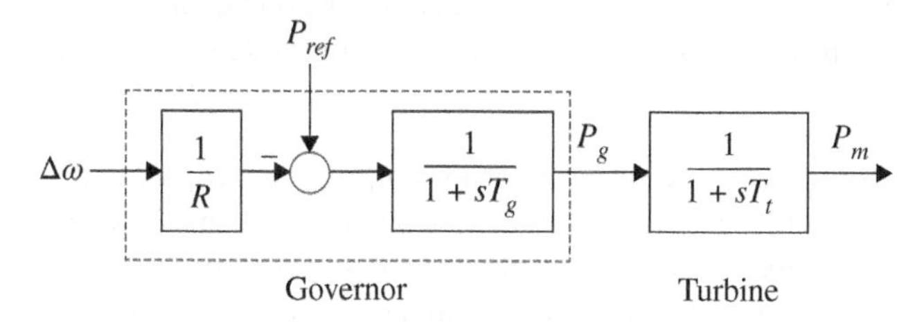

Figure 1.3 Governor and turbine models.

Table 1.2 Constants and variables of governor and turbine models.

Notation	Meaning
T_g (second)	Governor time constant
T_t (second)	Turbine time constant
P_{ref} (p.u.)	Setting point of the active power output determined by generation dispatch
$\Delta\overline{\omega}$ (p.u.)	Speed deviation in per unit, namely $(\omega - \omega_0)/\omega_0$
R (p.u.)	Speed regulation factor for the droop control
P_g (p.u.)	Output of the speed governor, i.e. the input to the turbine

The system consists of a speed governor and a first-order turbine model, represented by Eqs. (1.8) and (1.9), respectively.

$$T_g \dot{P}_g = -P_g + P_{ref} - \frac{1}{R} \cdot \Delta\overline{\omega} \tag{1.8}$$

$$T_t \dot{P}_m = -P_m + P_g \tag{1.9}$$

Table 1.2 provides a list of constants and variables used in the model. The speed governor, modeled by Eq. (1.8), regulates the opening of the turbine valve to maintain a constant rotational speed. The turbine model, given by Eq. (1.9), determines the active power output of the generator as a function of the valve position and the generator rotor speed deviation from the synchronous speed. The model assumes that the valve position directly controls the active power output of the generator. The model depicted in Figure 1.3 is a simplified linear approximation of an actual speed-governing system. It is employed in some chapters of the book for transient stability simulations of large-scale power systems.

A more practical speed-governing system model is presented in Figure 1.4, which is the IEEE Type-1 speed-governor model (IEEEG1) from the reference [4]. This model includes additional factors that are crucial for accurate representation of this type of real-world speed-governing system. These factors include:

- Upper and lower limits of the speed governor output P_g are considered.
- Speed deviation $\Delta\omega$ first goes through a deadband of, e.g., 30 mHz to avoid oversensitive speed regulation with small frequency variations.
- Nonlinearities, e.g. hysteresis and saturation, can be modeled with the turbine.

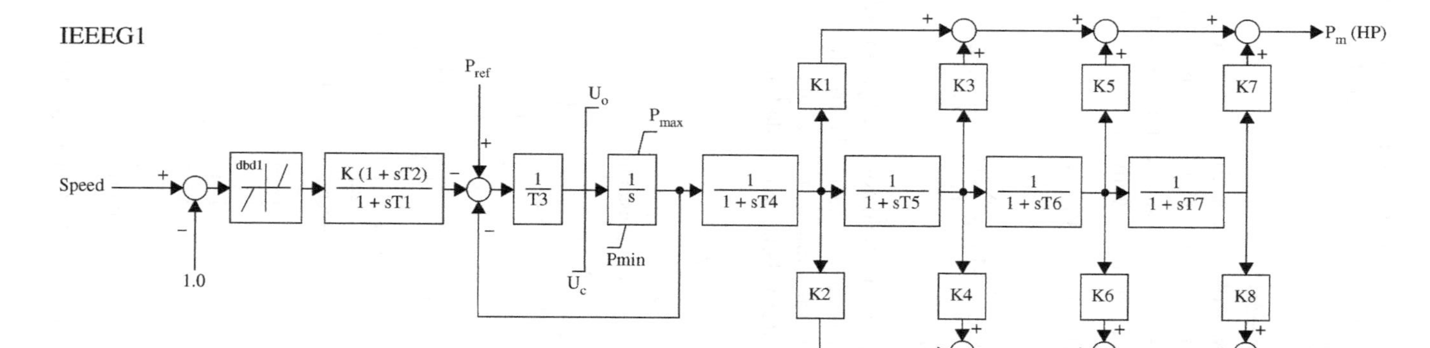

Figure 1.4 IEEE Type 1 speed-governor model. Source: [4].

- The first-order turbine model can be extended to multiple sequentially or parallelly connected first-order turbines to model, e.g., low-pressure to high-pressure turbines.

1.2.3.2 Excitation System Model

A simplified excitation control system for generators is illustrated in Figure 1.5 and modeled by Eq.s (1.10) to (1.13). It consists of four blocks: the exciter, amplifier, stabilizer, and transducer, and also receives the control signal V_s from a PSS for damping improvement.

$$T_A\dot{V}_R = -V_R + K_A(V_{ref} + V_s - V_F - V_c) \tag{1.10}$$

$$T_E\dot{E}_{FD} = -K_E E_{FD} + V_R \tag{1.11}$$

$$T_F\dot{V}_F = -V_F + K_F\dot{E}_{FD} = -V_F + \frac{K_F}{T_E}V_R - \frac{K_F}{T_E}K_E E_{FD} \tag{1.12}$$

$$T_R\dot{V}_c = -V_c + K_R V_t \tag{1.13}$$

The control system regulates the field voltage, E_{FD}, in order to automatically regulate the targeted voltage V_t measured through the transducer. The targeted voltage, V_t, can be the same as the voltage V in (1.5i) if the terminal voltage of the generator is being controlled; it can also be a remote

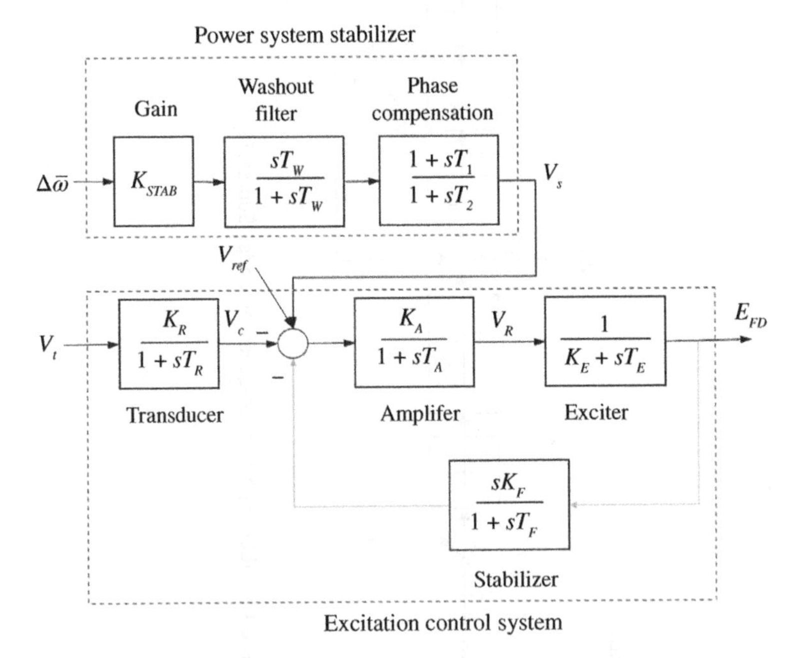

Figure 1.5 Linear excitation system with a power system stabilizer.

bus voltage if that is estimated via a load compensator based on the equivalent impedance from the terminal to the remote bus. The function of the stabilizer inside the excitation control system is responsible for ensuring the stability of the closed-loop transfer function on the exciter. To improve the damping of power system oscillation, the exciter also receives the output signal V_s from a PSS, which will be introduced later. It is worth noting that E_{FD} is the same as e_{fd} of (1.6c), with the exception that it is calculated using an alternative per-unit system. This amplifies the small per-unit value of e_{fd} to a value close to or bigger than 1 p.u. For more information, refer to Kundur [1].

Table 1.3 summarizes the variables and constants of the model. Time constants are given in seconds, and gains and voltage signals are expressed in per unit.

Figure 1.6 shows a practical simulation model using the IEEE Type-1 exciter (IEEET1) from reference [5]. The model includes additional factors compared to the model in Figure 1.5, such as:

- Upper and lower limits of the amplifier to ensure $V_{RMIN} \leq V_R \leq V_{RMAX}$.
- V_{OEL} and V_{UEL} are outputs from the over-excitation limiter and under-excitation limiter, respectively. These help prevent the generator from exceeding its reactive power capacity.
- Consideration of the saturation effect of the exciter. An exponential correction $A_E e^{B_E E_{FD}}$ is added to the constant K_E in (1.11) and (1.12) to account for this effect.

Table 1.3 Constants and variables of excitation system models.

Notation	Meaning
V_R, K_A, T_A	Output voltage signal, gain, and time constant of the amplifier.
E_{FD}, K_E, T_E	Field voltage and related parameters of the exciter.
V_F, K_F, T_F	Output voltage signal, gain, and time constant of the excitation stabilizer.
V_c, K_R, T_R	Output voltage signal, gain, and time constant of the voltage transducer.
V_{ref}, V_s	Voltage setting point and the control input from PSS.
K_{STAB}	Gain of PSS.
T_W, T_1, T_2	Time constants of the washout filter and phase compensator.

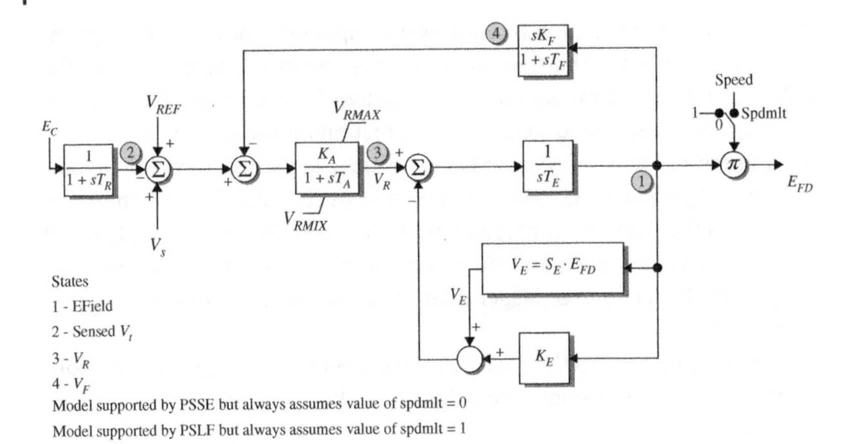

Figure 1.6 IEEE Type 1 exciter. Source: [5].

1.2.3.3 Power System Stabilizer

The excitation system of a synchronous generator may use a large gain K_A in its amplifier to improve the steady-state values of the controlled voltage, at the terminal bus or a remote bus, and the synchronizing torque of the generator. However, this can reduce the damping torque of the generator, making it more vulnerable to power system oscillation. To address this, a PSS, as shown earlier in Figure 1.5, is used to add damping against oscillation by controlling the exciter using a nonvoltage auxiliary signal, such as the frequency deviation $\Delta\bar{\omega}$ or its estimation.

Figure 1.5 has provided a simplified PSS model that includes three basic components: the gain, washout (high-pass) filter, and phase compensation, which can be modeled by:

$$T_W T_2 \ddot{V}_s + T_W \dot{V}_s + V_s = T_W T_1 K_{STAB} \Delta\ddot{\bar{\omega}} + T_W K_{STAB} \Delta\ddot{\bar{\omega}} \tag{1.14}$$

In this model, the gain K_{STAB} specifies the expected amount of damping to be added by the PSS. T_W is used to block low-frequency signals while keeping signals associated with power system oscillation unchanged. T_1 and T_2 together provide phase-lead compensation to the excitor for a range of frequencies centered at the frequency of generator oscillation that needs to be damped. The phase-lead compensation is necessary to compensate the phase-lag characteristic with the transfer function from the output signal V_s to the electrical torque T_e.

1.2.4 Load Models

In a bulk power system model, generators and the transmission and subtransmission facilities with voltages above a specific voltage level, such as 100 kV, are included, while distribution networks and electric loads in the area around each transmission substation are not modeled in detail. Instead, they are aggregated into an equivalent single-load device connected to the substation, which is represented as one bus in the bulk power system model. A load model is required to represent the equivalent load device on the bus.

1.2.4.1 Composite Load Model

One method for modeling the load at each bus is to use a composite load model. The devices to be reduced and merged into a bus for a composite load model may include substation step-down transformers, subtransmission and distribution feeders, distribution transformers, shunt capacitors, voltage regulators, customer wiring, etc. Figure 1.7 illustrates the structure of a composite load model, which is composed of static load components, where the active and reactive loads are dependent on the bus voltage, and dynamic load components, such as industrial motors and air conditioning motors, requiring ODEs to model.

1.2.4.2 ZIP Load Model

Load modeling typically employs static voltage-dependent loads to represent the majority of bus loads, with or without frequency dependency. Dynamic loads, such as motor loads, are typically only considered on selected buses where a significant industrial load is present or the air conditioning load is high. A widely used static load model in power systems is the ZIP load

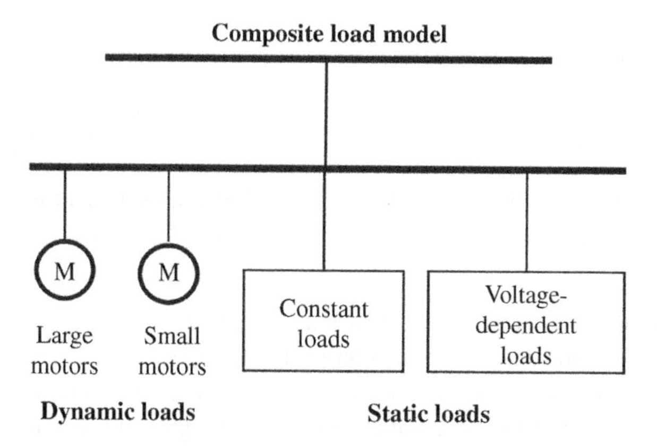

Figure 1.7 Composite load model.

model, which consists of three parallel load components for both active and reactive powers. These components, respectively, represent constant power, constant current, and constant impedance, as shown in Eq. (1.15):

$$P = P_0 \left[p_Z \left(\frac{|V|}{V_0} \right)^2 + p_I \left(\frac{|V|}{V_0} \right) + p_P \right] \cdot [1 + K_{pf}(f - f_0)],$$
$$(p_Z + p_I + p_P = 1)$$
$$Q = Q_0 \left[q_Z \left(\frac{|V|}{V_0} \right)^2 + q_I \left(\frac{|V|}{V_0} \right) + q_P \right] \cdot [1 + K_{qf}(f - f_0)],$$
$$(q_Z + q_I + q_P = 1) \tag{1.15}$$

Here, V is the voltage phasor at the load bus, P_0 and Q_0 are the nominal active and reactive powers, and V_0 is the nominal voltage magnitude at the bus. The parameters p_Z, p_I, and p_P define the percentages of the constant impedance, current, and power components in active power, while the parameters q_Z, q_I, and q_P do the same for reactive power. The parameters K_{pf} and K_{qf} represent the sensitivities of the active and reactive loads to deviations of the actual frequency f from the nominal, synchronous frequency f_0.

The static ZIP load model in Eq. (1.15) is also known as a polynomial load model since it maps the load characteristics onto a second-order polynomial function of the bus voltage magnitude. Alternatively, the model can be approximated by exponential functions in Eq. (1.16), where a and b are, respectively, the sensitivities of active and reactive power loads to changes in the bus voltage magnitude. When a (or b) is close to 0, 1, or 2, the active (reactive) load behaves similarly to a constant power, current, or impedance load, respectively.

$$P = P_0 \left(\frac{|V|}{V_0} \right)^a \cdot [1 + K_{pf}(f - f_0)]$$
$$Q = Q_0 \left(\frac{|V|}{V_0} \right)^b \cdot [1 + K_{qf}(f - f_0)] \tag{1.16}$$

In the absence of detailed load characteristics, the following assumptions are commonly made in industrial practice for transient stability simulation of bulk power systems:

- When $|V|$ is around V_0, $a = 1$ and $b = 2$ are assumed, treating P and Q as a constant current load and a constant impedance load, respectively.
- When $|V|$ becomes low, such as $|V| < 0.7\ V_0$, the entire bus load is converted to a constant impedance load with $a = b = 2$ to mitigate numerical issues in computations.

1.2.4.3 Motor Load Model

To provide an example of a dynamical load model, the third-order single-cage induction motor model is introduced below, which is often used to model a small motor of the composite load model shown in Figure 1.7. The model's ODEs and algebraic equations are listed in Eq. (1.17) and can be incorporated into the DAE model (1.4) for the entire grid.

$$2H_m\dot{s} = T_m - T_e \tag{1.17a}$$

$$\dot{v}'_d = -\frac{1}{T'_0}\left[v'_d + (X_s - X'_s)\, i_q\right] + s\omega_0 v'_q \tag{1.17b}$$

$$\dot{v}'_q = -\frac{1}{T'_0}\left[v'_q - (X_s - X'_s)\, i_d\right] - s\omega_0 v'_d \tag{1.17c}$$

$$T_m = \alpha s^2 + \beta s + \gamma \tag{1.17d}$$

$$T_e = v'_d i_d + v'_q i_q \tag{1.17e}$$

$$v_d + jv_q = (R_s + jX'_s)\,(i_d + ji_q) + (v'_d + jv'_q) \tag{1.17f}$$

Here, s is the slip, v'_d and v'_q are the d-axis and q-axis internal voltages, and $v_d + jv_q = V$ and $i_d + ji_q = I$ are the bus voltage phasor and current injection phasor, respectively. All parameters are in per unit except for the inertial time constant H_m and open-circuit transient time constant T'_0, which are in seconds.

1.2.5 Network Model

On the network of a power system, the power-flow equations belong to algebraic equations of the DAE model (1.4) and take the form of (1.18).

$$\mathbf{I_{bus}} = \mathbf{Y_{bus}}\mathbf{V_{bus}} \tag{1.18}$$

where $\mathbf{I_{bus}}$ is the complex vector composed of the phasors on current injections into all buses, $\mathbf{V_{bus}}$ is the complex vector having all bus voltage phasors, and $\mathbf{Y_{bus}}$ is the bus admittance matrix. For an n-bus network with an additional ground bus 0, the elements of $\mathbf{Y_{bus}}$ are defined by (1.19), where y_{ij} is the admittance between buses i and j.

$$\begin{aligned} Y_{ii} &= \sum_{j=0, j\neq i}^{n} y_{ij} \\ Y_{ij} &= -y_{ij} \quad \text{for } j \neq i \end{aligned} \tag{1.19}$$

$\mathbf{Y_{bus}}$ is a sparse matrix having nonzero elements only when a branch exists between the buses corresponding to the indices of the row and column. It is approximately symmetric, with asymmetries appearing only on a minority of branches that have a phase shift due to, for example, a phase-shifting transformer. If no branch has a phase shift, $\mathbf{Y_{bus}}$ is symmetric with each $y_{ij} = y_{ji}$.

From the definition of complex power at bus i

$$S_i = P_i + jQ_i = V_i I_i^* \tag{1.20}$$

the current injection I_i can be solved and plugged into (1.18) to rewrite it in terms of powers and voltages, as shown in (1.21).

$$S_i^* = P_i - jQ_i = V_i^* \sum_{j=1}^{n} Y_{ij} V_j \quad i = 1, \dots, n \tag{1.21}$$

Separating real and imaginary parts of (1.21) yields the power-flow equations for active and reactive powers at each bus i, as given in:

$$\begin{aligned} P_i &= \sum_{j=1}^{n} |V_j||V_i||Y_{ij}| \cos(\angle Y_{ij} - \angle V_i + \angle V_j) \\ Q_i &= -\sum_{j=1}^{n} |V_j||V_i||Y_{ij}| \sin(\angle Y_{ij} - \angle V_i + V_j) \end{aligned} \tag{1.22}$$

where the operator "$\angle$" takes the argument of a complex number. For an n-bus network, there are $2 \times n$ such real equations about four real quantities at each bus: active power P_i, reactive power Q_i, voltage magnitude $|V_i|$, and voltage angle $\angle V_i$. To solve these power-flow equations, two quantities need to be known at each bus, and the other two are solved. Power-flow analysis typically classifies all buses into several types, each freezing two quantities while relaxing the other two. For example, there are slack buses whose voltage magnitude and angle are specified, "PQ buses" such as load buses whose P_i and Q_i are given, and "PV buses" such as generator buses and other voltage-controlled buses whose P_i and $|V_i|$ are controlled.

1.2.6 Classical Power System Model

To qualitatively analyze transient stability for a power system, a simplified, classical multimachine model is often used as shown in Figure 1.8. This model involves several assumptions, such as:

- Each generator sis represented by an internal voltage phasor E' behind a synchronous reactance that is equal to its d-axis transient reactance x_d'. The armature resistance r_a is often ignored.
- Each load is represented by a constant admittance between its bus and the ground.

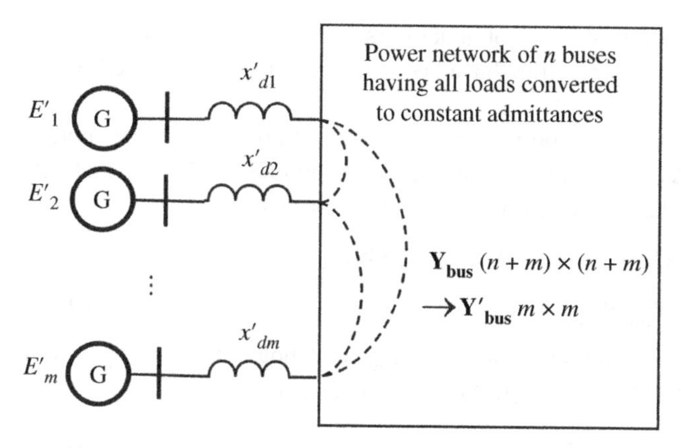

Figure 1.8 Classical power system model.

- The dynamics of each generator are modeled by its second-order classical model (1.7), in which the rotor angle coincides with the angle of E', the excitation control system and speed-governing system are ignored, and $|E'|$ and P_m remain constant.

As all loads are assumed to be constant impedance loads, the voltages of buses without a generator become dependent on the internal voltages of generators. Therefore, the algebraic equations with network model (1.18) can be reduced, leading to the classical power system model being rewritten into a purely ODE model.

To explain this, consider a power system with m generators at buses 1 to m and another n buses without a generator. Based on steady-state power-flow conditions, the internal voltage phasor E_i' behind the transient reactance x_{di}' is calculated for each bus connected to a generator, and load admittance y_{i0} is calculated for each bus connected to a load.

$$E_i' = V_i + jx_{di}' \frac{S_i^*}{V_i^*} \quad i = 1,2, \ldots, m \tag{1.23a}$$

$$y_{i0} = \frac{S_i^*}{|V_i|^2} \quad i = 1,2, \ldots, n \tag{1.23b}$$

The internal voltage phasors E_i' $(i = 1, \ldots, m)$ with m generators can be considered as m internal buses. Thus, by considering all $m + n$ buses in the entire system, the network model (1.18) becomes:

$$\begin{bmatrix} 0 \\ \mathbf{I}_{m\times 1} \end{bmatrix} = \begin{bmatrix} \mathbf{Y}_{n\times n} & \mathbf{Y}_{n\times m} \\ \mathbf{Y}_{m\times n} & \mathbf{Y}_{m\times m} \end{bmatrix} \begin{bmatrix} \mathbf{V}_{n\times 1} \\ \mathbf{E}'_{m\times 1} \end{bmatrix} \tag{1.24}$$

where only m internal buses of generators inject currents to the network while loads do not since they have become admittances embedded in the $\mathbf{Y_{bus}}$ matrix.

By eliminating $\mathbf{V}_{n\times 1}$, the vector of all external bus voltages, we obtain (1.25), where $\mathbf{Y'_{bus}}$ is an $m \times m$ bus admittance matrix on the reduced network that only keeps the internal buses of generators.

$$\begin{aligned}\mathbf{I}_{m\times 1} &= \left(\mathbf{Y}_{m\times m} - \mathbf{Y}_{m\times n}\mathbf{Y}_{n\times n}^{-1}\mathbf{Y}_{n\times m}\right)\mathbf{E}'_{m\times 1} \\ &\overset{\text{def}}{=} \mathbf{Y'_{bus}}\mathbf{E}'_{m\times 1}\end{aligned} \tag{1.25}$$

The ith row, jth column element of $\mathbf{Y'_{bus}}$ is denoted by $Y_{ij} = B_{ij} + jG_{ij}$. Electrical power P_{ei} satisfies (1.26), similar to (1.22).

$$P_{ei} = E'^2_i G_{ii} + \sum_{\substack{j=1 \\ j\neq i}}^{m} E'_i E'_j (B_{ij} \sin \delta_{ij} + G_{ij} \cos \delta_{ij}) \tag{1.26}$$

The swing equations are merged with (1.26), resulting in each generator being described by only two ODEs, with the network equations embedded.

$$\dot{\delta}_i = \Delta\omega_i \tag{1.27a}$$

$$\frac{2H_i}{\omega_0}\Delta\dot{\omega}_i = P_{mi} - D_i\frac{\Delta\omega_i}{\omega_0} - E'^2_i G_{ii} - \sum_{\substack{j=1 \\ j\neq i}}^{n} E'_i E'_j (B_{ij} \sin \delta_{ij} + G_{ij} \cos \delta_{ij}) \tag{1.27b}$$

As a result, the entire system is modeled by a simplified, purely ODE model, and its initial value problem can be solved through numerical integration for power system simulation.

1.3 Numerical Simulation

To simulate a power system accurately, it is essential to integrate the nonlinear ODEs of its model at small enough time steps for the desired period of simulation. This section provides a brief overview of the methods used to solve a power system model described by a set of ODEs like (1.27) or by a set of DAEs like (1.4).

Consider a nonlinear dynamical system, such as a classical power system model, represented by a set of nonlinear ODEs, as shown in:

$$\dot{\mathbf{x}} = \mathbf{f}(\mathbf{x}, t) \quad \mathbf{x} \in \mathbb{R}^n, t \in \mathbb{R} \tag{1.28}$$

To simulate the system's response to a disturbance, simulation needs to solve an initial value problem about (1.28), starting at time $t = t_0$ from an initial state $\mathbf{x} = \mathbf{x}_0$ deviated from its equilibrium. The goal is to calculate values of $\mathbf{x}$ at a series of time instants $t_k = t_0 + kh$ $(k = 1, 2, \dots)$ for the desired simulation period, where h is the size of the time step.

In practice, a contingency to be simulated involves a fault, switch, or perturbation that occurs in the system, leading to changes in the topology and parameters of its model. By considering, for instance, the occurrence, duration, and clearance of a fault, the entire simulation period is often partitioned into smaller subperiods, in which different initial value problems are solved using different models and initial states.

For example, a short-circuit fault can occur at $t = t_1 > t_0$ and then is cleared at $t_2 > t_1$ by itself as a temporary fault or by a protective action as a permanent fault. The simulation period, denoted by $[0, T]$, is partitioned into these three subperiods: $[t_0, t_1)$ for the prefault period, $[t_1, t_2)$ for the fault-on period, and $[t_2, T]$ for the postfault period.

In general, the system models are different in these subperiods, and the simulation on each subperiod is performed by solving the initial value problem with the initial state equal to the ending state of the previous subperiod. More specifically, at time t_1, the short-circuit fault adds a small impedance between the fault location and the ground, resulting in a change in the admittance matrix $\mathbf{Y}_{\text{bus}}$. Then, at time t_2, the protective system may trip one or multiple lines connected to the fault location, leading to a further change in $\mathbf{Y}_{\text{bus}}$. It should be noted that if the fault is temporary, $\mathbf{Y}_{\text{bus}}$ may be the same for both pre- and postfault periods. Simulations over these subperiods are performed sequentially until the entire simulation period is completed.

To solve each initial value problem, two types of numerical integration methods are used: explicit and implicit methods. Explicit methods, such as the *Forward Euler (FE) method* and *Runge–Kutta (RK) method*, use only values from past time steps in the computation at each time step, resulting in explicit formulas. In contrast, implicit methods, such as the *Backward Euler (BE) method* and *Trapezoidal-Rule (TRAP) method*, use values from future time steps as well, resulting in implicit equations that require iterative computations to solve. An explicit formula needs to be calculated only once, while an implicit equation needs to be solved by iterative computations.

1.3.1 Explicit Integration Methods

In an explicit integration method, t_k represents the current time, while t_{k+1} represents the next time instant at a time step of h (i.e. $t_{k+1} = t_k + h$). The method aims to calculate the state value $\mathbf{x}_{k+1} = \mathbf{x}(t_{k+1})$ at the future time t_{k+1} using an explicit formula based on t_k, h, and the already known state value $\mathbf{x}_k = \mathbf{x}(t_k)$ at time t_k.

1.3.1.1 Forward Euler Method

The FE method, a first-order method, calculates $\mathbf{x}_{k+1}$ from $\mathbf{x}_k$ by

$$\begin{aligned}\mathbf{x}_{k+1} &= \mathbf{x}_k + h\left.\frac{d\mathbf{x}}{dt}\right|_{t=t_k} \\ &= \mathbf{x}_k + h\mathbf{f}(\mathbf{x}_k, t_k)\end{aligned} \tag{1.29}$$

This method is based on the first two terms of the Taylor series of $\mathbf{x}(t)$ around $t = t_k$, and only the first-order derivative of $\mathbf{x}$ at t_k, $\mathbf{f}(\mathbf{x}_k, t_k)$, is used in approximation. Therefore, it is a first-order method with an error in the order of h^2.

1.3.1.2 Modified Euler Method

The Modified Euler (ME) method, a second-order method, estimates $\mathbf{x}_{k+1}$ at t_{k+1} using two steps in (1.30):

1. Predictor Step (1.30a): Use the FE method to calculate $\mathbf{x}^p{}_{k+1}$.
2. Corrector Step (1.30b): Apply the average of the calculated derivative at t_k and the predicted derivative at t_{k+1}.

$$\begin{aligned}k_1 &= \mathbf{f}(\mathbf{x}_k, t_k) \\ \mathbf{x}^p_{k+1} &= \mathbf{x}_k + hk_1\end{aligned} \tag{1.30a}$$

$$\begin{aligned}k_2 &= \mathbf{f}\left(\mathbf{x}^p_{k+1}, t_{k+1}\right) = \mathbf{f}(\mathbf{x}_k + hk_1, t_k + h) \\ \mathbf{x}_{k+1} &= \mathbf{x}_k + h\frac{k_1 + k_2}{2}\end{aligned} \tag{1.30b}$$

It can be shown from (1.30) that the ME method is a second-order method that applies the first three terms of Taylor's series of $\mathbf{x}(t)$ around $t = t_k$, with an error in the order of h^3 (see the following equation).

$$\begin{aligned}\mathbf{x}_{k+1} &= \mathbf{x}_k + h\frac{\mathbf{f}(\mathbf{x}_k, t_k) + \mathbf{f}\left(\mathbf{x}^p_{k+1}, t_{k+1}\right)}{2} \\ &= \mathbf{x}_k + \frac{h}{2}\mathbf{f}(\mathbf{x}_k, t_k) + \frac{h}{2}\mathbf{f}(\mathbf{x}_k + h\mathbf{f}(\mathbf{x}_k, t_k), t_k + h) \\ &= \mathbf{x}_k + \frac{h}{2}\mathbf{f}(\mathbf{x}_k, t_k) \\ &\quad + \frac{h}{2}\left[\mathbf{f}(\mathbf{x}_k, t_k) + h\left.\frac{\partial \mathbf{f}}{\partial t}\right|_{(\mathbf{x}_k, t_k)} + h\mathbf{f}\left(\mathbf{x}_k, t_k\right)\left.\frac{\partial \mathbf{f}}{\partial \mathbf{x}}\right|_{(\mathbf{x}_k, t_k)} + O(h^2)\right]\end{aligned}$$

$$= \mathbf{x}_k + h\mathbf{f}(\mathbf{x}_k, t_k) + \frac{h^2}{2}\left[\frac{\partial \mathbf{f}}{\partial t} + \frac{\partial \mathbf{f}}{\partial \mathbf{x}}\mathbf{f}\right]\Bigg|_{(\mathbf{x}_k, t_k)} + O(h^3)$$

$$= \mathbf{x}_k + h\left.\frac{d\mathbf{x}}{dt}\right|_{t=t_k} + \frac{1}{2!}h^2\left.\frac{d^2\mathbf{x}}{dt^2}\right|_{t=t_k} + O(h^3) \tag{1.31}$$

1.3.1.3 Runge–Kutta Methods

Formula (1.30) used in the ME method to calculate $\mathbf{x}_{k+1}$ is a special case of the second-order RK (RK-2) method, which is given by (1.32).

$$\begin{aligned} k_1 &= \mathbf{f}(\mathbf{x}_k, t_k) \\ k_2 &= \mathbf{f}(\mathbf{x}_k + \alpha h k_1, t_k + \alpha h) \\ \mathbf{x}_{k+1} &= \mathbf{x}_k + h\left(1 - \frac{1}{2\alpha}\right)k_1 + h\frac{1}{2\alpha}k_2 \end{aligned} \tag{1.32}$$

For a second-order method, α cannot be equal to 0.5, or else Eq. (1.32) will degrade to the first-order FE method. When $\alpha = 1$, Eq. (1.32) is the same as Eq. (1.30) used in the ME method. The RK-2 method, as compared to the FE method, generalizes the ME method by allowing for adjustment in the weight allocation between the gradient at $(\mathbf{x}_k, t_k)$ and the predicted gradient at $(\mathbf{x}_{k+1}, t_{k+1})$. The method is still based on a second-order approximation by Taylor's series unless the coefficients are adjusted by α. As a result, the error remains in order h^3.

To achieve a higher order approximation based on the Taylor's series, we can take more prediction steps using a higher order RK method. The widely used standard fourth-order RK (RK-4) method, shown in Eq. (1.33), has an error in the order h^5. While other alternative RK-4 methods exist for improved accuracy, their errors are still of the same order. Moreover, high-order RK methods are less common in applications since they may face increased computational burden and challenges in maintaining numerical stability.

$$\begin{aligned} k_1 &= \mathbf{f}(\mathbf{x}_k, t_k) \\ k_2 &= \mathbf{f}\left(\mathbf{x}_k + \frac{1}{2}hk_1, t_k + \frac{1}{2}h\right) \\ k_3 &= \mathbf{f}\left(\mathbf{x}_k + \frac{1}{2}hk_2, t_k + \frac{1}{2}h\right) \\ k_4 &= \mathbf{f}(\mathbf{x}_k + hk_3, t_k + h) \\ \mathbf{x}_{k+1} &= \mathbf{x}_k + h\left(\frac{k_1}{6} + \frac{k_2}{3} + \frac{k_3}{3} + \frac{k_4}{6}\right) \end{aligned} \tag{1.33}$$

In Figure 1.9, four methods are compared for simulation of a one-dimensional ODE $\dot{x} = f(x, t)$ $(x, t \in \mathbb{R})$ in a single time step, as shown by four straight dash lines from (x_k, t_k) to $(\tilde{x}_{k+1}, t_{k+1})$, i.e. the predicted (x_{k+1}, t_{k+1}).

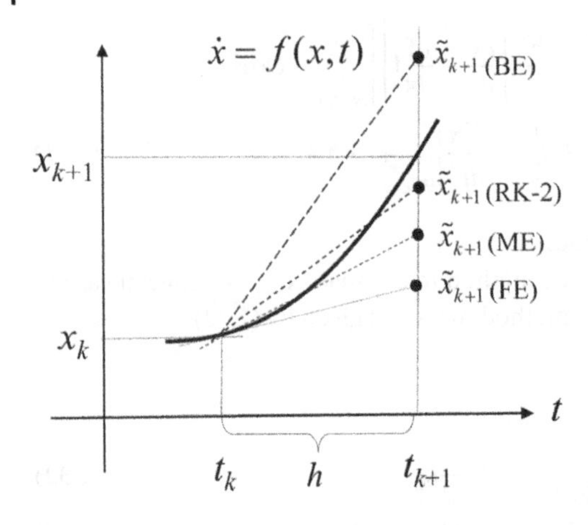

Figure 1.9 Comparison of the forward Euler method, modified Euler method, second-order Runge–Kutta method, and backward Euler method for one time step of integration.

These methods include the FE method, ME method, RK-2 method, and BE method, which is an implicit method and will be introduced later in this section. Their differences are the following:

1. The FE method predicts (x_{k+1}, t_{k+1}) using only the gradient calculated at the (x_k, t_k), resulting in a straight line that is tangent to the curve at (x_k, t_k), the starting point of the interval.
2. The ME method uses the average value of the gradient at (x_k, t_k) along with the predicted gradient at (x_{k+1}, t_{k+1}) to estimate (x_{k+1}, t_{k+1}), which is reflected by the slope of its straight line.
3. The RK-2 method generalizes the ME method and offers a more flexible approach by using a weighted average of the gradient at (x_k, t_k) and the predicted gradient at (x_{k+1}, t_{k+1}), with a user-defined weight allocation parameter α. Therefore, the slope of its straight line can vary with α. If α is appropriately chosen, the RK-2 method may achieve better accuracy than the ME method.
4. The BE method solves for the gradient at (x_{k+1}, t_{k+1}) together with x_{k+1}, which is reflected by the slope of its straight line being equal to that at the end point of the interval.

1.3.2 Implicit Integration Methods

1.3.2.1 Stiffness of ODEs

The stiffness of a system is a significant concern when using explicit methods for its simulations, as they have weak numerical stability when the set of ODEs modeling the system are highly stiff. Stiffness is measured by the ratio of the largest to smallest time constants, or more accurately by $|\lambda_{max}/\lambda_{min}|$, which is the absolute ratio of the largest eigenvalue to the smallest eigenvalue of the linearized model of the system. The stiffness increases as the ODEs characterize more detailed dynamics with smaller time constants.

For stiff systems, such as power systems, numerical integration with an explicit method can be time-consuming. Capturing the fastest mode as characterized by λ_{min} requires using a small time step in the order of less than $1/\lambda_{max}$, while assessing the system's stability regarding the slowest mode needs to run simulation for a number of its periods in the order of $1/\lambda_{min}$. During a simulation run, even after the fast modes characterized by large eigenvalues die out because of damping of the modes, the use of a small time step often needs to continue to maintain numerical stability. Thus, the total number of time steps is no fewer than the value of stiffness and, in fact, one or more orders higher. For instance, if the fastest and slowest modes of a power system model, respectively, have periods equal to 1 millisecond and 10 seconds, its stiffness is 10 000. To enable a high enough resolution of the waveforms in the fastest mode and ensure capturing a complete picture of the stability and damping process in the slowest mode, the simulation may need to integrate at a tiny step of 0.1 ms for, for example, 50 seconds, resulting in 500 000 steps of integration.

Let us use the FE method as an example to illustrate the numerical instability an explicit method can encounter. Suppose that during a simulation period, the dynamic of a power system is dominated by its largest eigenvalue, λ_{max}, associated with one state variable, x:

$$\dot{x} \approx \lambda_{max} x \tag{1.34}$$

To calculate x at t_k, there is

$$\begin{aligned} x_k &\approx x_{k-1} + h\lambda_{max} x_{k-1} \\ &= x_{k-1}(1 + h\lambda_{max}) \\ &\approx x_0(1 + h\lambda_{max})^k \end{aligned} \tag{1.35}$$

For numerical stability of variable x, there must be $|1 + h\lambda_{max}| < 1$, which means $h\lambda_{max}$ as a complex number staying inside a unit circle centered at -1. Thus, two necessary conditions are:

- λ_{max} has a negative real part, or in other words, a stable eigenvalue;
- the step size has to be small enough to satisfy.

$$h < \frac{2}{|\lambda_{max}|} \tag{1.36}$$

Other explicit methods encounter a similar numerical stability issue when using a step size that is not small enough for simulating the fastest mode of the system.

1.3.2.2 Backward Euler Method

The BE method is a simple implicit method that uses the following equation at each time step:

$$\begin{aligned}\mathbf{x}_{k+1} &= \mathbf{x}_k + h\left.\frac{d\mathbf{x}}{dt}\right|_{t=t_{k+1}} \\ &= \mathbf{x}_k + h\mathbf{f}(\mathbf{x}_{k+1}, t_k + h)\end{aligned} \tag{1.37}$$

Equation (1.37) is similar to Eq. (1.38) in form, but it uses the accurate gradient at $(\mathbf{x}_{k+1}, t_{k+1})$, which is unknown at time $t = t_k$. Thus, the method needs to solve a nonlinear algebraic equation (1.38) for $\mathbf{x}_{k+1}$ at each time step. Note that we cannot necessarily have an explicit formula for calculating $\mathbf{x}_{k+1}$ from $\mathbf{x}_k$ and t_k due to the nonlinear function $\mathbf{f}$.

$$\mathbf{x}_{k+1} - \mathbf{x}_k - h\mathbf{f}(\mathbf{x}_{k+1}, t_k + h) = 0 \tag{1.38}$$

The Gauss–Seidel method and the Newton–Raphson method can solve such an equation, but numerical iterations at each time step are usually needed, which increases the computation time compared to the FE method. However, the BE method avoids the numerical instability associated with the FE method. For the same system as (1.34), we have:

$$\begin{aligned}x_k &\approx x_{k-1} + h\lambda_{\max} x_k \quad \Leftrightarrow \\ x_k &\approx x_{k-1}\frac{1}{1 - h\lambda_{\max}} \\ &\approx x_0\left(\frac{1}{1 - h\lambda_{\max}}\right)^k\end{aligned} \tag{1.39}$$

For numerical stability, $|1 - h\lambda_{\max}| > 1$ is required, which means $h\lambda_{\max}$ as a complex number must stay outside of a unit circle centered at 1. To satisfy this condition, $h\lambda_{\max}$ must have a negative real part, or in other words, it must lie on the left-hand side of the complex plane. The BE method is said to have A-Stability [6], which the FE method does not have. Theoretically speaking, for a numerically stable simulation on x, the time step h can be an arbitrarily large positive number as long as $\lambda_{\max}$ itself is a stable eigenvalue with a negative real part.

1.3.2.3 Trapezoidal-Rule Method

The TRAP method is a widely used implicit method that conducts linear interpolation over each time interval between the starting point $(\mathbf{x}_k, t_k)$ and

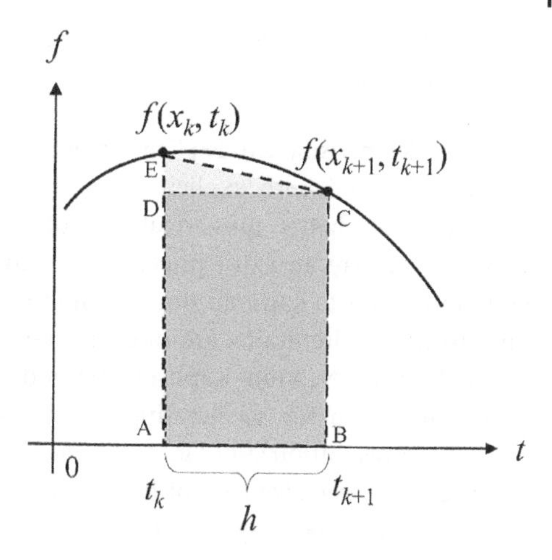

Figure 1.10 Illustration of the backward Euler and Trapezoidal-rule methods.

the ending point $(\mathbf{x}_{k+1}, t_{k+1})$ on the integrated expression **f**, as shown in (1.40). By contrast, both the FE method with Eq. (1.29) and the BE method with Eq. (1.37) use only the value of **f** at $(\mathbf{x}_k, t_k)$ or $(\mathbf{x}_{k+1}, t_{k+1})$.

$$\begin{aligned}\mathbf{x}_{k+1} &= \mathbf{x}_k + h\frac{\mathbf{f}(\mathbf{x}_k, t_k) + \mathbf{f}(\mathbf{x}_{k+1}, t_k + h)}{2} \\ &\approx \mathbf{x}_k + \int_{t_k}^{t_{k+1}} \mathbf{f}(\mathbf{x}_k, t_k)dt\end{aligned} \tag{1.40}$$

The TRAP method gets its name from the use of a trapezoid (as illustrated by ABCE in Figure 1.10 for a one-dimensional f) to approximate the integrated area under **f** for an interval between t_k and t_{k+1}. In comparison, the BE method uses a rectangle (i.e. ABCD in Figure 1.10) for this approximation. While the formula (1.40) for the TRAP method is similar to Eq. (1.31) for the ME method, it requires an accurate value of **f** for the future point $(\mathbf{x}_{k+1}, t_{k+1})$, which is unknown yet at $t = t_k$.

Like the BE method, the TRAP method needs to solve a nonlinear algebraic equation at each time step for $\mathbf{x}_{k+1}$ using numerical iterations with either the Gauss–Seidel method or the Newton–Raphson method. The equation for the TRAP method is (1.40), or equivalently (1.41):

$$\mathbf{x}_{k+1} - \mathbf{x}_k - h\frac{\mathbf{f}(\mathbf{x}_k, t_k) + \mathbf{f}(\mathbf{x}_{k+1}, t_k + h)}{2} = 0 \tag{1.41}$$

The stiffness of the system model being simulated using the TRAP method affects accuracy but not numerical stability of simulation. With larger time steps, fast transient dynamics or modes with frequencies higher than $\mathbf{f}(\mathbf{x}_k, t_k)$ and $\mathbf{f}(\mathbf{x}_{k+1}, t_{k+1})$ are filtered out since formula (1.40) assumes that the state of the system varies at a rate equal to their average, not faster.

However, slower modes than $\mathbf{f}(\mathbf{x}_k, t_k)$ or $\mathbf{f}(\mathbf{x}_{k+1}, t_{k+1})$ can still be captured by simulation results.

1.3.2.4 Comparison with Explicit Methods

When compared with explicit integration methods, implicit integration methods have better numerical stability when simulating stiff systems. However, to pay an extra price, they need to call a numerical algebraic equation solver to solve an implicit, nonlinear equation at each time step, and numerical iterations are often necessary for either the Gauss–Seidel method or the Newton–Raphson method. Implicit methods can adopt a larger time step for accelerating simulation than explicit methods, but when high-resolution simulation is required for fast modes to be accurately captured, the step size cannot be too large. Thus, the overall time cost by an implicit method may not necessarily be lower than that by an explicit method.

1.3.3 Solving Differential-Algebraic Equations

Simulations using the classical power system model (1.27) can use either an explicit method or an implicit method since the model has only ODEs. However, for simulations using a more accurate DAE model in the form of (1.4), particularly those that cannot be converted to a pure ODE model, ODEs need to be solved by an integration method, while algebraic equations are solved by a linear or nonlinear algebraic equation solver. The Gauss–Seidel method and the Newton–Raphson method are two widely used nonlinear algebraic equation solvers that conduct numerical iterations, with each iteration often requiring the solution of a linear equation, thereby necessitating an efficient linear solver.

To solve the ODEs and algebraic equations together for the solution of the DAE model, there are two approaches: the partitioned solution approach and the simultaneous solution approach.

1.3.3.1 Partitioned Solution Approach

The partitioned solution approach iteratively solves the ODEs using an explicit or implicit integration method and the algebraic equations using a numerical algebraic equation solver. For example, to solve an initial value problem for one time step from $(\mathbf{x}_k, t_k)$ to $(\mathbf{x}_{k+1}, t_{k+1})$ for (1.3), the algorithm alternatively assumes that all buses have constant voltages (namely, voltage sources) when solving ODEs for current injections and then assumes all constant current injections (namely, current sources) when solving the network equations for bus voltages. The algorithm performs the following steps:

Step 1: Assume $\mathbf{V}_{\mathbf{bus},k}$, i.e. the bus voltages $\mathbf{V}_{\mathbf{bus}}$ at $t = t_k$, with all dynamic devices (e.g. generators and motor loads) to be constant during the interval of $[t_k, t_{k+1}]$ so that their ODEs are decoupled from the network and each other;

Step 2: Integrate $\dot{\mathbf{x}} = \mathbf{f}(\mathbf{x}, \mathbf{V}_{\mathbf{bus},k})$ for $[t_k, t_{k+1}]$ to calculate $\mathbf{x}_{k+1}$ by using the integration method that can be conducted independently on each dynamic device.

Step 3: Update current injections by $\mathbf{I}_{\mathbf{bus},k+1}$ based on $\mathbf{V}_{\mathbf{bus},k}$ and the calculated $\mathbf{x}_{k+1}$ using the interfacing algebraic equations with dynamic devices, e.g. the algebraic equations in (1.5), and (1.6).

Step 4: With the updated $\mathbf{I}_{\mathbf{bus},k+1}$, solve algebraic equation (1.42) for $\mathbf{V}_{\mathrm{bus},k+1}$ with a linear solver. If iterations at steps 1–4 for one time interval have reached a threshold, let $k = k+1$ move forward by one time step. Go to step 1

$$\mathbf{G}(\mathbf{x}_{k+1}, \mathbf{V}_{\mathbf{bus},k+1}) \stackrel{\text{def}}{=} \mathbf{I}_{\mathbf{bus},k+1} - \mathbf{Y}_{\mathbf{bus}}\mathbf{V}_{\mathbf{bus},k+1} = 0 \tag{1.42}$$

The iterations at steps 1–4 improve the accuracy of calculated power flows by a linear solver in step 4. Power flows must satisfy nonlinear AC power-flow equations in (1.21). Buses are often characterized into different types, such as PV buses, PQ buses, and slack buses. Accordingly, the solution process can make use of the knowledge on which quantities tend to stay constant or change less than the others at a bus. Thus, to reduce iterations, an alternative method is to solve nonlinear equations (1.21) instead of (1.42) at step 4, making use of the types of buses. This requires a nonlinear algebraic equation solver such as the Gauss–Seidel method and the Newton–Raphson method.

1.3.3.2 Simultaneous Solution Approach

The simultaneous solution approach solves the entire DAE model by using an implicit integration method that converts all equations to algebraic equations and then solves them with an algebraic equation solver. An algorithm is illustrated as follows:

Step 1: Approximate the DAE model (1.4) by discretized algebraic equations (1.43) based on the trapezoidal rule on values at two consecutive steps:

$$\begin{aligned}\mathbf{F}(\mathbf{x}_{k+1}, \mathbf{V}_{\mathbf{bus},k+1}) &\stackrel{\text{def}}{=} \mathbf{x}_{k+1} - \mathbf{x}_k \\ &\quad - h\frac{\mathbf{f}(\mathbf{x}_k, \mathbf{V}_{\mathbf{bus}}) + \mathbf{f}(\mathbf{x}_{k+1}, \mathbf{V}_{\mathbf{bus},k+1})}{2} = 0 \\ \mathbf{G}(\mathbf{x}_{k+1}, \mathbf{V}_{\mathbf{bus},k+1}) &\stackrel{\text{def}}{=} \mathbf{I}_{\mathbf{bus}}(\mathbf{x}_{k+1}, \mathbf{V}_{\mathbf{bus},k+1}) - \mathbf{Y}_{\mathbf{bus}}\mathbf{V}_{\mathbf{bus},k+1} = 0\end{aligned} \tag{1.43}$$

Step 2: Solve (1.43) using a nonlinear solver.

1.4 Semi-Analytical Simulation

1.4.1 Drawbacks with Numerical Simulations

Exact, closed-form solutions are not generally available for nonlinear ODEs. Numerical methods are typically used to obtain their solutions. Two commonly used methods are explicit and implicit integration methods. An explicit integration method extrapolates the solution from past time instants, requiring small time steps, especially when multiple past time instants are used (a Kth-order explicit method) to ensure numerical stability and the accuracy of extrapolation. An implicit integration method interpolates the solution using both known and unknown values from past and future time instants, avoiding numerical instability but requiring iterative computations to solve a nonlinear algebraic equation for each time step.

When simulating a large-scale power system, a numerical solution to its model as a set of nonlinear ODEs or DAEs can be computationally burdensome, regardless of whether an explicit or implicit integration method is used. As a result, highly accurate numerical simulations can be slow due to the following reasons.

Implicit integration methods, which are implemented in a simultaneous or partitioned solution approach, can convert the overall ODEs or DAEs to discretized, nonlinear algebraic equations for each time step. To solve these algebraic equations, iterative computations are required, using either a linear solver or a nonlinear solver. While implicit integration can be performed at larger time steps than explicit integration and is numerically more stable, the benefits of an implicit method diminish when highly accurate simulation results are desired at small time steps. Additionally, nonlinear solvers such as the Newton–Raphson method often require Jacobians, which can be challenging to compute, especially when mathematical models of a power system contain nondifferentiable or black-box components, as is often the case for the models of many generators, controllers, and loads.

To accommodate a variety of equipment models, most commercialized power system simulators use a partitioned solution approach with an explicit integration method as the default option. However, this approach has a limitation when simulating large system models because the time step size must be small enough to avoid numerical instability. Although increasing the order of the explicit method can enlarge the step size, it can be challenging in practical applications. For example, while a first-order FE method or a second-order ME method can be extended to a fourth- or fifth-order RK method, extending the order further may not be easy due to some reasons. Firstly, new numerical issues can emerge with a high-order method using poorly designed k-values, such as k_1 to k_4 in (1.33), for the simulated system. Secondly, the computation burden at each time step will increase and offset the savings due to longer and fewer steps. Thirdly, these integration methods are generic methods for integrating a variety of types of nonlinear ODEs, and the differentials of different orders related to the k-values are estimated by locally interpolating the computed trajectory of the system without utilizing much knowledge inherent with the mathematical model to predict its trend. Eventually, these explicit integration methods will still be based on or equivalent to a lower order Taylor's series approximation.

Therefore, a natural question arises: *can we significantly increase the time step size by using a high-order approximant of $\mathbf{x}(t)$ derived directly from the mathematic model of the system?* This would require designing a custom integration method that is specific to the model at hand. While such a method may not be applicable to other methods, it can yield faster simulation times, which can be highly desirable in practical applications. Of course, there are tradeoffs to be made: designing a custom method requires additional effort, and the method may not be as generalizable as a generic integration method. Nonetheless, when speed is of the essence, one may be willing to accept these tradeoffs.

1.4.2 Emerging Methods for Semi-Analytical Power System Simulation

In recent years, several novel approaches have been developed and tested to expedite simulations of bulk power systems. Unlike traditional numerical simulation methods, these new methods are based on high-order approximants known as "semi-analytical solutions" for power systems modeled by ODEs or DAEs. A semi-analytical solution is characterized as "analytical" since it has an explicit, analytical expression in a series or recursive form that can adjust its order flexibly for increased accuracy or decreased complexity. On the other hand, it is also referred to as "semi-" because it represents a truncation of an infinite, analytical form that converges numerically to the actual solution. The degree of truncation, which is the order, determines the interval length of the solution's accuracy, based on a specified error tolerance. When using a high order, the time interval of accuracy is also extended compared to the step size of low-order integration methods. By consecutively applying the semi-analytical solution over extended time intervals, computations over the entire simulation period are completed.

Several semi-analytical simulation methods for power systems have been studied. One approach, proposed by Duan and Sun [7] and Duan and

Sun [8], uses a multistage Adomian Decomposition Method to decompose the ODE model of a power system into Adomian polynomials, which are used to find approximants on responses of the simulated system having all impedance loads. ([9, 10] studied the use of the Adomian Decomposition Method for simulating the DAE model of a power system by interfacing the semi-analytical solutions of ODEs on generators and controllers with the numerical solutions on the power network. Dinesha and Gurrala [11] obtained semi-analytical solutions using a multistage Homotopy Analysis Method, which involves an additional auxiliary parameter to adjust convergence region of the semi-analytical solution for better accuracy of simulation.

Abreut et al. [12] and Wang et al. [13] obtained power series-form semi-analytical solutions directly from the Taylor's series expansion of a true solution for a power system modeled by ODEs or DAEs. Liu et al. [14] and Liu et al. [15] utilized analytic continuation techniques to power series-form semi-analytical solutions, resulting in Padé approximants and continued fraction approximants that improve simulation accuracy and speed.

The Differential Transformation Method, Liu et al. [16], Liu and Sun [17], and Liu et al. [18] proposed systematic approaches for deriving high-order power series-form semi-analytical solutions for both the ODE model and the DAE model of a large-scale power system. They discovered that finding power series coefficients with a semi-analytical solution only requires solving formally linear equations recursively, from low-order to high-order terms.

Moreover, Holomorphic Embedding methods have been applied by Yao et al. [19], Yao et al. [20], and Yao et al. [21] to semi-analytical power system simulations, which consider QSS dynamics and electromechanical dynamics. Semi-analytical solutions, such as power series approximants or Padé approximants, were derived in the complex domain about an embedded variable. The parallelization of simulations using these semi-analytical solutions was also explored.

In the realm of parallel-in-time power system simulation, Park et al. [22] utilized semi-analytical solutions acquired through the Adomian Decomposition Method and Homotopy Analysis Method as coarse solutions within a Parareal framework. These semi-analytical solutions help connect many fine solutions obtained in parallel at successive time intervals. Further, Liu et al. [23] proposed an adaptive Parareal algorithm, which employs variable-order semi-analytical solutions derived from the Differential Transformation Method. This method enables simulations with variable time intervals.

1.4.3 Approaches to Semi-Analytical Solutions

Consider an n-dimensional nonlinear autonomous system in (1.44), where $\mathbf{f}(\mathbf{x}) = [f_1(\mathbf{x}), f_2(\mathbf{x}), \ldots, f_n(\mathbf{x})]^T$ is an analytical vector field.

$$\dot{\mathbf{x}} = \mathbf{f}(\mathbf{x}), \qquad t_0 = 0, \ \ \mathbf{x}(0) = \mathbf{a}_0 \tag{1.44}$$

A semi-analytical solution to its initial value problem can be derived through two mainstream approaches: the *analytical expansion* approach and *analytical homotopy* approach. Each approach involves specific ideas and steps but can lead to the same semi-analytical solution. Many existing semi-analytical methods are equivalent to one of these two approaches.

1.4.3.1 Analytical Expansion Approach

The *analytical expansion* approach selects either an infinite series form or an infinite nonseries recursive form that converges to $\mathbf{x}$ as the number of series terms or its order approaches infinity. The former involves summing an infinite number of terms by $\sum_{k=0}^{\infty} \mathbf{x}_k(t)$, while the latter involves relaxing the form to other nonseries mathematical structures such as fractions and growing its complexity by recursion on a particular operation.

In this section, we will use the series form as an example to explain the *analytical expansion* approach. A semi-analytical solution depends on time t, the initial state vector $\mathbf{x}(0)$, a vector $\boldsymbol{\mu}$ consisting of selected symbolized parameters, and an integer K that determines the order or the number of terms in the approximant. Thus, a semi-analytical solution is a function of t, $\mathbf{x}(0)$, and $\boldsymbol{\mu}$, represented by $\mathbf{x}_{SAS}^{K}(t; \mathbf{x}(0), \boldsymbol{\mu})$, or simply, $\mathbf{x}_{SAS}^{K}(t)$ in the rest of the section.

Equation (1.45) gives a series-form semi-analytical solution summing K terms of different orders:

$$\mathbf{x}_{SAS}^{K}(t) \stackrel{\text{def}}{=} \mathbf{x}_0(t) + \mathbf{x}_1(t) + \mathbf{x}_2(t) + \cdots + \mathbf{x}_K(t) \tag{1.45}$$

For example, if the power series form is adopted to be equal to Taylor's series of the exact solution expanded at $t = t_0$, the elements of each $\mathbf{x}_k$ are only k-degree monomials of time t:

$$\mathbf{x}(t) = \mathbf{a}_0 + \mathbf{a}_1 t + \mathbf{a}_2 t^2 + \ldots = \sum_{k=0}^{\infty} \mathbf{a}_k t^k \tag{1.46}$$

Elements of vector $\mathbf{a}_k$ on coefficients of the kth order term can be either constants according to specified values of $\mathbf{x}(0)$ and $\boldsymbol{\mu}$ for one operating condition or symbolic expressions with $\mathbf{x}(0)$ and $\boldsymbol{\mu}$ embedded to accommodate a variety of conditions.

Next, we calculate the derivative of $\mathbf{x}$ with respect to t by differentiating each term:

$$\frac{d\mathbf{x}(t)}{dt} = 0 + \mathbf{a}_1 + 2\mathbf{a}_2 t + 3\mathbf{a}_3 t^2 + \ldots = \sum_{k=0}^{\infty} (k+1)\mathbf{a}_{k+1} t^k \tag{1.47}$$

which approximates the left-hand side of the ODE. For the right-hand side, we expand $\mathbf{f}(\mathbf{x}(t))$ at $t = 0$ into its Taylor's series:

$$\begin{aligned}
\mathbf{f}(\mathbf{x}(t)) &= \mathbf{f}\left(\sum_{k=0}^{\infty} \mathbf{a}_k t^k\right) = \sum_{k=0}^{\infty} \mathbf{b}_k t^k \\
\text{where } \mathbf{b}_0 &= \mathbf{f}(\mathbf{a}_0), \\
\mathbf{b}_1 &= \frac{1}{1!}\mathbf{J}_\mathbf{f}(\mathbf{a}_0)\mathbf{a}_1 \\
\mathbf{b}_2 &= \frac{1}{2!}\left(\begin{bmatrix} \vdots \\ \mathbf{a}_1^T \mathbf{H}_{f_i}(\mathbf{a}_0)\mathbf{a}_1 \\ \vdots \end{bmatrix} + 2\mathbf{J}_\mathbf{f}(\mathbf{a}_0)\mathbf{a}_2\right) \\
&\cdots
\end{aligned} \tag{1.48}$$

where coefficient $\mathbf{b}_k$ depends on $\mathbf{a}_0, \ldots, \mathbf{a}_k$, and first three coefficients, $\mathbf{b}_0$, $\mathbf{b}_1$, and $\mathbf{b}_2$, are given. Here, $\mathbf{J}_\mathbf{f}(\mathbf{a}_0)$ is the Jacobian matrix of $\mathbf{f}$ at $\mathbf{x} = \mathbf{a}_0$, which consists of first-order derivatives with respect to $\mathbf{x}$. $\mathbf{H}_{fi}(\mathbf{a}_0)$ is the Hessian matrix of f_i, consisting of second-order derivatives with respect to $\mathbf{x}$. More details on the expanded terms can be found in Chapter 2.

Specifically, if $n = 1$, (1.48) becomes:

$$\begin{aligned}
f(x(t)) &= f(a_0 + a_1 t + a_2 t^2 + \cdots) \\
&= f(a_0) + D_x f(a_0) \cdot a_1 t + \frac{1}{2!}\left(D_x^2 f(a_0) \cdot a_1^2 + 2D_x f(a_0) \cdot a_2\right) t^2 + \cdots
\end{aligned} \tag{1.49}$$

We can then equate (1.47) and (1.48) and design a recursive formula to calculate $\mathbf{a}_k$ from $k = 0$ to any desired order K:

$$\begin{aligned}
0 + \mathbf{a}_1 + 2\mathbf{a}_2 t + 3\mathbf{a}_3 t^2 + \ldots &= \mathbf{b}_0 + \mathbf{b}_1 t + \mathbf{b}_2 t^2 + \ldots \\
\mathbf{a}_1 &= \mathbf{b}_0(\mathbf{a}_0) \\
\mathbf{a}_2 &= \tfrac{1}{2}\mathbf{b}_1(\mathbf{a}_0, \mathbf{a}_1) \\
\Rightarrow \mathbf{a}_3 &= \tfrac{1}{3}\mathbf{b}_2(\mathbf{a}_0, \mathbf{a}_1, \mathbf{a}_2) \\
&\cdots \\
\mathbf{a}_K &= \tfrac{1}{K}\mathbf{b}_{K-1}(\mathbf{a}_0, \mathbf{a}_1, \ldots, \mathbf{a}_{K-1})
\end{aligned} \tag{1.50}$$

Thus, a Kth order semi-analytical solution is given by

$$\mathbf{x}_{SAS}^K(t) \stackrel{\text{def}}{=} \mathbf{a}_0 + \mathbf{a}_1 t + \mathbf{a}_2 t^2 + \ldots + \mathbf{a}_K t^K \tag{1.51}$$

To simulate the system for a variety of operating conditions, such a semi-analytical solution is derived ahead of time before simulation, in which $\mathbf{a}_0$ to $\mathbf{a}_k$ are symbolized as vector-valued functions of variables or parameters that define the operating condition. When simulation is needed, the semi-analytical solution $\mathbf{x}_{SAS}^{K}(t)$ is calculated at sequential time intervals until the simulation period is completed. At each interval, $\mathbf{a}_0$, which is $\mathbf{x}_{SAS}(0)$, is equal to either the initial state of simulation for the first time interval or to the state vector at the end of the previous time interval for all subsequent intervals.

Alternatively, a semi-analytical solution can also be derived simultaneously with the simulation, in which case the values of all coefficients are calculated at the beginning of each time interval. Examples of such methods can be found in Chapter 3 on Differential Transformation-based semi-analytical solutions.

Besides the power series form, an exact solution can also be approximated by other series or recursive forms of semi-analytical solutions, which can achieve similar or higher accuracy by using fewer terms than a power series approximation. Examples of such methods can be found in Chapters 4 and 5.

A semi-analytical solution can incorporate complex-valued coefficients or even embed complex variables. The Holomorphic Embedding-based semi-analytical solution, introduced in Chapter 8, extends the power series expansion into the complex domain when $\mathbf{f}(\mathbf{x})$ is complex-analytic, or, in other words, a holomorphic function. This extension allows for the inclusion of complex variables and parameters related to voltage or current phasors, admittances, and complex powers into the power system model.

1.4.3.2 Analytical Homotopy Approach

When seeking a semi-analytical solution for an n-dimensional nonlinear system in (1.44), the *analytical expansion* approach, as introduced above, requires the preselection of an expansion form that converges to the exact solution. However, the *analytical homotopy* approach uses a different idea: it gradually deforms a known analytical solution of a simpler problem toward the solution of the original problem (1.44), as illustrated by Figure 1.11.

The *analytical homotopy* approach typically involves three steps. First, a simpler system of the same dimension is selected whose analytical solution $\mathbf{x}_0(t)$ exists. For example, the linear portion of the original system (1.44) can be chosen. Second, a homotopy is defined between the original system and the simpler system, such that their solutions are connected within a differentiable manifold of $n + 1$ dimensions. Third, the analytical solution of the simpler system is gradually modified, for instance, by adding new terms, to

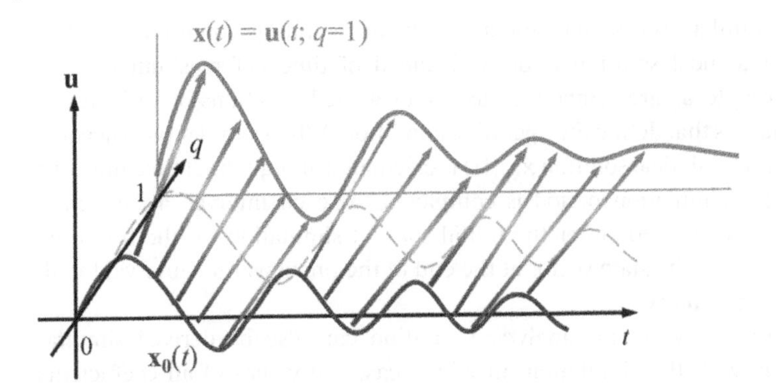

Figure 1.11 Illustration of the analytical homotopy approach.

approach the solution of the original system, as shown by the deformation toward the directions indicated by the arrows in Figure 1.11. Meanwhile, an intermediate approximation of moderate complexity can be used as a semi-analytical solution.

In this section, we introduce the *analytical homotopy* approach by utilizing a simple homotopy that deforms the solution of the nonlinear system in (1.44) from its linear portion toward its fully nonlinear form. This deformation is achieved through the application of Taylor's series expansion to the differentiable manifold. The idea of the analytical homotopy approach is introduced as follows.

We embed a real variable q that belongs to the interval [0, 1] into $\mathbf{x}$ to rewrite it as $\mathbf{u}(t; q)$, a function of both t and q. When q equals 1, $\mathbf{u}(t; q)$ reduces to $\mathbf{x}(t)$. We define a nonlinear operator $\mathcal{N}$ based on the original nonlinear system in (1.44):

$$\mathcal{N}(\mathbf{u}(t;q)) = \frac{\partial \mathbf{u}(t;q)}{\partial t} - \mathbf{f}(\mathbf{u}(t;q)) = 0, \qquad \mathbf{u}(0;q) = \mathbf{x}(0) \tag{1.52}$$

We also define another linear operator $\mathcal{L}$ that satisfies:

$$\mathcal{L}(\mathbf{u}(t;q) - \mathbf{x}_0(t)) = 0 \tag{1.53}$$

Here $\mathbf{x}_0(t)$ represents the initial guess for the original nonlinear equation (1.44), so that (1.53) is an approximation of (1.52). In homotopy theory, we consider the so-called zeroth-order deformation equation (1.54), which defines a family of equations.

$$\mathcal{H}(\mathbf{u}(t;q)) \stackrel{\text{def}}{=} (1-q)\mathcal{L}(\mathbf{u}(t;q) - \mathbf{x}_0(t)) + q\mathcal{N}(\mathbf{u}(t;q)) = 0 \tag{1.54}$$

When q equals 0, this equation corresponds to a linear equation that can be solvable analytically, whereas when q equals 1, it leads to the original nonlinear equation (1.44) that needs to be solved. As q increases continuously from 0 to 1, an analytical linear solution is deformed to an approximate solution, also known as a semi-analytical solution, of the original nonlinear equation.

Alternatively, a more general design for the homotopy can be adopted using the following deformation equation:

$$(1-q)\mathcal{L}(\mathbf{u}(t;q)-\mathbf{x}_0(t)) - qcH(t)\mathcal{N}(\mathbf{u}(t;q)) = 0 \tag{1.55}$$

This equation includes an additional constant, $-c$, and time-varying auxiliary parameter, $H(t)$, that multiply the nonlinear operator to improve the accuracy and convergence of the approximate solution. Chapter 6 provides further details on the Homotopy Analysis Method for more effective semi-analytical simulation.

To solve (1.55), we take two steps: first, we plug an infinite series that is assumed to approach $\mathbf{u}(t;\, q)$ into (1.55); second, we decompose equation (1.55) into a series of subequations to sequentially solve each term of the series up to a desired order. Since $\mathbf{u}(t;\, q)$ is analytic with respect to q, we can express it as a Maclaurin series about q that converges to $\mathbf{u}(t;\, q)$ at each time t:

$$\mathbf{u}(t;q) = \mathbf{x}_0(t) + \mathbf{x}_1(t)q + \mathbf{x}_2(t)q^2 + \cdots = \sum_{k=0}^{\infty} \mathbf{x}_k(t)q^k \tag{1.56}$$

Here, $\mathbf{x}_0(t)$ represents the initial guess of the solution. Based on Taylor's series expansion at $q = 0$, each coefficient can be expressed as:

$$\mathbf{x}_k(t) = \frac{1}{k!}\left.\frac{\partial^k \mathbf{u}(t;q)}{\partial q^k}\right|_{q=0} \tag{1.57}$$

Given already known $\mathbf{x}_0(t)$, $\mathbf{x}_1(t)$, $\mathbf{x}_2(t)$, …, $\mathbf{x}_{k-1}(t)$, we can solve for $\mathbf{x}_k(t)$ using the following kth-order deformation equation, which is typically a linear equation, to solve $\mathbf{x}_k(t)$:

$$\mathcal{L}(\mathbf{x}_k(t) - \chi_k \mathbf{x}_{k-1}(t)) = cH(t)\mathcal{R}_k(\mathbf{x}_0(t), \mathbf{x}_1(t), \ldots, \mathbf{x}_{k-1}(t)) \tag{1.58}$$

where $\mathcal{R}_k$ represents the kth term in the decomposition of $\mathcal{N}$ and depends only on the known values of $\mathbf{x}_0(t)$, …, $\mathbf{x}_{k-1}(t)$. There are the following equations:

$$\begin{aligned}
\mathcal{L}\mathbf{x}_1(t) &= cH(t)\mathcal{R}_1(\mathbf{x}_0(t)) \\
\mathcal{L}(\mathbf{x}_2(t) - \mathbf{x}_1(t)) &= cH(t)\mathcal{R}_2(\mathbf{x}_0(t), \mathbf{x}_1(t)) \\
\mathcal{L}(\mathbf{x}_3(t) - \mathbf{x}_2(t)) &= cH(t)\mathcal{R}_3(\mathbf{x}_0(t), \mathbf{x}_1(t), \mathbf{x}_2(t)) \\
&\cdots
\end{aligned}$$

In this section, we introduce and illustrate two methods for selecting the linear operator $\mathcal{L}$ in the deformation equation (1.54): $c{\cdot}H(t) = -1$ in (1.55). Let us assume that the system (1.44) has a stable equilibrium at the origin, i.e. $\mathbf{f}(0) = 0$. If not, we can move the coordinate origin to equilibrium.

- Method I involves defining $\mathcal{L} = \partial/\partial t$
- Method II involves defining $\mathcal{L} = \partial/\partial t - \mathbf{A}$, where $\mathbf{A}$ is the Jacobian matrix of $\mathbf{f}$ with respect to $\mathbf{x}$ at equilibrium $\mathbf{x} = 0$.

Method II linearizes $\mathcal{N}$ at equilibrium and incorporates linear information about the system in its linear operator, whereas the linear operator in Method I does not contain any intrinsic information about the system.

For Method I, the deformation equation (1.54) becomes (1.59):

$$(1-q)\frac{\partial}{\partial t}(\mathbf{u}(t;q) - \mathbf{x}_0(t)) + q\left(\frac{\partial}{\partial t}\mathbf{u}(t;q) - \mathbf{f}(\mathbf{u}(t;q))\right) = 0 \quad \Leftrightarrow$$
$$\frac{\partial}{\partial t}\mathbf{u}(t;q) - (1-q)\frac{\partial}{\partial t}\mathbf{x}_0(t) - q\mathbf{f}(\mathbf{u}(t;q)) = 0 \tag{1.59}$$

Applying Taylor's expansion to $\mathbf{f}$ with respect to q at $q = 0$, we obtain:

$$\mathbf{f}(\mathbf{u}(t;q)) = \sum_{k=0}^{\infty} q^k \mathcal{D}_k \mathbf{f}\left(\sum_{i=0}^{k} \mathbf{x}_i(t) q^i\right) \quad \text{where } \mathcal{D}_k = \left.\frac{1}{k!}\frac{\partial^k}{\partial q^k}\right|_{q=0} \tag{1.60}$$

When $q = 1$, $\mathbf{f}(\mathbf{x}(t))$ can be decomposed using Adomian polynomials, denoted by $\mathcal{A}_k$, as shown in

$$\mathbf{f}(\mathbf{x}(t)) = \mathbf{f}(\mathbf{u}(t;1)) = \left.\sum_{k=0}^{\infty} \mathcal{D}_k \mathbf{f}\left(\sum_{i=0}^{k} \mathbf{x}_i(t) q^i\right)\right|_{q=1} \stackrel{\text{def}}{=} \sum_{k=0}^{\infty} \mathcal{A}_k(\mathbf{x}_0(t), \mathbf{x}_1(t), \ldots \mathbf{x}_k(t)) \tag{1.61}$$

Each Adomian polynomial $\mathcal{A}_k$ is a function of $\mathbf{x}_0, \mathbf{x}_1, \ldots, \mathbf{x}_k$, and each term of the polynomial has subscripts that sum up to k. By using the decompositions for $\mathbf{u}$ as given in (1.55) and for $\mathbf{f}$ as given in (1.60) at $q = 1$, Eq. (1.59) can be expressed as:

$$\frac{d}{dt}\sum_{k=0}^{\infty} \mathbf{x}_k(t) = \sum_{k=0}^{\infty} \mathcal{A}_k(\mathbf{x}_0(t), \mathbf{x}_1(t), \ldots \mathbf{x}_k(t))$$
$$= \sum_{k=1}^{\infty} \mathcal{A}_{k-1}(\mathbf{x}_0(t), \mathbf{x}_1(t), \ldots \mathbf{x}_{k-1}(t)) \tag{1.62}$$

The next question is how to design an algorithm that sequentially solves for $\mathbf{x}_k(t)$ for $k > 1$ up to a desired $\mathbf{x}_K(t)$ to obtain a Kth-order semi-analytical solution $\mathbf{x}_{SAS}^K(t)$. By using Adomian polynomials to decompose $\mathbf{f}$, we can design the decomposition of $\mathcal{N}$ as shown in

$$\mathcal{R}_k = \frac{d}{dt}\mathbf{x}_{k-1}(t) - \mathcal{A}_{k-1}(\mathbf{x}_0(t), \mathbf{x}_1(t), \ldots \mathbf{x}_{k-1}(t)) \tag{1.63}$$

The kth-order deformation equation suggests the formulae in (1.64) to solve for $\mathbf{x}_k(t)$ $(k = 1, 2, \ldots)$.

$$\begin{cases} \dfrac{d}{dt}\mathbf{x}_1(t) = -\dfrac{d}{dt}\mathbf{x}_0(t) + \mathcal{A}_0(\mathbf{x}_0(t)) & \text{for } k = 1 \\ \dfrac{d}{dt}\mathbf{x}_k(t) = \mathcal{A}_{k-1}(\mathbf{x}_0(t), \mathbf{x}_1(t), \ldots, \mathbf{x}_{k-1}(t)) & \text{for } k \geq 2 \end{cases} \tag{1.64}$$

The selection of $\mathbf{x}_0(t)$ will influence the final form of $\mathbf{x}_{SAS}^K(t)$. For instance, if we let $\mathbf{x}_0(t)$ be constantly equal to the initial state $\mathbf{x}(0)$, the resulting $\mathbf{x}_{SAS}^K(t)$ will have a power series form equivalent to (1.51) from the *analytical expansion* approach. However, if $\mathbf{x}_0(t)$ is a known approximate solution, such as the solution of the linearization of (1.44), namely $d\mathbf{x}/dt = \mathbf{A}\mathbf{x}$, at equilibrium, a more complex semi-analytical solution is derived, which unnecessarily has a power series form.

For Method II, the deformation equation (1.54) becomes

$$0 = (1-q)\left(\frac{\partial}{\partial t} - \mathbf{A}\right)(\mathbf{u}(t;q) - \mathbf{x}_0(t)) + q\left(\frac{\partial}{\partial t}\mathbf{u}(t;q) - \mathbf{f}(\mathbf{u}(t;q))\right) \tag{1.65}$$

Using the same decomposition (1.63) for the nonlinear operator, the kth-order deformation equation suggests the formulae in (1.66), which each need to solve a linear differential equation on $\mathbf{x}_k(t)$:

$$\begin{cases} \left(\dfrac{d}{dt} - \mathbf{A}\right)\mathbf{x}_1(t) = -\dfrac{d}{dt}\mathbf{x}_0(t) + \mathcal{A}_0(\mathbf{x}_0(t)) & \text{for } k = 1 \\ \left(\dfrac{d}{dt} - \mathbf{A}\right)\mathbf{x}_k(t) = -\mathbf{A}\mathbf{x}_{k-1}(t) + \mathcal{A}_{k-1}(\mathbf{x}_0(t), \mathbf{x}_1(t), \ldots, \mathbf{x}_{k-1}(t)) & \text{for } k \geq 2 \end{cases} \tag{1.66}$$

Because it uses $\mathbf{A}$ in addition, the formulae in (1.66) can generate a semi-analytical solution $\mathbf{x}_{SAS}^K(t)$ that is not in a power series form and typically converges faster to the true solution than (1.64).

In the rest of the book, Chapter 5 discusses power system simulations using the multistage Adomian Decomposition Method, which belongs to the family of *analytical homotopy* approach for semi-analytical solutions. Chapter 6 further introduces a more general method also in this approach, the multistage Homotopy Analysis Method for semi-analytical power system simulation. Both methods are applied in the parallel-in-time framework in Chapter 7. The Holomorphic Embedding Method introduced in Chapter 8 can also be considered an *analytical homotopy* approach, with its embedding variable taking a complex number so that the homotopy in complex domain is realized by holomorphic functions. The resulting semi-analytical solution has a power series form and can further be transformed into a fractional form by Padé approximation for better accuracy.

1.4.4 Forms of Semi-Analytical Solutions

Semi-analytical solutions of power series forms are most commonly used, as illustrated with the *analytical expansion* approach and *analytical homotopy* approach. In fact, a Kth-order semi-analytical solution $\mathbf{x}_{SAS}^{K}(t; \mathbf{x}(0), \boldsymbol{\mu})$ can adopt a variety of forms as long as it is a function of time t, the initial state vector $\mathbf{x}(0)$, and a vector of symbolized parameters $\boldsymbol{\mu}$.

1.4.4.1 Power Series Form

The power series form given in (1.67) is the most commonly used for a semi-analytical solution since the true solution of the initial value problem describing a system response in time is generally expendable into a power series in time at a starting time, e.g. $t = 0$.

$$\mathbf{x}_{SAS}^{K}(t; \mathbf{x}(0), \boldsymbol{\mu}) = \sum_{k=0}^{K} \mathbf{a}_k(\mathbf{x}(0), \boldsymbol{\mu}) \cdot t^k \tag{1.67}$$

If taking the *analytical expansion* approach, one can directly assume this form up to order K, plug it into (1.44) or the DAE (1.4), and then equate both sides of each equation to solve for coefficients $\mathbf{a}_k(\mathbf{x}(0), \boldsymbol{\mu})$ from $k = 0$ to K by using recursive formulae.

If the *analytical homotopy* approach is used, the decomposition of the nonlinear operator can also lead to a power series form. For instance, when the Adomian decomposition is used, each decomposed Adomian polynomial $\mathcal{A}_k$ is a polynomial about $\mathbf{x}_0, \mathbf{x}_1, \ldots, \mathbf{x}_k$ whose subscripts sum up to k. Since terms $\mathbf{x}_0, \mathbf{x}_1, \ldots, \mathbf{x}_k$ are solved sequentially, if $\mathbf{x}_0$ is selected as a polynomial of time (including a constant), each $\mathbf{x}_k$ is a polynomial of time, too. Thus, the resulting semi-analytical solution will be a power series in time.

1.4.4.2 Other Series Forms

Other series forms that summate terms of increasing complexities can also be adopted in the *analytical expansion* approach if a preknowledge or preference on accurate approximation of the true solution is available. In the *analytical homotopy* approach, the decomposition of the nonlinear operator, the initial guess $\mathbf{x}_0(t)$ on the solution, and the deformation equation together determine the final form of the semi-analytical solution. Especially if $\mathbf{x}_0(t)$ is not a power series in time, then neither is $\mathbf{x}_k(t)$. Another example is that even if $\mathbf{x}_0(t)$ is a power series in time, the deformation equation, e.g. (1.66), can generate an $\mathbf{x}_k(t)$ in a different form from $\mathbf{x}_{k-1}(t)$. Semi-analytical solutions in a nonpower series form can have the coefficient of each term be time-dependent as well (see an example in Chapter 5). On one hand, it can increase the complexity of each term; on the other hand, it may speed up convergence of a semi-analytical solution using fewer terms.

1.4.4.3 Fractional Forms

Semi-analytical solutions in series form can be transformed into other forms, such as fractional forms, for better convergence. Padé approximants and Continued Fractions are two examples of nonseries forms for approximation. They can be derived based on series-form semi-analytical solutions. Examples are given in Chapters 4 and 8 of the book.

1.4.5 Schemes on Semi-Analytical Power System Simulation

After deriving a semi-analytical solution $\mathbf{x}_{SAS}^{K}(t; \mathbf{x}(0), \boldsymbol{\mu})$ for Eq. (1.44) to approximate its true solution, the simulation only needs to evaluate the values of the solution over consecutive time intervals, maintaining the accuracy of the semi-analytical solution. There are two semi-analytical simulation schemes: the one-stage and two-stage schemes, depending on whether the solution $\mathbf{x}_{SAS}^{K}(t; \mathbf{x}(0), \boldsymbol{\mu})$ needs to be derived simultaneously with the simulation or ahead of time:

- **One-stage scheme**: This scheme derives a semi-analytical solution when simulation is needed. The solution is specific to the system condition and contingency of simulation, which are given per the requirement of simulation, so only the initial state and time need to have symbolic representations. If the simulation allows for the solution to be repeatedly derived at each time interval using the end state from the previous interval, then the solution is only dependent on time, whose coefficients are all constant for each interval. As a result, the solution can take a simple form without any additional symbolic variables. Because each solution is only valid for a specific simulation, it does not require significant storage on computers. The use of the semi-analytical solution is similar to that of a numerical, explicit integration method, but the order can be much higher.
- **Two-stage scheme**: This scheme derives a semi-analytical solution ahead of time in the solution stage (called "stage 1"). The solution's applicability depends on how many parameters are symbolized in $\boldsymbol{\mu}$. The more symbolic parameters, the more complex the solution, but the more conditions or contingencies it can be applied to. The derived solution needs to be saved in storage for later use in the simulation stage (called

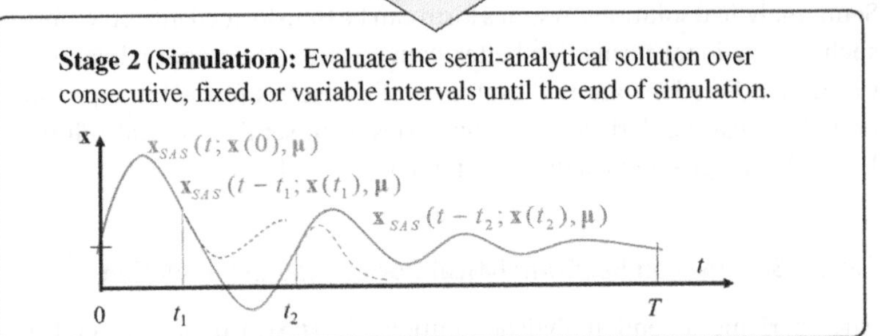

Stage 1 (Solution): Find the semi-analytical solution with symbolic representations of time t, initial state x_ini, and selected parameters μ
ẋ = f(x,y,t)
0 = g(x,y,t)
x(t; x_ini, μ) = Σ_{k=0}^∞ x_k(t; x_ini, μ)
≈ Σ_{k=0}^K x_k(t; x_ini, μ) =def x^K_SAS(t; x_ini, μ)
Stage 2 (Simulation): Evaluate the semi-analytical solution over consecutive, fixed, or variable intervals until the end of simulation.
x
x_SAS(t; x(0), μ)
x_SAS(t − t_1; x(t_1), μ)
x_SAS(t − t_2; x(t_2), μ)
t
0
t_1
t_2
T

Figure 1.12 Illustration of a semi-analytical solution approach.

"stage 2"), where it will be evaluated to generate a simulation result. The storage needed to save the solution can be huge if the simulated system is large and many parameters need to be symbolized to cover a variety of simulation scenarios.

Figure 1.12 illustrates a two-stage scheme for online power system simulation using a Kth-order series-form semi-analytical solution $\mathbf{x}_{SAS}^{K}(t; \mathbf{x}_{ini}, \boldsymbol{\mu})$. The solution stage can be offline, ahead of simulation. The vector $\boldsymbol{\mu}$ includes all parameters that become known when simulation starts. Then, at the online simulation stage, the solution is repeatedly evaluated based on the system condition and the contingency for consecutive time intervals of h. In each interval, values are simply plugged into all symbolic parameters until the simulation period T is completed. In series, the computed state at the end of each time interval is then used as the initial state $\mathbf{x}_{ini}$ of the next interval. Power system simulation using a semi-analytical solution derived ahead of time needs to do such a calculation for T/h times sequentially without back-and-forth numerical iterations. Such a two-stage scheme using the semi-analytical solution more easily enables faster-than-real-time simulations, in which a simulation run needs to be completed in a clock time earlier than the end of the simulation time period. Here, we only need evaluation of the solution for each time interval to be shorter than h.

For example, a semi-analytical solution $\mathbf{x}_{SAS}^{K}$ can be used to simulate a practical contingency that has a short-circuit fault near bus i on the line from bus i to bus j at $t = t_1 > t_0$, cleared at $t_2 > t_1$ by tripping the faulted line. Assume that all entries of $\mathbf{Y}_{\text{bus}}$ are symbolized as variables of the solution, namely $\mathbf{x}_{SAS}^{K} = \mathbf{x}_{SAS}^{K}(t; \mathbf{x}_{ini}, \mathbf{Y}_{\text{bus}})$. This allows for the semi-analytical solution to be flexible enough to accommodate change in the network topology due to a fault or control. The entire simulation period, denoted by $[0, T]$, is partitioned into the following subperiods, and a semi-analytical simulation using a fixed length h of time intervals is conducted as follows:

- **Prefault** ($t = t_0$ to t_1): Starting from the equilibrium state $\mathbf{x}_{eq}$ of the system at $t = t_0$ for the first interval, evaluate $\mathbf{x}_{SAS}^{K}$ at consecutive intervals of h until $t = t_1$, using $\mathbf{x}_{ini} = \mathbf{x}_{eq}$ as the initial state.
- **During fault** ($t = t_1$ to t_2): Update the admittance matrix $\mathbf{Y}_{\text{bus}}$ by updating Y_{ii}, Y_{jj}, and Y_{ij} for the fault-on period. Using the values of $\mathbf{x}_{SAS}^{K}$ at $t = t_1$ from the prefault period as the initial state, evaluate $\mathbf{x}_{SAS}^{K}$ at intervals of h until $t = t_2$.
- **Postfault** ($t = t_2$ to T): Update $\mathbf{Y}_{\text{bus}}$ again to reflect the admittance matrix after the fault is cleared. Using the values of $\mathbf{x}_{SAS}^{K}$ at $t = t_2$ as the initial state, evaluate $\mathbf{x}_{SAS}^{K}$ at intervals of h until $t = T$.

Alternatively, we can use a numerical integration method to simulate the system's responses during the prefault and fault-on periods, which are generally short in duration. This will allow us to determine the initial state for a semi-analytical simulation that covers the next postfault period until the end of simulation.

1.5 Parallel Power System Simulation

Parallel computing can significantly accelerate the power system simulation process by parallelizing the solution approach for the power system model in DAEs or ODEs across multiple processors. However, conventional numerical integration methods for ODEs and nonlinear solvers for algebraic equations were not designed for parallel computing. Yet, approaches to parallel power system simulation have been developed by parallelizing computational tasks that are inherently or intentionally decoupled.

In the literature, three directions of parallelizing power system simulation have been explored: parallelization across solutions, parallelization in space, and parallelization in time. Parallelization across solutions conducts concurrent simulations on parallel computers when many independent contingencies need to be simulated, with each computer still conducting a sequential simulation run. Parallelization in space and parallelization in time aim to parallelize each simulation run onto multiple computers or processors and, therefore, require parallel algorithms.

1.5.1 Parallelization in Space

In general, a power system has a loosely coupled structure as its power plants and load areas are located spatially apart. The changes in many power system variables only influence others that are in the same local area or directly connected with them by either electrical coupling or a designed control law. Therefore, solutions for many equations of a power system model are inherently parallelizable on different processors and then integrated via their loose coupling for a system-wide solution. A parallel-in-space algorithm can be designed based on either the mathematical model or the network topology to accelerate power system simulation.

1.5.1.1 Natural Decoupling

The mathematical model can generally indicate which variables have numerical loose coupling at each iteration of the whole set of equations. These variables are often associated with devices separated by long electrical distances. Therefore, when a change happens first on a variable, it may not have instantaneous impacts on values of all variables in the rest of the system.

Specifically, when a simulation needs to solve a set of linear or nonlinear algebraic equations, such as power-flow equations, many equations are decoupled at least for one time step or for one numerical iteration. Thus, they can be solved simultaneously by multiple processors. For example, the computation solves $\mathbf{Y}_{n\times n}\mathbf{A}_{n\times m} = \mathbf{B}_{n\times m}$ for an unknown matrix $\mathbf{A}_{n\times m}$, where $\mathbf{Y}_{n\times n}$ is the bus admittance matrix of a power network. Obviously, the solutions to different columns of $\mathbf{A}_{n\times m}$ can be parallelized onto different processors.

When using the partitioned solution approach to solve the DAE model for one time step, solutions to the ODEs of different dynamic devices, such as generators and motors, are parallelizable because their buses are treated as constant voltage sources in one time interval. For each dynamic device, integrations of its different first-order ODEs can be further parallelized.

1.5.1.2 Network Partitioning

With topological information available on a power system, spatial parallelization can be designed among existing control regions of its network.

Those control regions defined by power system planning engineers are weakly interconnected via few tie lines or boundary buses. When a disturbance occurs in one region, it will propagate first to the nearest neighboring regions and then to the rest of the system. Simulations of these regions can be parallelized onto different processors. In each time step of simulation, all regions assume their boundary buses to have either constant voltages or constant currents so that their respective simulations are performed in parallel. Mismatches in voltages or currents at boundary buses are corrected by iterating such parallel local simulations. At the end, parallelization on multiple processors can still accelerate the overall simulation, as long as the number of iterations remains moderate.

1.5.2 Parallelization in Time

Parallelization of a power system simulation in time is a complex task compared to parallelization in space. This section introduces the Parareal algorithm as a parallel-in-time simulation method that will be utilized in Chapter 7 as a framework for semi-analytical solutions to accelerate power system simulation through parallel computers.

As illustrated in Figure 1.13, the Parareal algorithm is based on the decomposition of the simulation period T into separate, consecutive time intervals of ΔT called "coarse" intervals. Within each coarse interval, fine intervals of δt far shorter than ΔT are defined. The Parareal algorithm requires two solvers that, respectively, solve two mathematical models, which usually have different levels of detail. The high-level coarse solver integrates a reduced model of the system at large time steps equal to coarse intervals of ΔT through the entire period T for prediction of the overall system response. The concurrent low-level fine solvers, as many as coarse intervals, integrate a detailed model at small steps of δt within coarse intervals for correction of the coarse solution.

The initial state of each fine solution at the beginning of the coarse interval is taken from the coarse solution. Then, in the next iteration, it is corrected based on the end state of the fine solution of the preceding coarse interval. The Parareal algorithm stops when gaps between the predicted and correct values at all coarse intervals can be tolerated. Thus, the Parareal algorithm introduces a two-level hierarchy to solve the problem of simulation: the high-level, approximate, coarse solution predicts at large intervals in a sequential manner, while the low-level, accurate, fine solutions are linked by the coarse simulation but correct it for concurrent coarse intervals using parallel processors. For accuracy of simulation, the algorithm performs a number of prediction-correction iterations between the sequential coarse

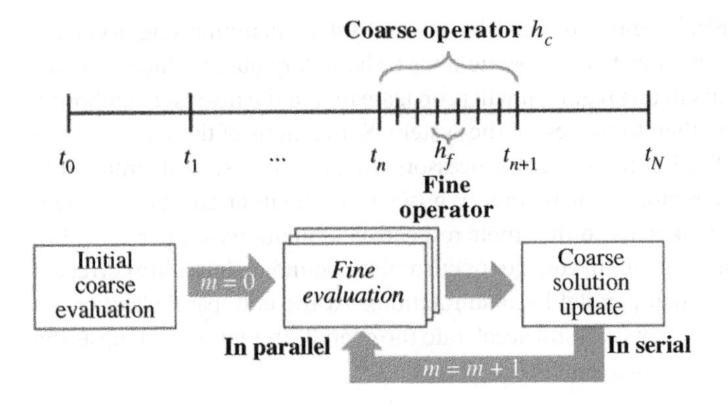

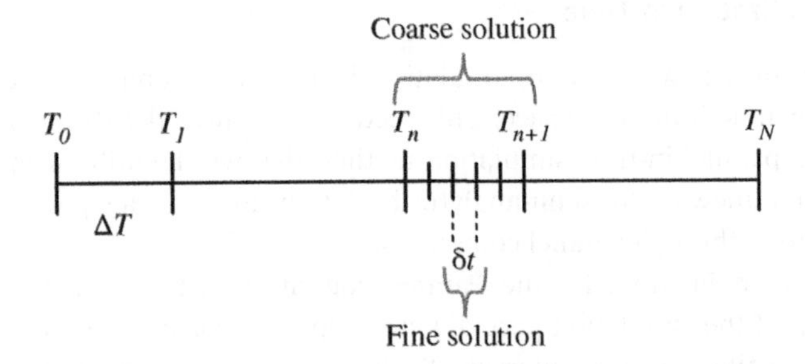

Figure 1.13 Illustration of the parareal algorithm.

solution and concurrent fine solutions until all fine simulations of sequential coarse intervals are connected.

The overall speedup of the Parareal algorithm depends on several factors, such as the time cost and accuracy of a coarse solution, the number of coarse intervals, the number of prediction-correction iterations, and communications between coarse and fine solutions. For example, assume the following to estimate the speedup:

- There are totally $N = T/\Delta T$ coarse intervals parallelized onto N worker processors.
- The coarse solver and fine solver, respectively, take constant times τ_c and τ_f to finish the simulation for each coarse interval ΔT.
- There are totally k prediction–correction iterations, or, in other words, k times sequential coarse solutions.

A sequential simulation purely by the fine solver at steps of δt takes $N{\cdot}\tau_f$ time. Each iteration of a Parareal simulation takes a runtime of $N{\cdot}\tau_c$ for the coarse solution by one processor and a runtime of τ_f by concurrent fine solutions on N processors, which totals $N{\cdot}\tau_c + \tau_f$ if communication time is ignored. Thus, an approximate estimate of the speedup by using a Parareal algorithm is:

$$\text{Speedup} \approx \frac{N}{k}\left[1 + \frac{N\tau_c}{\tau_f}\right]^{-1} < \frac{N}{k} \tag{1.68}$$

An ideal speedup is the ratio of the number of coarse intervals (namely, the number of processors) to the number of iterations. Using a large number of parallel processors with a small number of iterations, the speedup can be significant. In practice, a high speedup is achieved when, first, the number of prediction-correction iterations k is much less than the number of coarse intervals N and, second, the time $N{\cdot}\tau_c$ spent on the coarse solution is negligible compared to τ_f. However, it is rare for both conditions to be simultaneously satisfied in practice when simulating a nonlinear system such as a power system.

If a reduced model of the system is used by the coarse solver, aiming to reduce its time cost τ_c, the inaccuracy in the solution can lead to an increase in k, i.e. the number of iterations, which in turn decreases the speedup. Moreover, to avoid divergence, the coarse solution may have to take a shorter integration step than ΔT, which, however, increases τ_c and decreases the speedup.

A more accurate speedup estimation depends on the detailed design of the Parareal simulation. While some time costs can be saved, others that were neglected have to be considered. For instance, the simulation periods of k coarse solutions do not necessarily have to be the entire period T. With fine solutions available and connected at earlier intervals, coarse solutions are necessary thereafter.

The actual speedup achieved by the Parareal algorithm on parallel processors also has to consider the communication between coarse and fine solutions. In practice, there are two Parareal algorithms that depend on memory systems of computers: the master–workers algorithm and the distributed algorithm. In the master–workers algorithm, the initial coarse solution is performed by a master processor, and the results are sent to the worker processors for fine solutions on respective coarse intervals. The fine solutions are then sent back to the master processor to correct the coarse solution for the next iteration. This algorithm is suited for shared memory systems where all processors share the same global memory. In the distributed algorithm, the coarse solution and correction are distributed across all processors. This algorithm is suited for distributed memory systems where each processor uses its own local memory.

More details about the Parareal method and its integration of semi-analytical solutions as coarse solutions will be introduced in Chapter 7.

1.5.3 Parallelization of Semi-Analytical Solutions

Semi-analytical solutions provide structural information about a power system solution that numerical solutions cannot reveal. Thus, there is potential to develop more methods for parallel power system simulation based on semi-analytical methods. Here are some examples:

First, the analytical form of a semi-analytical solution makes it clear how symbolic variables and coefficients are associated by operations and how they may be decoupled at each interval. While a more accurate semi-analytical solution is typically more complex and of a higher order, parallel computing can decompose the evaluation of a complex semi-analytical solution into concurrent subtasks. For example, when a series-form semi-analytical solution is used, evaluations of all terms that need to be summed can be distributed to different processors. Thus, the complexity of the solution due to an ultrahigh order is mitigated when multiple processors are available.

Second, because semi-analytical solutions are approximations based directly on the mathematical model, they can vary their order over a much larger range than numerical solutions. They can either adopt a concise, lower order approximation for easier computation or a more accurate, ultrahigh-order approximation for longer intervals. The flexibilities in the accuracy of approximation and the length of time intervals make semi-analytical solutions better candidates for the coarse and fine solvers with the Parareal algorithm than numerical solutions.

1.6 Final Remark

Numerical and analytical methods are two legs of power system analysis and simulation. While numerical methods are commonly used in power system simulation programs due to rapid advancement of computer technology, they may not always be efficient for power system simulation since they are built upon general numerical integrators and solvers designed to solve a large variety of nonlinear differential and algebraic equations.

As the mathematical model of a power grid becomes more complex, nonlinear, and stiff, the computational burden with numerical simulations can significantly increase, making it difficult to meet the requirement for real-time applications where the simulation of a contingency needs to be fast enough to trigger an early warning or proactive control action before a stability threat impacts the grid.

In contrast, analytical methods can uncover valuable structural information intrinsic to a power system model and its possible solutions, simplifying computations required by power system simulation. It is true that such structural information may be less general and system- or model-dependent. When the model, its operating condition, or the contingency changes, the solution and its analytical approximant will also be different. Moreover, since finite approximation is used, the error from a true solution can only be tolerated within a certain time interval of accuracy, whose length depends on the complexity or order of the approximant. All these mean that moderate numerical computations are still needed to extend the accuracy of an analytical approximant by, for example, a multistage scheme that repeatedly uses the same approximant for a series of time intervals. Thus, a semi-analytical–semi-numerical simulation approach is introduced, combining both numerical and analytical methods.

In the rest of the book, a variety of semi-analytical methods are introduced for finding semi-analytical solutions of power system ODE and DAE models, and for accelerating power system simulations for real-time applications. The accuracies and time performances of these methods are demonstrated on realistic bulk power system models with comparisons against traditional numerical methods.

References

1 Kundur, P. (1994). *Power System Stability and Control*. McGraw-Hill.

2 Padiyar, K.R. (2008). *Power System Dynamics: Stability and Control*, 2e. BS Publications.

3 Anderson, P.M. and Fouad, A.A. (2002). *Power System Control and Stability*. Wiley-IEEE Press.

4 IEEE PES Task Force on Turbine-Governor Modeling (2013). Dynamic Models for Turbine-Governors in Power System Studies, PES-TR1.

5 Demetriou, P. et al. (2017). Dynamic IEEE test systems for transient analysis. *IEEE Systems Journal* 11 (4): 2108–2117.

6 Dahlquist, G. (1963). A special stability problem for linear multistep methods. *BIT* 3 (1): 27–43. https://doi.org/10.1007/BF01963532, hdl:10338.dmlcz/103497, S2CID 120241743.

7 Duan, N. and Sun, K. (2015). Application of the adomian decomposition method for semi-analytic solutions of power system differential algebraic equations. In: *Proceedings of the Powertech Conference*, Eindhoven, Netherlands, 29 June-2 July.

8 Duan, N. and Sun, K. (2017). Power system simulation using the adomian decomposition method. *IEEE Transactions on Power Systems* 32 (1): 430–441.

9 Gurrala, G., Dimitrovski, A., Sreekanth, P. et al. (2015). Application of adomian decomposition for multi-machine power system simulation. *Proceedings of the IEEE PES General Meeting*, Denver, CO, 26-30 July.

10 Gurunath, G., Dinesha, D.L., Dimitrovski, A. et al. (2017). Large multi-machine power system simulations using multi-stage adomian decomposition. *IEEE Transactions on Power Systems* 32 (5): 3594–3606.

11 Dinesha, D.L. and Gurrala, G. (2018). Application of multi-stage homotopy analysis method for power system dynamic simulations. *IEEE Transactions on Power Systems* 34 (3): 2251–2260.

12 Abreut, E., Wang, B., and Sun, K. (2017). Semi-analytical fault-on trajectory simulation and its application in direct methods. In: *Proceedings of the IEEE PES General Meeting*, Chicago, Illinois, 16-20 July.

13 Wang, B., Duan, N., and Sun, K. (2019). A time-power series based semi-analytical approach for power system simulation. *IEEE Transactions on Power Systems* 34 (2): 841–851.

14 Liu, C., Wang, B., and Sun, K. (2017). Fast power system simulation using semi-analytical solutions based on pade approximants. In: *Proceedings of the IEEE PES General Meeting*, Chicago, Illinois, 16-20 July.

15 Liu, C., Wang, B., and Sun, K. (2018). Fast power system dynamic simulation using continued fractions. *IEEE Access* 6 (1): 62687–62698.

16 Liu, Y., Sun, K., Yao, R., and Wang, B. (2019). Power system time domain simulation using a differential transformation method. *IEEE Transactions on Power Systems* 34 (5): 3739–3748.

17 Liu, Y. and Sun, K. (2020). Solving power system differential algebraic equations using differential transformation. *IEEE Transactions on Power Systems* 35 (3): 2289–2299.

18 Liu, Y., Sun, K., and Dong, J. (2020). A dynamized power flow method based on differential transformation. *IEEE Access* 8: 182441–182450.

19 Yao, R., Sun, K., Shi, D., and Zhang, X. (2019). Voltage stability analysis of power systems with induction motors based on holomorphic embedding. *IEEE Transactions on Power Systems* 34 (2): 1278–1288.

20 Yao, R., Sun, K., and Qiu, F. (2019). Vectorized efficient computation of pade approximation for semi-analytical simulation of large-scale power systems. *IEEE Transactions on Power Systems* 34 (5): 3957–3959.

21 Yao, R., Liu, Y., Sun, K. et al. (2020). Efficient and robust dynamic simulation of power systems with holomorphic embedding. *IEEE Transactions on Power Systems* 35 (2): 938–949.

22 Park, B., Sun, K., Dimitrovski, A. et al. (2021). Examination of semi-analytical solution methods in the coarse operator of parareal algorithm for power system simulation. *IEEE Transactions on Power Systems* 36 (6): 5068–5080.

23 Liu, Y., Park, B., Sun, K. et al. (2022). Parallel-in-time power system simulation using a differential transformation based adaptive parareal method. *IEEE Open Access Journal of Power and Energy* 10: 61–72.

2

Power System Simulation Using Power Series-Based Semi-Analytical Methods

Bin Wang

Department of Electrical and Computer Engineering, University of Texas at San Antonio, San Antonio, TX, USA

2.1 Power Series-Based SAS for Simulating Power System ODEs

2.1.1 Power Series-Based SAS for ODEs

This section derives power series-based SAS for general ODEs [1], including power system ODEs. The resulting SAS of each state variable is expected to be a function of time, system initial states, and parameters.

Consider a general dynamical system represented by ODEs

$$\begin{cases} \dot{\mathbf{x}}(t) = \mathbf{f}(\mathbf{x}(t)) \\ \mathbf{x}(t_0) = \mathbf{x}_0 \end{cases} \tag{2.1}$$

where $\mathbf{x}$ is the state vector of dimension $N \times 1$, $\mathbf{x}_0$ is the initial condition and $\mathbf{f}$ is a smooth vector field.

The existence and uniqueness of the exact solution $\mathbf{x}(t)$ to (2.1) is guaranteed by Caratheodory's existence theorem [2]. Also, assume that $\mathbf{x}(t)$ is differentiable with respect to t up to any desired order. Thus, $\mathbf{x}(t)$ can be expanded to the power series in (2.2) with unknown coefficients in vector $\mathbf{a}_k$ to be determined. The time derivative of $\mathbf{x}(t)$ is given in (2.3). Note that $\mathbf{a}_0$ equals the initial state $\mathbf{x}_0$, and $\mathbf{a}_k$ ($k \geq 1$) depends on system parameters, denoted by vector $\mathbf{p}$, and $\mathbf{a}_0$.

$$\mathbf{x}(t) = \sum_{k=0}^{\infty} \mathbf{a}_k (t - t_0)^k \tag{2.2}$$

$$\dot{\mathbf{x}}(t) = \sum_{k=0}^{\infty} (k+1)\mathbf{a}_{k+1} (t - t_0)^k \tag{2.3}$$

Power System Simulation Using Semi-Analytical Methods, First Edition. Edited by Kai Sun.

Published 2024 by John Wiley & Sons, Inc.

Then, substitute (2.2) into the right-hand side of the first equation in (2.1) and expand each element of the vector field $\mathbf{f}$ about $\mathbf{t}$ at $\mathbf{t}_0$ to give (2.4), where $\mathbf{b}$ depends on $\mathbf{a}$.

$$\mathbf{f}(\mathbf{x}(t)) = \mathbf{f}\left(\sum_{k=0}^{\infty}\mathbf{a}_k(t-t_0)^k\right) = \sum_{k=0}^{\infty}\mathbf{b}_k(t-t_0)^k \tag{2.4}$$

Substitute (2.3) and (2.4) into the first equation of (2.1) and obtain (2.5), or equivalently (2.6), for solving unknown coefficients $\mathbf{a}_k$.

$$\sum_{k=0}^{\infty}(k+1)\mathbf{a}_{k+1}(t-t_0)^k = \sum_{k=0}^{\infty}\mathbf{b}_k(t-t_0)^k \tag{2.5}$$

$$(k+1)\mathbf{a}_{k+1} = \mathbf{b}_k \quad \text{for } k = 0, 1, 2, \ldots \tag{2.6}$$

Theorem 2.1 The coefficients $\mathbf{a}$ in the power series-based SAS, i.e. $\mathbf{a}_k$ for $k = 1, 2, \ldots$ in (2.2), can be solved from Eq. (2.6) recursively.

Proof: The idea of the proof is to expand the expression of $\mathbf{b}$ in (2.4) and find out how it depends on $\mathbf{a}$, and it will be observed that the recursive solution of $\mathbf{a}$ is given by (2.6).

Since $\mathbf{x}(t_0) = \mathbf{x}_0$, then the Taylor expansion of $\mathbf{f}(\mathbf{x}(t))$ at t_0 is equivalent to the Taylor expansion of $\mathbf{f}(\mathbf{x})$ at $\mathbf{x}_0$, i.e.

$$\mathbf{f}(\mathbf{x}(t)) = \mathbf{f}(\mathbf{x}) \Rightarrow \sum_{k=0}^{\infty}\mathbf{b}_k(t-t_0)^k = \mathbf{f}(\mathbf{x}_0) + \sum_{k=1}^{\infty}\mathbf{q}_k\Delta\mathbf{x}^k \tag{2.7}$$

$$\mathbf{q}_k = \left.\frac{\partial^k\mathbf{f}}{\partial\mathbf{x}^k}\right|_{\mathbf{x}=\mathbf{x}_0} \tag{2.8}$$

$$\Delta\mathbf{x} = \mathbf{x}(t) - \mathbf{x}_0 = \sum_{k=0}^{\infty}\mathbf{a}_k(t-t_0)^k - \mathbf{x}_0 = \sum_{k=1}^{\infty}\mathbf{a}_k(t-t_0)^k \tag{2.9}$$

where $\Delta\mathbf{x}^k$ consists of all kth order homogeneous polynomials in $\Delta\mathbf{x}$; $\mathbf{q}_k$ is the corresponding coefficient matrix. Note that $\mathbf{q}_k$ does not depend on t or any $\mathbf{a}_k$ with $k > 0$.

According to (2.9), how the coefficient of $(t-t_0)^m$ in the expression of $\Delta\mathbf{x}_n$ depends on the unknown $\mathbf{a}$ can be determined, as shown in Table 2.1. For example, the coefficient of term $(t-t_0)^2$ in the expression $\Delta\mathbf{x}^1$ only depends on $\mathbf{a}_0$ and $\mathbf{a}_2$. Thus, considering (2.7), Table 2.1 shows that $\mathbf{b}_k$, i.e. the coefficient of $(t-t_0)^k$ on the left side of (2.7), only depends on $\mathbf{a}_0, \mathbf{a}_1, \ldots, \mathbf{a}_k$, for any $k = 0, 1, 2, \ldots$. Thus, (2.10) holds, where $\mathbf{g}_k$ is a vector function depending only on $\mathbf{a}_0, \mathbf{a}_1, \ldots, \mathbf{a}_k$.

Table 2.1 Coefficients of $(t-t_0)^m$ in $\Delta\mathbf{x}^n$.

	$\Delta\mathbf{x}^0$	$\Delta\mathbf{x}^1$	$\Delta\mathbf{x}^2$	$\Delta\mathbf{x}^3$	...	$\Delta\mathbf{x}^k$	...
$(t-t_0)^0$	$\mathbf{a}_0$	0	0	0	...	0	...
$(t-t_0)^1$	0	$\mathbf{a}_0, \mathbf{a}_1$	0	0	...	0	...
$(t-t_0)^2$	0	$\mathbf{a}_0, \mathbf{a}_2$	$\mathbf{a}_0, \mathbf{a}_1$	0	...	0	...
$(t-t_0)^3$	0	$\mathbf{a}_0, \mathbf{a}_3$	$\mathbf{a}_0, \mathbf{a}_1, \mathbf{a}_2$	$\mathbf{a}_0, \mathbf{a}_1$	...	0	...
$\vdots$	$\vdots$	$\vdots$	$\vdots$	$\vdots$	$\ddots$	$\vdots$	...
$(t-t_0)^k$	0	$\mathbf{a}_0, \mathbf{a}_k$	$\mathbf{a}_0 \ \cdots \ \mathbf{a}_{k-1}$	$\mathbf{a}_0 \ \cdots \ \mathbf{a}_{k-2}$	...	$\mathbf{a}_0, \mathbf{a}_1$	...
$\vdots$	$\vdots$	$\vdots$	$\vdots$	$\vdots$	...	$\vdots$	$\ddots$

Finally, substitute (2.10) into (2.6) to give (2.11), which is a recursive solution for coefficients $\mathbf{a}$. Note that by substitution, $\mathbf{a}_{k+1}$ only depends on $\mathbf{a}_0$ and $\mathbf{p}$. □

$$\mathbf{b}_k = \mathbf{g}_k(\mathbf{a}_0, \mathbf{a}_1, \ldots, \mathbf{a}_k) \ \text{for}\ k = 0, 1, 2, \ldots \tag{2.10}$$

$$\mathbf{a}_{k+1} = \frac{\mathbf{g}_k\left(\mathbf{a}_0, \mathbf{a}_1(\mathbf{p}), \ldots, \mathbf{a}_k(\mathbf{p})\right)}{k+1} = \frac{\tilde{\mathbf{g}}_k(\mathbf{a}_0, \mathbf{p})}{k+1} \ \text{for}\ k = 0, 1, 2, \ldots \tag{2.11}$$

Remark 2.1 Theorem 3.1 shows that the power series-based SAS can always be achieved in a recursive way for ODEs, where the solution contains an infinite number of polynomial terms in t. In a numerical implementation, only a finite number of terms can be handled in (2.2), say n, which will lead to $N \times n$ equations in (2.6) used for solving $N \times n$ unknown coefficients $\mathbf{a}_0, \mathbf{a}_1, \ldots, \mathbf{a}_n$. Denote the corresponding power series-based SAS as (2.12), which is a truncated approximation of the true solution $\mathbf{x}(t)$, where n is the order of the SAS $\mathbf{x}_{\text{sas}}^{\langle n\rangle}(t)$.

$$\mathbf{x}(t) \approx \mathbf{x}_{\text{sas}}^{\langle n\rangle}(t) = \sum_{k=0}^{n} \mathbf{a}_k (t-t_0)^k \tag{2.12}$$

Note that the approximate SAS in (2.12) is accurate at the initial point and gradually deviates from the true solution over time. If the system dynamics in a short period of time are focused, then the error of SAS (2.12) can be small and ignorable. In Section 2.1.2, a use case for simulating the fault-on trajectory of power systems will be presented. If a longer period of time is simulated, then the error can become too large to be acceptable. To this end, the multistage approach can be adopted to achieve an accurate solution for the entire time period by sequentially plugging the system condition at the end time of the previous time step into the SAS to simulate system dynamics in the current time step. How to select the length of the time step, and how to deal with power system models represented by DAEs will be respectively covered in Sections 2.2 and 2.3.

2.1.2 SAS-Based Fault-On Trajectory Simulation and Its Application in Direct Methods

This section first introduces the application of the power series-based SAS on simulating fault-on trajectories of power systems represented by ODEs [3], where the simplified, classical power system model is adopted. Then, the endpoint of the fault-on trajectory is fed to the decoupling-based direct method to determine the transient stability of the post-fault power system without conducting the time-domain simulation. Examples are presented on two test systems in this section: IEEE 9-bus system and the simplified WECC 179-bus system [4].

2.1.2.1 SAS-Based Simulation of Fault-On Trajectories

Short-circuit faults in power systems are always cleared by the protection system typically within 5–6 cycles, i.e. 0.08–0.1 seconds, upon the occurrence of a fault. Such short fault-on periods naturally admit the potential application of SAS for simulating the fault-on trajectories.

Consider N-machine power system ODEs in (2.13), where all machines are represented by a classical model and all loads by constant impedance.

$$\ddot{\delta}_i + \frac{D_i}{2H_i}\dot{\delta}_i + \frac{\omega_s}{2H_i}(P_{ei} - P_{mi}) = 0 \tag{2.13}$$

$$P_{ei} = E_i^2 G_i + \sum_{j=1, j\neq i}^{m} \left(C_{ij}\sin(\delta_i - \delta_j) + D_{ij}\cos(\delta_i - \delta_j) \right) \tag{2.14}$$

where $i \in \{1, 2, \dots, m\}$, δ_i, P_{mi}, P_{ei}, E_i, H_i, and D_i respectively represent the absolute rotor angle, mechanical power, electrical power, field voltage, the inertia constant, and damping constant of machine i, and G_i, C_{ij}, and D_{ij} represent network parameters including loads represented by impedance; ω_s represents the synchronous angular frequency.

According to the above modeling and the derivations in Section 2.1.1, the 3rd order power series-based SAS of the IEEE 9-bus system (see its one-line diagram in Figure 2.1) is shown in (2.15), where a_{i0} and a_{i1} are respectively the initial conditions δ_i^0 and $\dot{\delta}_i^0$ of generator i, where $i \in \{1, 2, 3\}$, and a_{i2} and

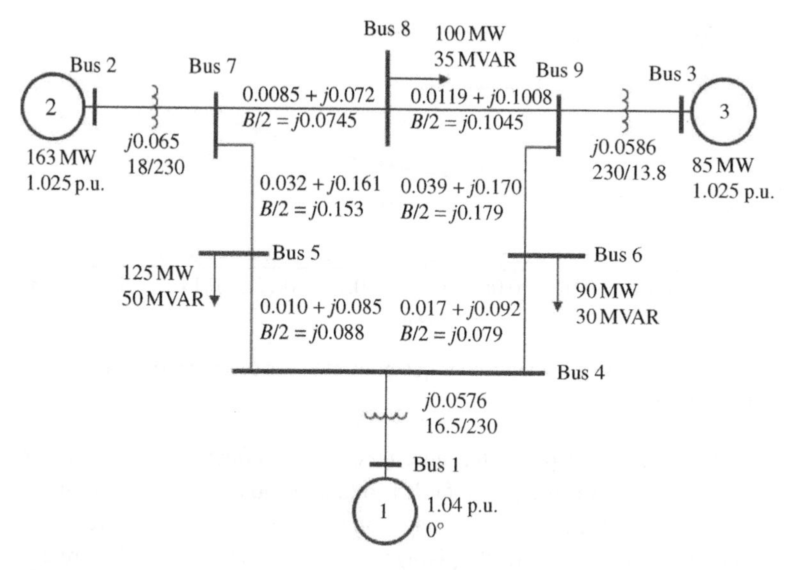

Figure 2.1 One-line diagram of the IEEE 9-bus system. Source: Bin Wang.

a_{i3} are shown in (2.16).

$$\delta(t) = \begin{pmatrix} \delta_1(t) \\ \delta_2(t) \\ \delta_3(t) \end{pmatrix} = \begin{pmatrix} a_{10} + a_{11}t + a_{12}t^2 + a_{13}t^3 \\ a_{20} + a_{21}t + a_{22}t^2 + a_{23}t^3 \\ a_{30} + a_{31}t + a_{32}t^2 + a_{33}t^3 \end{pmatrix} \tag{2.15}$$

$$\begin{cases} a_{12} = \dfrac{D_1\dot{\delta}_1^0}{4H_1} + \dfrac{\omega_s(G_1E_1^2 + C_{12}\sin\delta_{12}^0 + C_{13}\sin\delta_{13}^0 + D_{12}\cos\delta_{12}^0 + D_{13}\cos\delta_{13}^0 - P_{m1})}{4H_1} \\ a_{13} = \dfrac{a_{12}D_1\dot{\delta}_1^0}{6H_1} + \dfrac{\omega_s(C_{12}\delta_{12}^0\cos\delta_{12}^0 + C_{13}\delta_{13}^0\cos\delta_{13}^0 - D_{12}\delta_{12}^0\sin\delta_{12}^0 - D_{13}\delta_{13}^0\cos\delta_{13}^0)}{12H_1} \\ a_{22} = \dfrac{D_2\dot{\delta}_2^0}{4H_2} + \dfrac{\omega_s(G_2E_2^2 + C_{21}\sin\delta_{21}^0 + C_{23}\sin\delta_{23}^0 + D_{21}\cos\delta_{21}^0 + D_{23}\cos\delta_{23}^0 - P_{m2})}{4H_2} \\ a_{23} = \dfrac{a_{22}D_2\dot{\delta}_2^0}{6H_2} + \dfrac{\omega_s(C_{21}\delta_{21}^0\cos\delta_{21}^0 + C_{23}\delta_{23}^0\cos\delta_{23}^0 - D_{21}\delta_{21}^0\sin\delta_{21}^0 - D_{23}\delta_{23}^0\cos\delta_{23}^0)}{12H_2} \\ a_{32} = \dfrac{D_3\dot{\delta}_3^0}{4H_3} + \dfrac{\omega_s(G_3E_3^2 + C_{31}\sin\delta_{31}^0 + C_{32}\sin\delta_{32}^0 + D_{31}\cos\delta_{31}^0 + D_{32}\cos\delta_{32}^0 - P_{m3})}{4H_3} \\ a_{33} = \dfrac{a_{32}D_1\dot{\delta}_3^0}{6H_3} + \dfrac{\omega_s(C_{31}\delta_{31}^0\cos\delta_{31}^0 + C_{32}\delta_{32}^0\cos\delta_{32}^0 - D_{31}\delta_{31}^0\sin\delta_{31}^0 - D_{32}\delta_{32}^0\cos\delta_{32}^0)}{12H_3} \end{cases} \tag{2.16}$$

where $\delta_{ij}^0 = \delta_i^0 - \delta_j^0$.

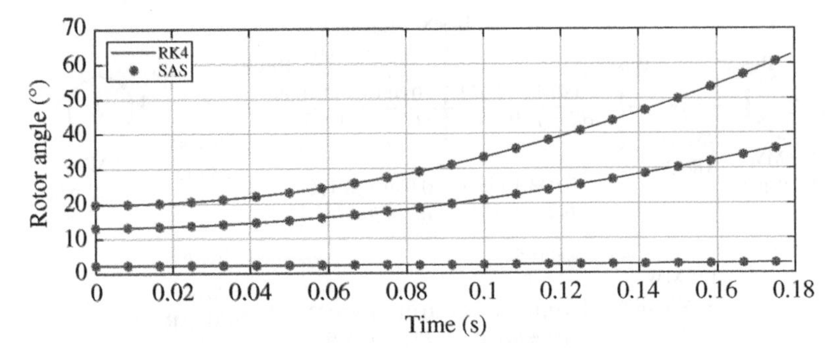

Figure 2.2 Comparison of fault-on trajectories of 9-bus system: SAS vs. RK4. Source: Bin Wang.

Three examples are presented here to show the effectiveness of SAS. In the first example, a three-phase fault is added at bus 7 and cleared after 5 cycles by tripping the line 5–7. The fault-on trajectories by the 3rd order SAS and the 4th order Runge–Kutta (RK4) method are compared in Figure 2.2, where the star marks are the results simulated by SAS, while the solid lines are by RK4. Other faults have been checked and found to give similar comparisons, whose figures are omitted here. This example illustrates that SAS can accurately simulate the fault-on period for all faults in the IEEE 9-bus system.

In the second example, the fault-on trajectories of the IEEE 9-bus system are simulated for an extended period of time, i.e. >0.4 seconds. This fault duration is not realistic but would be useful to demonstrate how much longer the SAS can be accurate. The result in Figure 2.3 illustrates that the 3rd order SAS can be accurate for up to 0.2–0.3 seconds, and would result in noticeable errors for 0.3–0.4 seconds or longer.

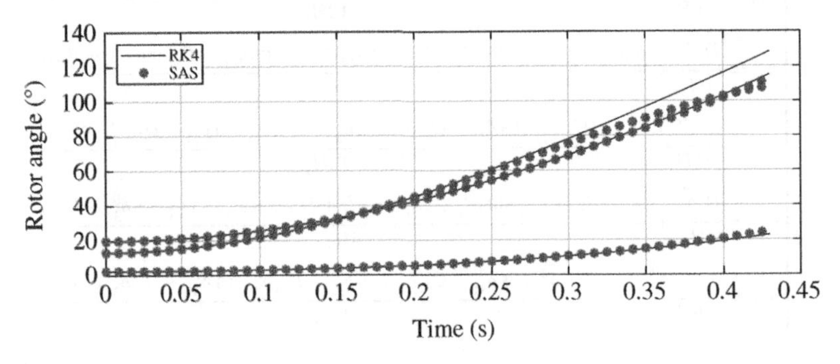

Figure 2.3 Comparison of extended fault-on trajectories of 9-bus system. Source: Bin Wang.

The third example is on the large 179-bus system, whose one-line diagram is shown in Figure 2.4. Figure 2.5 contains the trajectories simulated by 4th-order SAS and RK4, respectively, which are fairly close to each other. This example is included here to show that the SAS can be applied to

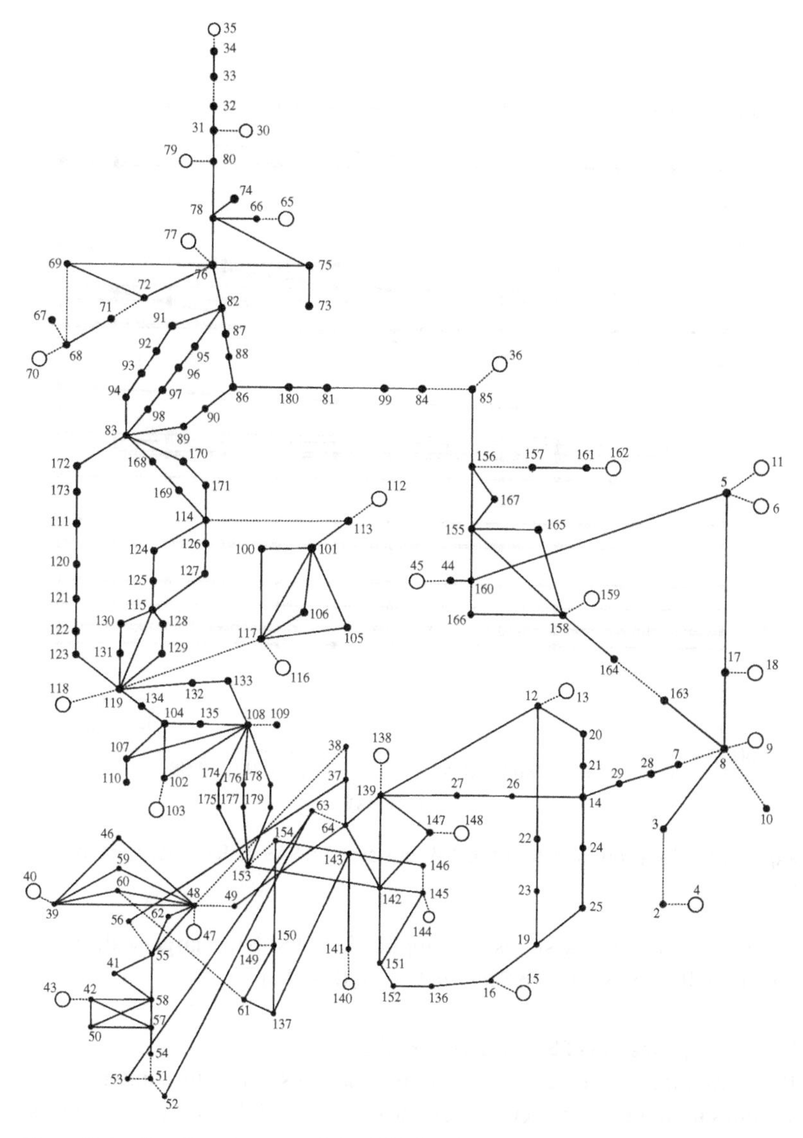

Figure 2.4 One-line diagram of the WECC 179-bus system. Source: [4] (©[2016] IEEE).

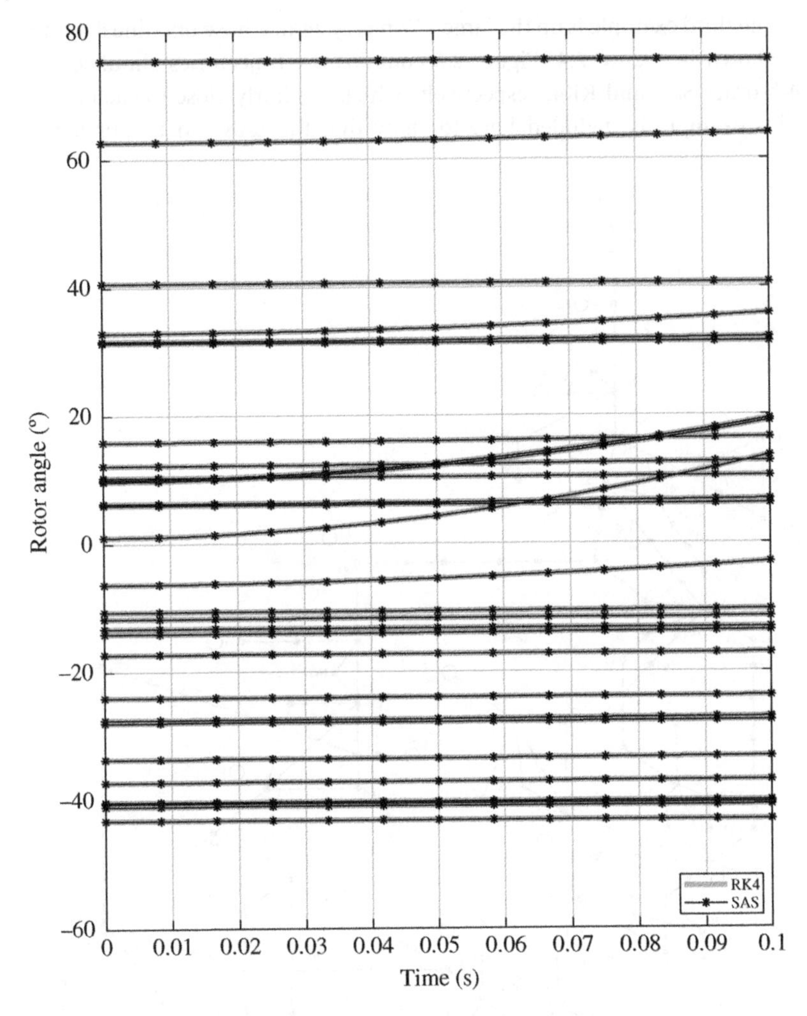

Figure 2.5 Fault-on trajectories of 179-bus system: SAS vs. RK4. Source: Bin Wang.

medium-sized power systems. The application of SAS to a large-scale power system will be presented in Section 2.4.3 of this chapter.

2.1.2.2 Application of SAS in Direct Methods

Direct methods evaluate power system transient stability by comparing the system's total transient energy upon the clearance of the fault, i.e. at the last point of the fault-on trajectory, to the critical transient energy. According to the results in Section 2.1.2.1, SAS can predict an accurate estimate of the system state at the time of the fault clearance based on the analytical form of the SAS without simulating any trajectory in between. With the last point from the fault-on trajectory estimated by SAS, any direct methods can be adopted to assess the transient stability of the post-fault system. Here, the decoupling-based direct method [5] is employed for the demonstration purpose. Details about the decoupling-based direct method are omitted here. Interested readers are referred to [5] for more details.

The above application is demonstrated using different faults with a common fault duration fixed at five cycles. The result is summarized in Table 2.2. The stability margins w.r.t. the two oscillatory modes are respectively shown in columns three and four, where the smaller one is selected as the stability margin of the system, because (i) if the system is stable, then the two modes should both have positive stability margins; and (ii) if any of the two modes has a negative stability margin, the system will be unstable. Column five gives the severity ranking of these faults by the decoupling-based direct method while the last two columns give the ranking by the critical clearing time (CCT) obtained by running time-domain simulations multiple times. It can be seen that the two rankings roughly align with each other, while the difference is mainly caused by the error of the direct method, instead of the small error caused by SAS.

To show a quantified error of the SAS, the largest angle difference between trajectories by SAS and the corresponding ones by RK4 is selected as the error index. Table 2.3 shows the errors of the SAS with a fault duration of 5/60 seconds and a fault duration equal to CCT, respectively, for each fault considered in Table 2.2. It can be seen that (i) if faults are cleared within

Table 2.2 Stability analysis of 10 faults on the IEEE 9-bus system.

Tripped line	Faulty bus	Mode 1	Mode 2	Ranking by ΔV_n	ΔV_n	Ranking by CCT	CCT (s)
5–7	7	2.09	4.0×10^3	1	2.09	1	0.179
7–8	7	5.81	169.91	2	5.81	2	0.195
5–7	5	8.72	4.5×10^3	3	8.72	8	0.353
6–9	9	8.99	145.52	4	8.99	3	0.231
7–8	8	13.02	80.50	5	13.02	5	0.297
8–9	8	20.55	67.29	6	20.55	6	0.324
8–9	9	34.97	21.11	7	21.11	4	0.249
6–9	6	25.90	1.2×10^3	8	25.90	9	0.430
4–6	4	33.19	7.4×10^3	9	33.19	7	0.329
4–6	6	50.98	998.22	10	50.98	10	0.493

Table 2.3 Largest error of angle differences (in degree) by SAS.

Tripped line	Faulty bus	CCT (s)	SAS error when fault duration = 5 cycles	SAS error when fault duration ≈ CCT
5–7	7	0.179	0.001203	0.20
7–8	7	0.195	0.001203	0.34
5–7	5	0.353	0.000486	0.49
6–9	9	0.231	0.000486	0.77
7–8	8	0.297	0.001325	4.35
8–9	8	0.324	0.001325	7.21
8–9	9	0.249	0.000270	1.38
6–9	6	0.430	0.000721	6.69
4–6	4	0.329	0.000474	19.94
4–6	6	0.493	0.000474	41.57

five cycles (by the protection system), the error in SAS is fairly small, e.g. $<2\times10^{-3}$ degree; (ii) the error in SAS is small for faults with small CCTs, while the error is large for faults with large CCTs. This is reasonable since the SAS will be less accurate over a longer time window.

The following gives two examples from Table 2.3. In example one, a three-phase fault is added at bus 9 and cleared after 0.231 seconds by tripping line 6–9 and the fault-on trajectories are shown in Figure 2.2. In example two, a three-phase fault is added at bus 6 and cleared after 0.430 seconds by tripping line 6–9 and the resulting trajectories are shown in Figure 2.3. Although the error increases over a longer time window, that should not discourage the use of the SAS in direct methods because the faults with small CCTs are more severe and more interesting, and their fault-on trajectories can be accurately simulated by SAS. For those faults with large CCTs, they are less severe such that errors from SAS may not negatively impact the stability analysis.

The SAS-based direct method is applied to the 179-bus WECC system to demonstrate its application to a medium-sized power system. Twenty faults used in this example were selected from all $N-1$ line-tripping contingencies by evenly picking cases with low to high CCTs, where the fault duration takes 0.05 seconds for all faults. As a matter of fact, any fault with a CCT lower than 0.05 seconds causes instability, while any fault with a CCT greater than 0.05 seconds does not cause instability. The stability analysis results are summarized in Table 2.4, which shows that all unstable faults are captured by the SAS-based direct method. This demonstrates the applicability of the SAS-based direct method to this medium-sized 179-bus power system.

Table 2.4 Stability analysis of 20 selected faults on the WECC 179-bus system.

Tripped line	Faulty bus	Ranking by ΔV_n	ΔV_n	Ranking by CCT	CCT (s)
31–80	80	1	−1.00	3	0.043
24–25	24	2	−0.98	1	0.026
22–23	23	3	−0.79	2	0.035
114–171	171	4	−0.42	4	0.049
115–127	127	5	−0.01	5	0.071
130–131	130	6	1.81	6	0.095
108–133	108	7	5.71	15	0.392
14–21	21	8	5.92	10	0.225
19–25	19	9	14.15	11	0.259
83–172	172	10	15.98	7	0.102
104–135	104	11	16.35	17	0.511
48–55	55	12	16.94	18	0.584
136–152	136	13	19.61	16	0.428
41–58	41	14	29.30	19	0.757
49–64	49	15	33.30	20	1.228
69–72	69	16	33.54	8	0.138
82–87	82	17	69.04	9	0.178
115–127	115	18	129.19	12	0.286
111–173	173	19	176.95	14	0.376
82–91	91	20	247.32	13	0.315

2.2 Power Series-Based SAS for Simulating Power System DAEs

2.2.1 Power Series-Based SAS for Power System DAEs

A realistic power system model is usually represented by a large set of DAEs, including differential equations (DEs) describing, e.g. generators, associate controllers and dynamic loads, etc., and algebraic equations (AEs) describing the network and static loads. Directly finding the symbolized power series-based SAS for each state variable of such systems is difficult, though not theoretically impossible, mainly because of the difficulty in calculating the inverse of high-dimensional symbolic matrices.

This section introduces a dynamic bus method that can extend the semi-analytical approach to simulating large power systems represented by DAEs. The idea of the dynamic bus method is to express the voltage of the selected bus by an explicit function of time with unknown parameters to be continuously estimated and updated over the simulation. Such a time function should be selected to accurately represent the dynamics of the bus voltage over a period of time. In this section, the power series in time is chosen as the function for each bus voltage. The dynamic bus method is introduced in Section 2.2.1 while detailed steps of the SAS-based simulation involving such dynamic buses will be presented in Section 2.2.2.

For a general g-generator, b-bus, l-load, m-motor power system, denote the sets of all generator, load, and motor buses respectively as $\mathcal{B}$, $\mathcal{G}$, $\mathcal{L}$, and $\mathcal{M}$. Obviously, there are $g \leq b$, $l \leq b$, $m \leq l$. Without loss of generality, consider a 6th order generator model, a 1st order exciter, a 1st order governor and the load represented by ZIP model plus a 3rd order motor model, which are respectively shown in (2.17)–(2.23) [6, 7]. The network equation is shown in (2.24). The initialization of these models can be found in [7].

$$\begin{cases} \dot{\delta}_i = \omega_s \Delta\omega_i \\ \Delta\dot{\omega}_i = (P_{mi} - P_{ei} - D_i\Delta\omega_i)/2H_i \\ \dot{e}'_{qi} = \left(-\dfrac{X_{di} - X''_{di}}{X'_{di} - X''_{di}} e'_{qi} + \dfrac{X_{di} - X'_{di}}{X'_{di} - X''_{di}} e''_{qi} + e_{fdi} \right) / T'_{d0i} \\ \dot{e}'_{di} = \left(-\dfrac{X_{qi} - X''_{qi}}{X'_{qi} - X''_{qi}} e'_{di} + \dfrac{X_{qi} - X'_{qi}}{X'_{qi} - X''_{qi}} e''_{di} \right) / T'_{q0i} \\ \dot{e}''_{qi} = \left(e'_{qi} - e''_{qi} - (X'_{di} - X''_{di}) I_{di} \right) / T''_{d0i} \\ \dot{e}''_{di} = \left(e'_{di} - e''_{di} - (X'_{qi} - X''_{qi}) I_{qi} \right) / T''_{q0i} \end{cases} \tag{2.17}$$

$$\dot{e}_{fdi} = \left(K_i (V_{refi} - |V_i|) - e_{fdi} \right) / T_{ei} \tag{2.18}$$

$$\dot{P}_{mi} = \left(P_{refi} - P_{mi} - \Delta\omega_i / R_i \right) / T_{gi} \tag{2.19}$$

$$\begin{cases} V_{di} + jV_{qi} = V_i e^{-j(\delta_i - \pi/2)} \\ I_{di} = \left((e''_{di} - V_{di}) R_{ai} + (e''_{qi} - V_{qi}) X''_{qi} \right) / (R_{ai}^2 + X''_{di} X''_{qi}) \\ I_{qi} = \left(-(e''_{di} - V_{di}) X''_{di} + (e''_{qi} - V_{qi}) R_{ai} \right) / (R_{ai}^2 + X''_{di}) \\ I_{di} + jI_{qi} = I_{gi} e^{-j(\delta_i - \pi/2)} \\ P_{ei} = V_{di} I_{di} + V_{qi} I_{qi} \end{cases} \tag{2.20}$$

where $i \in \mathcal{G}$; δ_i, $\Delta\omega_i$, e'_{qi}, e'_{di}, e''_{qi}, e''_{di}, e_{fdi}, and P_{mi} are state variables, respectively representing rotor angle, rotor speed deviations, q- and d-axes transient and sub-transient voltages, field voltage and the mechanical power of generator i; V_{di}, V_{qi}, I_{di}, I_{qi}, and P_{ei} are d- and q-axes voltages and currents and the electrical power; V_i is the terminal bus voltage, I_{gi} is the current injection from generator i to the terminal bus; all other symbols are machine parameters.

$$\begin{cases} \dot{s}_i = (T_{\text{load}i} - T_{\text{motor}i})/2H_{\text{m}i} \\ \dot{v}'_{\text{d}i} = -\omega_{\text{s}} r_{\text{r}i}\left((X_{\text{s}i} - X'_{\text{s}i})I_{\text{qm}i} + v'_{\text{d}i}\right)/X_{\text{r}i} + \omega_{\text{s}} s_i v'_{\text{q}i} \\ \dot{v}'_{\text{q}i} = \omega_{\text{s}} r_{\text{r}i}\left((X_{\text{s}i} - X'_{\text{s}i})I_{\text{dm}i} - v'_{\text{q}i}\right)/X_{\text{r}i} - \omega_{\text{s}} s_i v'_{\text{d}i} \end{cases} \tag{2.21}$$

$$\begin{cases} y_i = s_i X_{\text{r}i}/r_{\text{r}i} \\ z_{\text{re}i} = r_{\text{s}i} + y_i(X_{\text{s}i} - X'_{\text{s}i})/(1 + y_i^2) \\ z_{\text{im}i} = Z'_{\text{s}i} + (X_{\text{s}i} - X'_{\text{s}i})/(1 + y_i^2) \\ I_{\text{dm}i} + jI_{\text{qm}i} = V_i/(z_{\text{re}i} + jz_{\text{im}i}) \\ T_{\text{motor}i} = v'_{\text{d}i} I_{\text{dm}i} + v'_{\text{q}i} I_{\text{qm}i} \\ T_{\text{load}i} = f_1 s_i^{i_1} + f_2(1 - s_i)^{i_2} \end{cases} \tag{2.22}$$

where $i \in \mathcal{M}$; s_i, v'_{qi}, and v'_{di} are state variables, representing motor slip, q- and d-axes transient voltages, respectively, of motor i; all other symbols are motor parameters.

$$I_{\text{cc}i} + I_{\text{cp}i} = \frac{p_{i2}P_{i0}\dfrac{|V_i|}{|V_{i0}|} - jq_{i2}Q_{i0}\dfrac{|V_i|}{|V_{i0}|} + p_{i3}P_0 - jq_{i3}Q_0}{V_i^*} \tag{2.23}$$

where $i \in \mathcal{L}$; I_{cci} and I_{cpi} respectively represent the current injections from constant-current load and constant-power load at bus i, while the constant-impedance load is included into the network admittance matrix; V_{i0}, P_{i0}, and Q_{i0} are the initial bus voltage, active and reactive load powers at bus i at the initialization of the dynamic simulation; p_{i2}, q_{i2}, p_{i3}, and q_{i3} are the percentages of constant-current and constant-power components of the total active and reactive load powers at bus i. Note that for load buses with and without motors, we respectively have $p_{i1} + p_{i2} + p_{i3} + p_{im} = 1$ and $p_{i1} + p_{i2} + p_{i3} = 1$ for active load, where p_{i1} and p_{im} represent the percentages of constant-impedance load and motor load, respectively. We have similar equations in q for the reactive load.

$$\mathbf{I}_{\mathcal{B}} = \mathbf{Y}_{\mathcal{B}}\mathbf{V}_{\mathcal{B}} \tag{2.24}$$

where $\mathbf{I}_{\mathcal{B}}$ and $\mathbf{V}_{\mathcal{B}}$ are bus injection current vector and bus voltage vector, $\mathbf{Y}_{\mathcal{B}}$ is the admittance matrix including the constant-impedance load. In this work, since each generator is interfaced with the network as a current source in parallel with an admittance, $\mathbf{Y}_B$ in (2.24) should also contain the admittance of the shunt branch from the equivalent circuit of each generator.

The generator equations (2.17)–(2.20), motor equations (2.21) and (2.22) and the ZIP load equation (2.23) are linked through the network equation in (2.24). Seen from each dynamic element, the impacts from all other elements of the system are fully reflected by the dynamics of its terminal bus voltage, i.e. $V_i(t)$. If the terminal bus is considered a dynamic bus whose voltage can be expressed as a function $V_i(t)$ at least for a short period of time, each individual dynamic element can be completely decoupled from the rest of the system and hence they can all be modeled by ODEs with exogenous control $V_i(t)$ that only depends on time. The resulting system can be solved for its power series-based SAS recursively, similar to that in Section 2.1. The above is the main idea of the dynamic bus method [1] for extending the semi-analytical approach for power system DAEs.

In this chapter, $V_i(t)$ is assumed to be in the form of a power series in time. The polar form is adopted since it is reported to outperform the rectangular form in terms of the extrapolation accuracy [8].

$$V_i(t) = V_{\mathrm{m}i}(t)e^{jV_{\mathrm{a}i}(t)} \approx \left(\sum_{k=0}^{n_v} V_{\mathrm{m}ik}t^k\right)e^{j\left(\sum_{k=0}^{n_v} V_{\mathrm{a}ik}t^k\right)} \tag{2.25}$$

To sum up, with the dynamic bus method, the SAS of the element j on bus i can be written as (2.26), where element j can be a generator or a motor.

$$\mathbf{x}_{ij}(t, \mathbf{x}_{ijt0}, V_i, \mathbf{p}_{ij}) = \mathbf{x}_{ijt0} + \sum_{k=1}^{n} a_k(t - t_0)^k \tag{2.26}$$

where $\mathbf{x}_{ijt0}$ is the initial condition, $\mathbf{p}_{ij}$ represents the parameter vector of the element j on bus i, a_k is the coefficient, which is a function of $\mathbf{x}_{ijt0}$, $\mathbf{p}_{ij}$, $V_{\mathrm{m}i}$ and $V_{\mathrm{a}i}$.

2.2.2 SAS-Based Simulation of Power System DAEs

With the dynamic bus method introduced in Section 2.2.1, the semi-analytical approach can be extended to simulate large power systems modeled by DAEs following the flow chart shown in Figure 2.6. More details about each key step are explained below:

- **Step 1**: All dynamic elements are initialized at the stable equilibrium point of the pre-disturbance system.
- **Step 2**: Disturbances can be added or cleared at any time during the simulation, such as the disconnection or closure of a line, generation trip, and

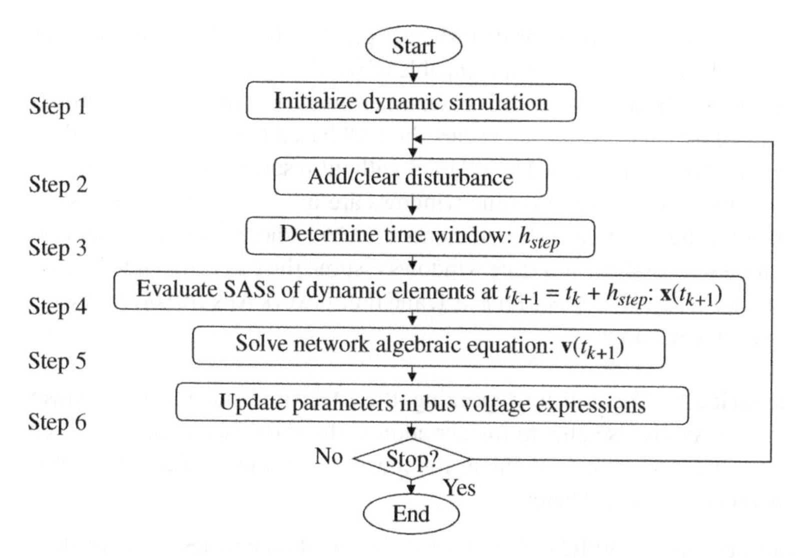

Figure 2.6 Flow chart of SAS-based simulation of power system DAEs using the dynamic bus method. Source: Bin Wang.

load shedding. Note that different system conditions can be represented by the same set of symbolic DAEs with only differences in parameter values. For example, any 3-phase fault or any change in network topology will only change the admittance matrix, such that exactly the same SAS can be used to simulate those disturbances by using proper parameters.

- **Step 3**: The length of the time step for SAS evaluation can be either fixed, e.g. 1 ms, or adaptively selected over the simulation. An adaptive time-stepping method will be introduced in Section 2.3.
- **Step 4**: By the dynamic bus method, the SAS evaluations for all dynamic elements connected to either the same dynamic bus or different dynamic buses become independent of each other and hence their computations are ready for parallel computing.
- **Step 5**: The network equation is solved through a few iterations to achieve a match between the bus current injection in the algebraic power flow model and the net current injection from the generators and loads on the same bus. In each iteration, $\mathbf{V}_B$ has to be solved from (2.24). In this work, the column approximate minimum degree permutation and the LU factorization are adopted to speed up the solution by making use of the sparsity of $\mathbf{Y}_B$.
- **Step 6**: The coefficients of all bus voltages in (2.25) are updated by solving a number of low-order linear least square error problems using several voltage samples from previous time steps. Note that at the beginning of

the simulation or after any disturbance, previous bus voltage samples may either be not available or not capable of predicting bus voltage dynamics to be used in the SAS. In this case, other numerical integration methods can be adopted to start or restart the SAS-based simulation. To achieve a self-starting SAS-based simulation without resorting to other methods, a number of shorter sub-time windows are used to create enough samples for the estimation of coefficients in (2.26), where the SAS evaluations on these very short sub-time windows assume the bus voltages to be constant, i.e. $V_{\text{m}ik} = V_{\text{a}ik} = 0$ for $k \geq 1$, and the network AEs are solved in each sub-time window.

Remark 2.2 Theoretically speaking, the SAS-based simulation of power system DAEs is also able to further address the three factors below, which makes it a promising candidate for practical, online applications. Some remarks are provided below.

More complex models: The dynamic bus method enables the derivation of the power series-based SAS for any dynamic element by adopting a smooth function in time to represent the dynamics of the terminal bus voltage, including wind and solar generation and the saturation effect with a generator. This is because, with the terminal bus voltage expressed as a function in time, the dynamics of any element connected to that bus will be fully determined, and the underlying equations become a set of DEs. Therefore, its power series-based SAS can always be derived. In addition, simulations considering stochastic effects of load and renewable generations can also be achieved by incorporating stochastic processes into the SAS thanks to its symbolic form [9].

Changes in network topology: Any changes in the network topology, e.g. the disconnection or closure of a transmission line, is reflected by using a different set of values for a small number of entries in the admittance matrix $\mathbf{Y}_{\mathcal{B}}$ in (2.24). This does not have any impact on the symbolic format of the derived SAS, since $\mathbf{Y}_{\mathcal{B}}$ is symbolized and ready to take any values without changing the form [10].

Limits in controllers: The limits in generator exciters or other controllers can be considered in the SAS-based simulation by symbolizing necessary intermediate variables whose limits need to be addressed in the SAS expression and then adding a limit-checking procedure after the evaluation at each time step. Once any of such variables meets its upper or lower limit, its value will be frozen at the limit in the following time windows unless it is determined that the variable should be released by the limit-checking procedure.

2.3 Adaptive Time-Stepping Method for SAS-Based Simulation

From Section 2.1.2, we learned that an SAS can be considered accurate when the time step is short enough. As an analytical, approximate solution, the SAS also provides important information to determine the largest time step for simulations [10], given an error tolerance. This section presents an adaptive time-stepping method to find the largest time step satisfying any prespecified error tolerance by leveraging the error-rate upper bound. In addition, it is worth mentioning that the adopted adaptive time-stepping method is not limited to the power series-based SAS, it can also be applied to other SASs introduced in Chapters 3–6 of this book.

2.3.1 Error-Rate Upper Bound

"Error rate" represents the error of an approximate solution accumulated per unit of time. Suppose that an SAS has been obtained, e.g. (2.12). Denote the time derivative of the SAS to be (2.27) and define function $r^{\langle n\rangle}(h)$ as in (2.28), where $h \geq t_0$. Then, we have Theorem 3.2.

$$\dot{\mathbf{x}}_{\text{sas}}^{\langle n\rangle}(t) = \sum_{k=1}^{n} k\mathbf{a}_k(t-t_0)^{k-1} \tag{2.27}$$

$$r^{\langle n\rangle}(h) \triangleq \sup_{t\in[t_0,h]} \|\dot{\mathbf{x}}(t) - \dot{\mathbf{x}}_{\text{sas}}^{\langle n\rangle}(t)\| = \sup_{t\in[t_0,h]} \|\mathbf{f}(\mathbf{x}(t)) - \dot{\mathbf{x}}_{\text{sas}}^{\langle n\rangle}(t)\| \tag{2.28}$$

where $\|\cdot\|$ represents a certain norm, e.g. the infinity norm.

Theorem 2.2 Function $r^{\langle n\rangle}(h)$ defined in (2.28) is an error-rate upper bound of SAS $\mathbf{x}_{\text{sas}}^{\langle n\rangle}(t)$ to the ODE problem defined in (2.1).

Proof: The error of SAS $\mathbf{x}_{\text{sas}}^{\langle n\rangle}(t)$ accumulated from t_0 to the time instance h, is

$$\begin{aligned}\left\|\int_{t_0}^{h}\left(\dot{\mathbf{x}}(\tau) - \dot{\mathbf{x}}_{\text{sas}}^{\langle n\rangle}(\tau)\right)d\tau\right\| &= \left\|\int_{t_0}^{h}\left(\mathbf{f}(\mathbf{x}(\tau)) - \dot{\mathbf{x}}_{\text{sas}}^{\langle n\rangle}(\tau)\right)d\tau\right\| \\ &\leq \int_{t_0}^{h}\left\|\mathbf{f}(\mathbf{x}(\tau)) - \dot{\mathbf{x}}_{\text{sas}}^{\langle n\rangle}(\tau)\right\| d\tau \\ &\leq \int_{t_0}^{h} r^{\langle n\rangle}(h)d\tau \\ &= (h-t_0)r^{\langle n\rangle}(h)\end{aligned} \tag{2.29}$$

Therefore, (2.29) indicates $r^{\langle n\rangle}(h)$ to be an error-rate upper bound by definition. □

$$\frac{\left\| \int_{t_0}^{h} \left(\dot{\mathbf{x}}(\tau) - \dot{\mathbf{x}}_{\text{sas}}^{\langle n\rangle}(\tau) \right) d\tau \right\|}{h - t_0} \leq r^{\langle n\rangle}(h) \tag{2.30}$$

2.3.2 Adaptive Time-Stepping for SAS-Based Simulation

This section will present an adaptive time-stepping method for the SAS-based simulation based on the error-rate upper bound. The SAS is accurate at the initial condition and its error is assumed to monotonically increase over time within a small neighborhood of the initial point, as illustrated in Figure 2.7, one can always find a time instance, say $h_{\max}$, such that the error-rate upper bound is equal to a prespecified error-rate tolerance, say ϵ. Then, the adaptive time window h_{step} for the current simulation step is determined by $h_{\text{step}} = h_{\max} - t_0$. For simplicity, $r^{\langle n\rangle}(h)$ is denoted as $r(h)$ for the rest of this chapter, where the order of the associate SAS will be clearly mentioned.

In the numerical implementation, the time cost for accurately finding $h_{\max}$ can be expensive due to the fact that the evaluation of $r(h)$ has to be done multiple times before time advances. This may undermine the benefit of using an adaptive time window for increasing the simulation speed. Thus, the following approximate but computationally efficient approach is employed for any given error-rate tolerance ϵ.

- **Step 1**: Let the time step determined in the previous step be h_{pre}. For the initial simulation step, h_{pre} can take a small value, e.g. $h_{\text{pre}} = 1$ ms is used in this chapter. In the current time step, calculate the SAS at $t_0 + \alpha h_{\text{pre}}$ and

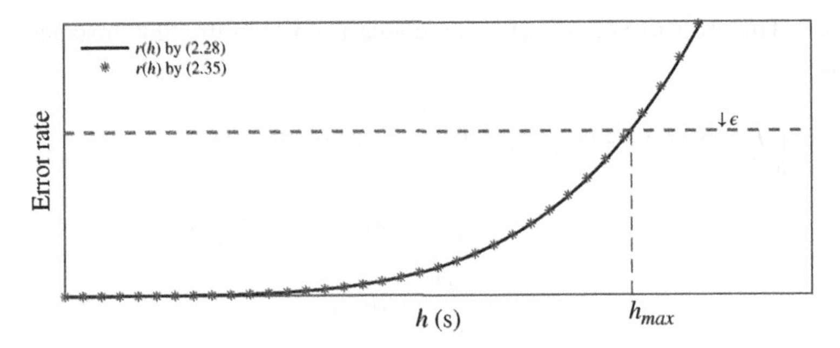

Figure 2.7 Typical curve of the error-rate upper bound. Source: Bin Wang.

the error-rate upper bound $r(t_0 + \alpha h_{\text{pre}})$, where α is a safety factor adopted to maintain the conservativeness. $\alpha = 0.95$ is used in this paper.

- **Step 2**: Approximate $r(h)$ by $\hat{r}(h)$ in (2.31). Use the point from step 1 to solve for μ by (2.32).
- **Step 3**: Approximate $h_{\max}$ by $\hat{h}_{\max}$ in (2.33).
- **Step 4**: Calculate the adaptive time step h_{step} by (2.34) to be used by the next time step.

$$\hat{r}(h) = \mu(e^{h-t_0} - 1) \tag{2.31}$$

$$\mu = \frac{r(t_0 + \alpha h_{\text{pre}})}{e^{\alpha h_{\text{pre}}} - 1} \tag{2.32}$$

$$\hat{h}_{\max} = t_0 + \ln\left(\epsilon \frac{e^{h-t_0} - 1}{r(t_0 + \alpha h_{\text{pre}})} + 1\right) \tag{2.33}$$

$$h_{\text{step}} = \hat{h}_{\max} - t_0 = \ln\left(\epsilon \frac{e^{h-t_0} - 1}{r(t_0 + \alpha h_{\text{pre}})} + 1\right) \tag{2.34}$$

Remark 2.3 There are several important remarks for the error-rate upper bound:

Practical error-rate upper bound: The error-rate upper bound is applicable to any SAS with the help of the assumed exact solution. However, the exact solution does not exist for most dynamical systems requiring simulations. To make the upper bound applicable in practice, the exact solution $\mathbf{x}(t)$ used in the definition (2.28) is replaced by the SAS $\mathbf{x}_{\text{sas}}^{\langle n \rangle}(t)$, leading to an approximate error-rate upper bound in (2.35).

$$r(h) \approx \sup_{t \in [t_0, h]} \|\mathbf{f}\left(\mathbf{x}_{\text{sas}}^{\langle n \rangle}(t)\right) - \dot{\mathbf{x}}_{\text{sas}}^{\langle n \rangle}(t)\| \tag{2.35}$$

Conservativeness: The error-rate upper bound describes the upper bound of the error in SAS accumulated over a time interval. Within the time interval, since the actual error rate is not always equal to the upper bound, the actual accumulated error is often much smaller than that predicted by the upper bound. Thus, using the upper bound may usually result in a conservative, i.e. shorter than the potentially largest, adaptive time step. The conservativeness can be reduced by making use of the observed characteristics of the curve of error-rate upper bound.

Flexibility: The error-rate upper bound can be applied either to each single state variable to make sure the error of a specific quantity, e.g. rotor angle, within a prespecified error rate tolerance, e.g. 0.01°/s, or to the whole state vector to guarantee some overall accuracy.

Self-sufficiency: The approximate error-rate upper bound in (2.35) is self-sufficient, which means that for any derived SAS, its error rate can be bounded without resorting to any other SAS or numerical integration methods.

2.4 Numerical Examples

This section first presents examples using a simple linear dynamical system having a closed-form solution, where the SAS-based simulation with two selected popular numerical integration methods, i.e. RK4 and the backward differentiation formula (BDF) methods are compared with the true solution in terms of the accuracy and adaptive time steps. Then, the application of SAS-based simulation to the New England 39-bus system [10] and the Polish 2383-bus power system [11] are presented to show the accuracy and speedup when simulating large power systems modeled by DAEs.

2.4.1 SAS vs. RK4 and BDF

The linear dynamical system in (2.36) is considered which has a closed-form solution $x_{\text{true}}(t)$ as shown in (2.37). The following examples will compare the SAS-based simulation to simulation results from the RK4 and BDF. The Matlab functions "ode45" and "ode15s" are properly configured to respectively represent RK4 and BDF. The power series-based Nth order SAS is shown in (2.38), where initial state determines a_0 and a_1 by $a_0 = x(0)$, $a_1 = \dot{x}(0)$, and a_k is determined recursively by (2.39) for $k = 2, 3, \ldots, N$. Let the numerical solutions by the RK4 and BDF be $x_{\text{rk4}}(t)$ and $x_{\text{bdf}}(t)$, respectively. The error at time t is defined in (2.40), where $x(t)$ could be any solution by SAS, RK4 or BDF. For all results in this section, the parameters and initial conditions are taken as $\omega = \pi$, $\sigma = -0.1$, $x(0) = 0$, $\dot{x}(0) = \pi$.

$$\ddot{x} - 2\sigma\dot{x} + (\omega^2 + \sigma^2)x = 0 \tag{2.36}$$

$$x_{\text{true}}(t) = e^{\sigma t} \sin \omega t \tag{2.37}$$

$$x_{\text{sas}}^{\langle N \rangle}(t) = \sum_{k=0}^{N} a_k t^k \tag{2.38}$$

$$a_k = \frac{2\sigma(k-1)a_{k-1} - (\omega^2 + \sigma^2)a_{k-2}}{k(k-1)} \tag{2.39}$$

$$e(t) = x(t) - x_{\text{true}}(t) \tag{2.40}$$

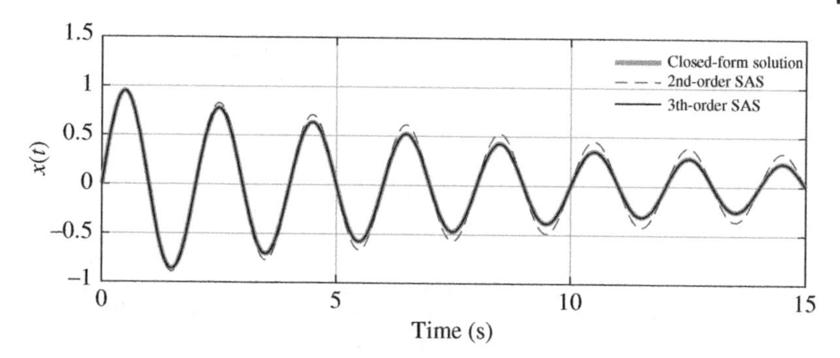

Figure 2.8 Closed-form solution compared with 2nd and 3th order SASs. Source: Bin Wang.

The first example shows the relationship between the accuracy and the order of the SAS when a fixed time window is used for evaluation, e.g. 0.01 seconds is used here. Figure 2.8 shows the 2nd and 3rd order SASs and their comparisons to the closed form solution, where the largest absolute errors are about 0.1 and 0.96×10^{-3}, respectively. Note that under the same condition, the largest absolute error of the RK4 is about 1.3×10^{-3}, comparable to that of the 3rd order SAS. Errors with higher-order SASs become even smaller as shown in Table 2.5.

The second example illustrates the adaptive time step used in the SAS-based simulation. To make a fair comparison to the RK4, the error tolerances with the BDF and SAS are carefully adjusted to make their errors comparable to the default Matlab solver "ode45." For the BDF, it is found that setting option "AbsTol" to be 4×10^{-5} with solver "ode15s" can achieve this. For simulations using the 3rd order to 8th order SASs, the tolerances of error rate, i.e. ϵ, need to take 0.0025, 0.004, 0.006, 0.007, 0.009, and 0.015, respectively. For example, the absolute errors calculated by (2.40) for three methods are shown in Figure 2.9, which are all around the order of 0.001. Then, the fixed time step used by RK4 and the adaptive time steps by BDF and SASs are shown in Figure 2.10, which shows: (i) the BDF and the 4th order SAS have a comparable size of adaptive time steps; (ii) SASs with orders higher than 4 can adopt longer adaptive time steps than BDF.

Table 2.5 Largest absolute SAS errors in time domain.

N	4	5	6	7	8
$\max \lvert e_{\text{sas}}^{\langle N \rangle}(t) \rvert$	7.4×10^{-6}	4.7×10^{-8}	2.4×10^{-10}	1.3×10^{-12}	2.0×10^{-13}

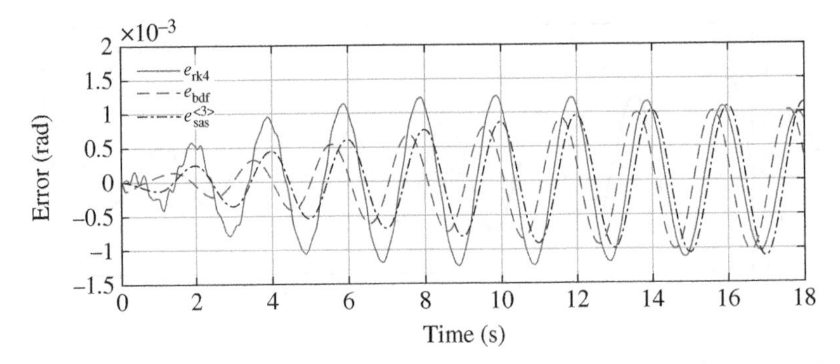

Figure 2.9 Error curves of RK4, BDF, and 3rd order SAS. Source: Bin Wang.

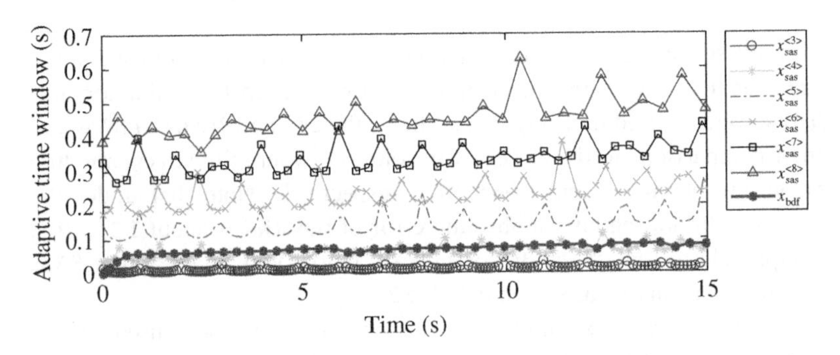

Figure 2.10 Adaptive time steps by BDF and SASs with orders from 3 to 8. Source: Bin Wang.

2.4.2 SAS Derivation

This section shows the offline performance in deriving the power series-based SAS in terms of the time consumption and required memory space for storage. To achieve a fair comparison with Adomian decomposition method (ADM) [10], in this section, (i) a power system model only containing 6th order generators is used; (ii) the dynamic bus approach, which further reduces the complexity of the SAS, is not used for the power series-based SAS; and (iii) all variables and parameters are symbolized when deriving the SAS, including time t, initial state variables, and the reduced admittance matrix.

The time cost for the ADM to offline derive the 2nd order SAS of the New England 39-bus 10-machine power system is about 16 339.7 seconds and this solution stored in the text without compression takes about 15 380 KB memory space [10]. On an Inter CoreTM i7-6700 3.4 GHz desktop computer using Symbolic Math Toolbox in Matlab, deriving the 2nd order power

Table 2.6 Time cost for deriving higher order SASs of 39-bus system.

Order of SAS	3	4	5	6	7	8
CPU time (s)	10.1	22.5	52.6	116.3	246.6	486.0

series-based SAS only costs about 4.4 seconds. The solution stored in the text is much more compact which only takes about 307 KB space, i.e. 2.0% of the space required by ADM. In addition, the time costs for deriving higher-order power series-based SASs are summarized in Table 2.6.

2.4.3 Application of SAS-Based Simulation on Polish 2383-Bus Power System

The Polish power system model adopted in this section has 327 machines, 2383 buses, and 1826 loads [11]. All generators use the model in (2.17)–(2.20). 366 buses, i.e. 20% of all load buses, have motor loads that consume 40% of the total active power load on those buses. The non-motor loads on those 366 buses and the rest 1460 buses are represented by a ZIP load model with 20% constant impedance, 30% constant current, and 50% constant power components. The adopted contingency is a three-phase temporary fault on generator bus 10, the same fault investigated by Gurrala et al. [12]. The fault is cleared after four cycles without disconnecting any line, and the post-fault system is simulated for 10 seconds to cover the period of transient dynamics. To have a reference for comparison, the forward Euler (FE) numerical integration method with fixed time windows is adopted as a fast simulation method, and two cases are created below. The SAS with $n = 2$ and $n_v = 1$ is used for the simulation with adaptive time windows. The error-rate tolerances for all rotor angles, all voltages, and all mechanical powers are respectively selected as 2°/s, 0.01 pu/s, and 0.001 pu/s.

- **Case 1**: A small time step, i.e. 0.2 ms, is used in the FE method to produce a reference result for showing the accuracy of the SAS-based simulation.
- **Case 2**: A 1-ms time step is used to provide a typical time cost to show the improvement achieved by the SAS-based approach.

Figure 2.11a shows the simulated angles in case 1 of five generators near the fault location. Considering all 327 generators, the largest angle difference between the SAS-based approach and the FE method is less than 0.7°, which is much less than the maximum error determined by the error-rate

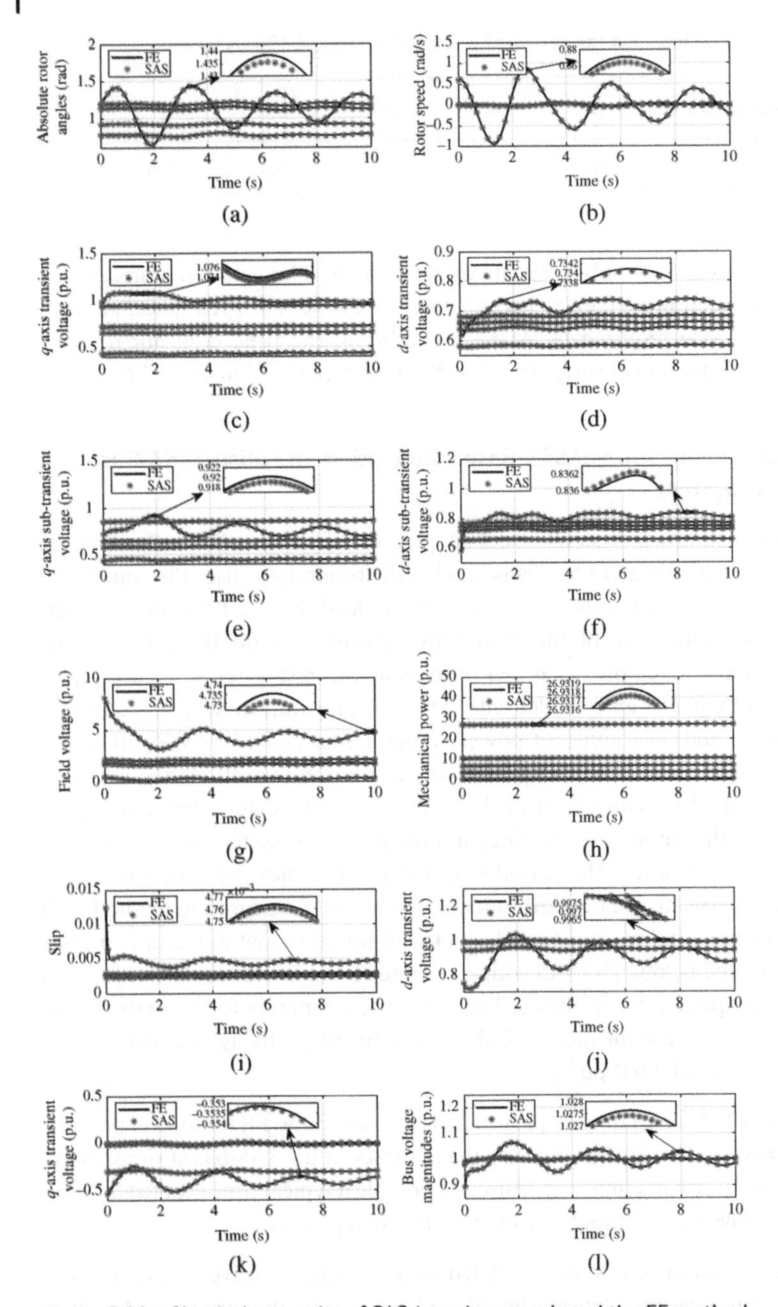

Figure 2.11 Simulation results of SAS-based approach and the FE method. (a) Generator rotor angle, (b) generator rotor speed, (c) generator q-axis transient voltage, (d) generator d-axis transient voltage, (e) generator q-axis sub-transient voltage, (f) generator d-axis sub-transient voltage, (g) generator field voltage, (h) generator mechanical power, (i) motor slip, (j) motor q-axis transient voltage, (k) motor d-axis transient voltage, and (l) bus voltage. Source: Bin Wang.

Table 2.7 Comparison of time cost between SAS and FE on polish 2383-bus system (unit: seconds).

Simulation routine	FE	SAS-sequential	SAS-Ideal parallelization
Generator DEs	5.1	65.9	0.20
Motor DEs	Included to above	88.0	0.24
Network AEs	49.5	6.11	6.11
Adaptive time stepping	N/A	0.36	0.02
Total	54.6	166.9	6.93
# of time steps	10 000	781	781

tolerance, i.e. 20°, reflecting the conservativeness of the adaptive time step. The other seven states of these five generators, the three states of five motors, and their bus voltages near the fault location are respectively illustrated in Figure 2.11b–l. These results demonstrate the SAS-based approach for simulating a power system in DAEs with adaptive time steps while satisfying the given error-rate tolerances.

Table 2.7 summarizes the comparison between the SAS-based approach and the FE method when simulating case 2. For the FE method, most of the time is spent on solving the network AEs, which has to be performed at each time step. Since the fixed time step is small, i.e. 1 ms, then the AEs have to be solved many times, leading to a total of 54.6 seconds time cost for simulating the 10-second dynamics of the system. In the SAS-based simulation without considering parallel computing, denoted as "SAS-Sequential," the time spent on solving network AEs is significantly reduced to 6.11 seconds since the AEs are solved for much fewer times with the longer adaptive time steps used by the SAS of DEs, as shown in Figure 2.12. If all SAS evaluations are performed by a single processor in a sequential manner, that takes as much as 153.9 seconds. Thanks to the fact that the SAS evaluations can be parallelized among multiple processors, if we can afford one processor for each dynamic element, i.e. 693 processors in total for all generators and all motors, and envisage an ideal parallelization, the total time cost of SAS-based simulation of generators and motors will decrease from 153.9 seconds to 0.44 seconds and the overall time cost of the SAS-based simulation will be less than 7 seconds, which is much less than the 54.6 seconds by the FE method. In addition, the SAS evaluation of each dynamic element can be further parallelized if using an ideally tremendous number of processors, i.e. 12.6 million, the overall time cost can be further reduced to about

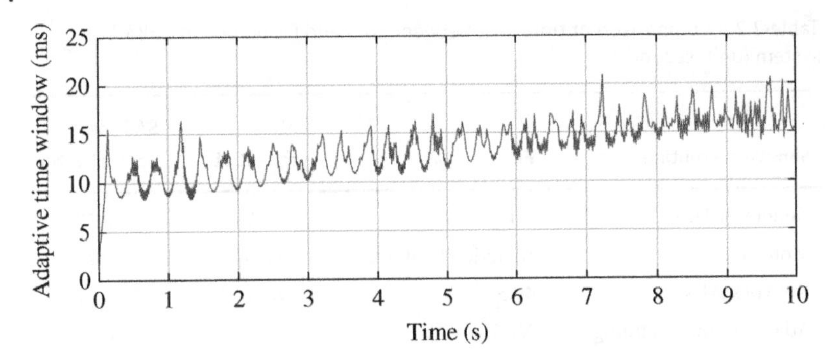

Figure 2.12 Adaptive time windows determined in the SAS-based simulation. Source: Bin Wang.

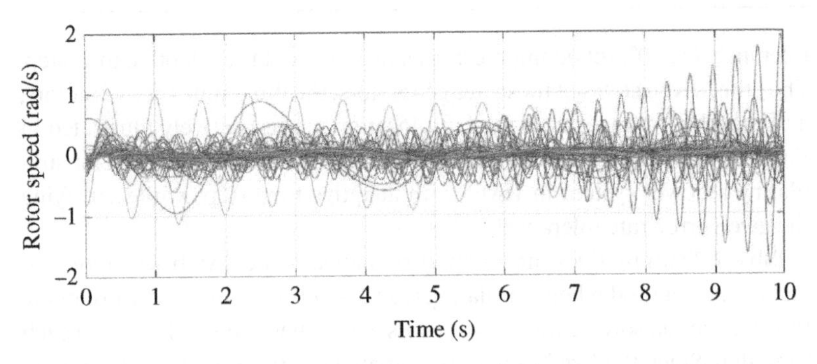

Figure 2.13 Simulated rotor speed by FE with 2 ms fixed time window. Source: Bin Wang.

6.5 seconds in theory. Since that incremental improvement is too expensive, parallelization is recommended for the element level to achieve a high payback.

An argument on the above comparison results could be that the network AEs can also be solved fewer times using the FE method on a longer fixed time step. Thus, the FE method is additionally simulated with longer time windows. It is found that starting from a 2-ms time step, the numerical instability occurs to the FE method, as illustrated in Figure 2.13.

References

1 Wang, B., Duan, N., and Sun, K. (2019). A time-power series-based semi-analytical approach for power system simulation. *IEEE Transactions on Power Systems* 34 (2): 841–851.

2 Coddington, E.A. and Levinson, N. (1955). *Theory of Ordinary Differential Equations*. New York, McGral-Hill.

3 Abreut, E., Wang, B., and Sun, K. (2017). Semi-analytical fault-on trajectory simulation and its application in direct methods. *IEEE Power & Energy Society General Meeting*, 1–5.

4 Maslennikov, S., Wang, B., Zhang, Q. et al. (2016). A test cases library for methods locating the sources of sustained oscillations. *IEEE Power & Energy Society General Meeting*, 1–5.

5 Wang, B., Sun, K., and Su, X. (2015). A decoupling based direct method for power system transient stability analysis. *IEEE Power & Energy Society General Meeting*, 1–5.

6 Milano, F. (2010). *Power System Modelling and Scripting*. London: Springer.

7 Chow, J.H. and Cheung, K.W. (1992). A toolbox for power system dynamics and control engineering education and research. *IEEE Transactions on Power Systems* 7 (4): 1559–1564.

8 Stott, B. (1979). Power system dynamic response calculations. *Proceedings of the IEEE* 67 (2): 219–241.

9 Duan, N. and Sun, K. (2017). Stochastic power system simulation using the adomian decomposition method. https://arxiv.org/abs/1710.02415.

10 Duan, N. and Sun, K. (2017). Power system simulation using the multistage adomian decomposition method. *IEEE Transactions on Power Systems* 32 (1): 430–441.

11 Zimmerman, R.D., Murillo-Sánchez, C.E., and Thomas, R.J. (2011). MATPOWER: steady-state operations, planning, and analysis tools for power systems research and education. *IEEE Transactions on Power Systems* 26 (1): 12–19.

12 Gurrala, G., Dinesha, D.L., Dimitrovski, A. et al. (2017). Large multi-machine power system simulations using multi-stage adomian decomposition. *IEEE Transactions on Power Systems* 32 (5): 3594–3606.

3

Power System Simulation Using Differential Transformation Method

Yang Liu[1] *and Kai Sun*[2]

[1] *Division of Energy Systems and Infrastructure Analysis, Argonne National Laboratory, Lemont, IL, USA*
[2] *Department of Electrical Engineering & Computer Science, University of Tennessee, Knoxville, TN, USA*

3.1 Introduction to Differential Transformation

The section illustrates the concept of the differential transformation (DT) as per Liu et al. [1–3]. The differential transformation method is a powerful mathematical tool for approximating the solutions to nonlinear differential equations. Initially utilized by researchers in the fields of applied mathematics and physics for low-dimensional nonlinear dynamic systems [4, 5], the method was introduced to the power system research by Liu et al. [1–3] to evaluate its effectiveness in addressing real-life, high-dimensional nonlinear networked dynamical systems. Since its introduction, the method has been applied to study various power system problems, including the efficient computation of loading limit in the static voltage stability assessment problem [3], the energy flow in heat and electricity-integrated systems [6], the frequency control of wind turbine generators [7], and the temporal parallelization of high-performance power system simulations [8].

The differential transformation method, like many other semi-analytical methods in power system simulation, approximates solutions of power system models in the form of a polynomial function of time up to a desired order. However, the differential transformation method offers unique advantages over other semi-analytical solution methods.

First, the differential transformation method allows for arbitrary high-order approximations, which are difficult to achieve with other methods. This is because the differential transformation method does not require explicit calculations of high-order derivatives to obtain the power

Power System Simulation Using Semi-Analytical Methods, First Edition. Edited by Kai Sun.

Published 2024 by John Wiley & Sons, Inc.

series coefficients, thanks to the various differential transformation rules for generic nonlinear functions.

Second, the transformation rules provided by the differential transformation methods enable a nonlinear algebraic equation to be converted to a formally linear equation in terms of the power series coefficients, thereby avoiding the iterative calculation required by conventional numerical solvers. The high-order noniterative features of the differential transformation make it a promising approach for solving various power system problems, which are modeled as nonlinear ordinary differential equations (ODEs), differential-algebraic equations (DAEs), or algebraic equations.

In the remainder of this section, the definition and widely used transformation rules of the differential transformation method will be presented in detail.

Definition 3.1 Consider a function $x(t)$ of a real continuous variable t. The differential transformation of $x(t)$ is defined by Eq. (3.1), and the inverse differential transformation of $X(k)$ is defined by Eq. (3.2), where $k \in \mathbb{N}$ is the order.

$$X(k) = \frac{1}{k!}\left[\frac{d^k x(t)}{dt^k}\right]_{t=0} \tag{3.1}$$

$$x(t) = \sum_{k=0}^{\infty} X(k)\, t^k \tag{3.2}$$

The differential transformation method offers a set of transformation rules, as outlined in Propositions 3.1–3.5. Propositions 3.3–3.5 are accompanied by proof examples. To distinguish them from original functions, differential transformations of a function are denoted using capital letters. Besides, the time "t" of original functions is sometimes omitted for simplicity.

Proposition 3.1 Denote $x(t)$, $y(t)$, and $z(t)$ as the original functions and $X(k)$, $Y(k)$, and $Z(k)$ as their differential transformations, respectively. The following propositions (a)–(f) hold, where c is a constant, n is a nonnegative integer, and ϱ is the Kronecker delta function defined in the discrete domain.

(a) $X(0) = x(0)$

(b) $y(t) = cx(t) \rightarrow Y(k) = cX(k)$

(c) $z(t) = x(t) \pm y(t) \rightarrow Z(k) = X(k) \pm Y(k)$

(d) $z(t) = x(t)y(t) \rightarrow Z(k) = \sum_{m=0}^{k} X(m)Y(k-m)$

(e) $$y(t) = t^n \rightarrow Y(k) = \varrho(k-n) = \begin{cases} 1, & k = n \\ 0, & k \neq n \end{cases}$$

(f) $$y(t) = c \rightarrow Y(k) = c\varrho(k) = \begin{cases} c, & k = 0 \\ 0, & k \neq 0 \end{cases}$$

(g) $$y(t) = \frac{dx(t)}{dt} \rightarrow Y(k) = (k+1)X(k+1)$$

Proposition 3.2 If $\phi(t) = \sin\delta(t)$, $\psi(t) = \cos\delta(t)$, and $\Phi(k)$,$\Psi(k)$, and $\Delta(k)$ are the differential transformations of $\phi(t)$, $\psi(t)$, and $\delta(t)$, respectively, then $\Phi(k)$ and $\Psi(k)$ are calculated by:

$$\Phi(k) = \sum_{m=0}^{k-1} \frac{k-m}{k} \Psi(m)\Delta(k-m)$$

$$\Psi(k) = -\sum_{m=0}^{k-1} \frac{k-m}{k} \Phi(m)\Delta(k-m) \tag{3.3}$$

Proposition 3.3 Given function $y_1(t) = e^{x(t)}$, if $Y_1(k)$ and $X(k)$ are the differential transformations of $y_1(t)$ and $x(t)$, respectively, then $Y_1(k)$ is calculated by:

$$Y_1(k) = \frac{1}{k}\sum_{m=0}^{k-1} (k-m)Y_1(m)X(k-m) \tag{3.4}$$

Proposition 3.4 Given function $y_2(t) = \sqrt{x}$, $Y_2(k)$, and $X(k)$ are the differential transformations of $y_2(t)$ and $x(t)$, respectively, and

$$Y_2(k) = \frac{1}{2Y_2(0)}X(k) - \frac{1}{2Y_2(0)}\sum_{m=1}^{k-1} Y_2(m)Y_2(k-m) \tag{3.5}$$

Proposition 3.5 Given function $z(t) = x(t)/y(t)$, if $X(k)$, $Y(k)$, and $Z(k)$ are the differential transformations of $x(t)$, $y(t)$, and $z(t)$, respectively, then $Z(k)$ is calculated by:

$$Z(k) = \frac{1}{Y(0)}X(k) - \frac{1}{Y(0)}\sum_{m=0}^{k-1} Z(m)Y(k-m) \tag{3.6}$$

Proof of Proposition 3.3 From $\dot{y}_1 = y_1\dot{x}$, there is

$$(k+1)Y_1(k+1) = \sum_{m=0}^{k} Y_1(m) \cdot (k+1-m)X(k+1-m)$$

For convenience, it can be written as:

$$Y_1(k+1) = \frac{1}{(k+1)} \sum_{m=0}^{k} (k+1-m) Y_1(m) X(k+1-m)$$

Replacing k by $k-1$ will lead to Eq. (3.4). ■

Proof of Proposition 3.4 From $y_2^2 = x$, there is

$$\sum_{m=0}^{k} Y_2(m) Y_2(k-m) = X(k)$$

$$2Y_2(0)Y_2(k) + \sum_{m=1}^{k-1} Y_2(m) Y_2(k-m) = X(k)$$

Then, it is easy to obtain Eq. (3.5). ■

Proof of Proposition 3.5 Observe that $x = zy$, we have

$$\sum_{m=0}^{k} Z(m) Y(k-m) = X(k)$$

$$Y(0)Z(k) + \sum_{m=0}^{k-1} Z(m) Y(k-m) = X(k)$$

Then, it is easy to obtain Eq. (3.6). ■

In a power system model, currents and voltages are typically expressed using rectangular coordinates in matrix form. To simplify the differential transformation expression, the existing transformation rules for scalar-valued functions can be extended to vectorized transformation rules. This allows for direction application to vector-valued functions without the need to expand them into multiple scalar-valued functions first.

Proposition 3.6 provides six vectorized transformation rules for common vector or matrix operations that appear in power system DAE models. These rules can be easily obtained by applying the existing transformation rules to each element of the vector-valued function. Their proofs are omitted for brevity.

Proposition 3.6 Given $\boldsymbol{x}(t)$ and $\boldsymbol{y}(t)$ as vector-valued functions having differential transformations as $\boldsymbol{X}(k)$ and $\boldsymbol{Y}(k)$, $h(t)$ and $H(k)$ as a scalar function and its differential transformation, and $\boldsymbol{c}$ and $\boldsymbol{d}$ are constant matrices, the transformation rules (a)–(f) hold.

(a) $\boldsymbol{x}(t) \pm \boldsymbol{y}(t) \rightarrow \boldsymbol{X}(k) \pm \boldsymbol{Y}(k)$

(b) $\boldsymbol{x}(t)^T \rightarrow \boldsymbol{X}(k)^T$

(c) $\boldsymbol{cx}(t) \rightarrow \boldsymbol{cX}(k)$

(d) $\boldsymbol{x}(t)\boldsymbol{d} \rightarrow \boldsymbol{X}(k)\boldsymbol{d}$

(e) $\boldsymbol{x}(t)\boldsymbol{y}(t) \rightarrow \boldsymbol{X}(k) \otimes \boldsymbol{Y}(k) = \sum_{m=0}^{k} \boldsymbol{X}(m)\boldsymbol{Y}(k-m)$

(f) $\frac{1}{h(t)}\boldsymbol{y}(t) \rightarrow \frac{1}{H(0)}\left(\boldsymbol{Y}(k) - \sum_{m=0}^{k-1} H(k-m)\boldsymbol{Z}(m)\right)$

3.2 Solving the Ordinary Differential Equation Model

This section focuses on the use of the differential transformation method for solving ODEs in power system modeling. We will begin by explaining the derivation process for typical power system models. Next, we will introduce a solution algorithm for solving ODEs using the differential transformation method. Finally, we will showcase the effectiveness of this method through several case studies.

3.2.1 Derivation Process

3.2.1.1 Governor Model

The equation of a first-order governor model is given in Eq. (3.7), where p_{sv} is the governor output power, p_{ref} is the electrical power set point, and s_m is rotor slip; T_{sv} and R_d are governor time constant and the droop coefficient, respectively.

$$T_{sv}\dot{p}_{sv} = -p_{sv} + p_{ref} - \frac{1}{R_d}s_m \tag{3.7}$$

For illustration, the differential transformations of the governor model in (3.7) is derived using Proposition 3.1, as explained in detail in the following:

Step 1: The left-hand side term $T_{sv}\dot{p}_{sv}$ is a product of a constant and the derivative of p_{sv}. The differential transformation of $\dot{p}_{sv}$ can be obtained from Proposition 3.1(g):

$$\dot{p}_{sv} \rightarrow (k+1)P_{sv}(k+1)$$

Then the differential transformation of $T_{sv}\dot{p}_{sv}$ can be obtained from Proposition 3.1(b):

$$T_{sv}\dot{p}_{sv} \rightarrow (k+1)T_{sv}P_{sv}(k+1)$$

Step 2: Similarly, the differential transformations of three right-hand-side terms can be obtained using Proposition 3.1(b):

$$\begin{aligned} -p_{sv} &\rightarrow -P_{sv}(k) \\ p_{ref} &\rightarrow p_{ref}\varrho(k) \\ -\frac{1}{R_d}s_m &\rightarrow -\frac{1}{R_d}S_m(k) \end{aligned}$$

Step 3: Finally, the original equation becomes

$$(k+1)T_{sv}P_{sv}(k+1) = -P_{sv}(k) + p_{ref}\varrho(k) - \frac{1}{R_d}S_m(k) \tag{3.8}$$

3.2.1.2 Turbine Model

Consider the first-order turbine model given in Eq. (3.9) about mechanical power p_m with a time constant T_{ch}:

$$T_{ch}\dot{p}_m = -p_m + p_{sv} \tag{3.9}$$

Similar to the above procedure for the governor model, apply the differential transformation to both sides:

$$(k+1)T_{ch}P_m(k+1) = -P_m(k) + P_{sv}(k) \tag{3.10}$$

3.2.1.3 Power System Stablizer Model

A PSS Type I model is given in Eq. (3.11), where v_s is the PSS output voltage that is used to modify the exciter reference voltage; the input signal is the rotor speed w, electrical power p_e, and bus voltage magnitude v_t; T_w is the time constant and K_w is the stabilizer gain.

$$\begin{aligned} T_w\dot{v}_1 &= -(K_w w + K_p p_e + K_v v_t + v_1) \\ v_s &= K_w w + K_p p_e + K_v v_t + v_1 \end{aligned} \tag{3.11}$$

Similar to the procedure for the governor and turbine model, the differential transformation of the PSS model is written in Eq. (3.12).

$$\begin{aligned} (k+1)T_wV_1(k+1) &= -(K_wW(k) + K_pP_e(k) + K_vV_t(k) + V_1(k)) \\ V_s(k) &= K_wW(k) + K_pP_e(k) + K_vV_t(k) + V_1(k) \end{aligned} \tag{3.12}$$

3.2.1.4 Synchronous Machine Model

A detailed sixth-order synchronous machine model is given in Eqs. (3.13)–(3.16), including a coordinate transform at the terminal bus for the network interface under the constant impedance load assumption.

$$T'_{q0}\dot{e}'_d = -\frac{x_q - x''_q}{x'_q - x''_q}e'_d + \frac{x_q - x'_q}{x'_q - x''_q}e''_d$$

$$T'_{d0}\dot{e}'_q = -\frac{x_d - x''_d}{x'_d - x''_d}e'_q + \frac{x_d - x'_d}{x'_d - x''_d}e''_q + e_{fd}$$
$$T''_{q0}\dot{e}''_d = e'_d - e''_d + \left(x'_q - x''_q\right) i_q$$
$$T''_{d0}\dot{e}''_q = e'_q - e''_q - \left(x'_d - x''_d\right) i_d$$
$$\dot{\delta} = \omega_s s_m$$
$$2H\dot{s}_m = p_m - p_e - Ds_m \tag{3.13}$$

$$p_e = v_d i_d + v_q i_q$$
$$\begin{bmatrix} i_d \\ i_q \end{bmatrix} = \begin{bmatrix} r_a & -x''_q \\ x''_d & r_a \end{bmatrix}^{-1} \left(\begin{bmatrix} e''_d \\ e''_q \end{bmatrix} - \begin{bmatrix} v_d \\ v_q \end{bmatrix} \right) \tag{3.14}$$

$$i = \boldsymbol{Y_r} v \tag{3.15}$$

$$\begin{bmatrix} i_x \\ i_y \end{bmatrix} = \boldsymbol{R} \begin{bmatrix} i_d \\ i_q \end{bmatrix}, \begin{bmatrix} v_x \\ v_y \end{bmatrix} = \boldsymbol{R} \begin{bmatrix} v_d \\ v_q \end{bmatrix}, \quad \text{where } \boldsymbol{R} = \begin{bmatrix} \sin\delta & \cos\delta \\ -\cos\delta & \sin\delta \end{bmatrix} \tag{3.16}$$

State variables δ, ω, e'_q, e'_d, e''_q, and e''_d, are, respectively, the rotor angle, rotor speed, q-axis, and d-axis transient voltages and sub-transient voltages; p_e, i_d, and i_q are the electrical power, d-axis stator current, and q-axis stator currents, respectively; v_d and v_q are the d-axis and q-axis terminal voltages; i_x and i_y are the x-axis and y-axis terminal currents; v_x and v_y are the x-axis and y-axis terminal voltages, respectively; p_m is the mechanical power; e_{fd} is field voltage; H is the inertia and D is the damping constant; T'_{d0}, T'_{q0}, T''_{d0}, and T''_{q0} are the open circuit transient time constants and sub-transient time constants in d-axis and q-axis; x_d, x_q, x'_d, x'_q, x''_d, and x''_q are the d-axis and q-axis synchronous reactances, transient reactances, and transient reactances; $\omega_s = 2\pi \times 60$ is the nominal frequency and $\boldsymbol{Y_r}$ is the reduced network admittance matrix.

The differential transformations of Eqs. (3.13)–(3.16) are given in Eqs. (3.17)–(3.20), where $\Phi(m)$ and $\Psi(m)$ denote the differential transformations of $\sin\delta$ and $\cos\delta$ obtained by Proposition 3.2.

$$(k+1)T'_{q0}E'_d(k+1) = -\frac{x_q - x''_q}{x'_q - x''_q}E'_d(k) + \frac{x_q - x'_q}{x'_q - x''_q}E''_d(k)$$
$$(k+1)T'_{d0}E'_q(k+1) = -\frac{x_d - x''_d}{x'_d - x''_d}E'_d(k) + \frac{x_d - x'_d}{x'_d - x''_d}E''_q(k) + E_{fd}(k)$$
$$(k+1)T''_{q0}E''_d(k+1) = E'_d(k) - E''_d(k) + \left(x'_q - x''_q\right) I_q(k)$$
$$(k+1)T''_{d0}E''_q(k+1) = E'_q(k) - E''_q(k) - \left(x'_d - x''_d\right) I_d(k)$$
$$(k+1)\Delta(\mathrm{k}+1) = \omega_s S_m(k)$$

$$2(k+1)HS_m(k+1) = P_m(k) - P_e(k) - DS_m(k) \tag{3.17}$$

$$\begin{bmatrix} I_d(k) \\ I_q(k) \end{bmatrix} = \begin{bmatrix} r_a & -x_q'' \\ x_d'' & r_a \end{bmatrix}^{-1} \left(\begin{bmatrix} E_d''(k) \\ E_q''(k) \end{bmatrix} - \begin{bmatrix} V_d(k) \\ V_q(k) \end{bmatrix} \right)$$

$$P_e(k) = \sum_{m=0}^{k} [V_d(m)I_d(k-m) + V_q(m)I_q(k-m)] \tag{3.18}$$

$$I(k) = Y_r V(k) \tag{3.19}$$

$$\begin{bmatrix} I_x(k) \\ I_y(k) \end{bmatrix} = \sum_{m=0}^{k} \begin{bmatrix} \Phi(m) & \Psi(m) \\ -\Psi(m) & \Phi(m) \end{bmatrix} \begin{bmatrix} I_d(m-k) \\ I_q(m-k) \end{bmatrix}$$

$$\begin{bmatrix} V_x(k) \\ V_y(k) \end{bmatrix} = \sum_{m=0}^{k} \begin{bmatrix} \Phi(m) & \Psi(m) \\ -\Psi(m) & \Phi(m) \end{bmatrix} \begin{bmatrix} V_d(m-k) \\ V_q(m-k) \end{bmatrix} \tag{3.20}$$

3.2.1.5 Exciter Model

The IEEE Type I exciter model is given in Eqs. (3.21)–(3.23), where e_{fd}, v_f, v_{ts}, and v_r are the field voltage, feedback voltage, sensed terminal voltage, and regulator voltage, respectively; T_e, T_f, T_r, and T_a are exciter time constant, feedback time constant, the filter time constant, and regulator time constant, respectively; K_e, K_f, and K_a are exciter constants related to the self-excited field, feedback gain, and regulator gain, respectively; s_e is the exciter saturation function, which is a nonlinear exponential function determined by the two constants a_e and b_e; v_t is the bus voltage, which is a nonlinear square root function; and v_{rmax} and v_{rmin} are the maximum and minimum voltage regulator outputs.

$$\begin{aligned} T_e \dot{e}_{fd} &= v_r - s_e e_{fd} - K_e e_{fd} \\ T_f \dot{v}_f &= -v_f + K_f \dot{e}_{fd} \\ T_r \dot{v}_{ts} &= -v_{ts} + v_t \\ T_a \dot{v}_r &= \begin{cases} -v_r + K_a(v_{ref} + v_s - v_{ts} - v_f), \\ \quad if\, v_{rmin} < v_r < v_{rmax} \\ 0, if\, (v_r = v_{rmax}, \dot{v}_r > 0), OR \\ \quad (v_r = v_{rmin}, \dot{v}_r < 0) \end{cases} \end{aligned} \tag{3.21}$$

$$s_e = a_e e^{b_e e_{fd}} \tag{3.22}$$

$$v_t = \sqrt{v_x^2 + v_y^2} \tag{3.23}$$

The differential transformation of Eq. (3.21) is given in Eq. (3.24). Since the differential transformation can handle the discrete events in Eq. (3.21) by deriving the differential transformation for both expressions as shown in the last two equations in (3.24), they can be switched in the simulation as the traditional method does.

$$
\begin{aligned}
&(k+1)T_eE_{fd}(k+1) = V_r(k) - K_eE_{fd}(k) - \sum_{m=0}^{k} S_e(m)E_{fd}(k-m) \\
&(k+1)T_fV_f(k+1) = -V_f(k) + (k+1)K_fE_{fd}(k+1) \\
&(k+1)T_rV_{ts}(k+1) = -V_{ts}(k) + V_t(k) \\
&(k+1)T_aV_r(k+1) = -V_r(k+1) \\
&\qquad\qquad + K_a(V_{ref}\delta(k) + V_s(k) - V_{ts}(k) - V_f(k)) \\
&(k+1)T_aV_r(k+1) = 0
\end{aligned} \tag{3.24}
$$

Both Eqs. (3.22) and (3.23) are composite nonlinear functions. To obtain their differential transformations, Propositions 3.3 and 3.4 for the composite exponential function and composite square root function are applied. From Proposition 3.3, the differential transformation of the saturation function is given in Eq. (3.25).

$$
S_e(k) = a_e\frac{1}{k}\sum_{m=0}^{k-1}(k-m)S_e(m)E_{fd}(k-m) \tag{3.25}
$$

From Proposition 3.4, the differential transformation of terminal voltage is given in Eqs. (3.26) and (3.27), where $u = v_t^2 = v_x^2 + v_y^2$, and $U(k)$ is its differential transformation.

$$
U(k+1) = \sum_{m=0}^{k} V_X(m)V_X(k-m) + \sum_{m=0}^{k} V_Y(m)V_Y(k-m) \tag{3.26}
$$

$$
V_t(k+1) = \frac{1}{2U(0)}V_t(k) - \frac{1}{2U(0)}\sum_{m=1}^{k-1} U(m)U(k-m) \tag{3.27}
$$

3.2.2 Solution Algorithm

The variables of the above power system ODE model are given in Eq. (3.28).

$$
\begin{aligned}
&x(t) = \left[p_{sv}, p_m, v_s, e'_d, e'_q, e''_d, e''_q, \delta, s_m, e_{fd}, v_f, v_r, v_{ts}\right] \\
&y(t) = [i_d, i_q, p_e, i_x, i_y, v_d, v_q, v_x, v_y, s_e, v_t]
\end{aligned} \tag{3.28}
$$

In the differential transformation-based dynamic simulation scheme, these variables are approximated by the Kth-order power series. After obtaining the coefficients, the trajectory can be obtained by evaluating

Eq. (3.29). Thus, the key is to calculate the coefficients $X(k)$, $Y(k)$, $k = 0, 1, 2, \ldots, K$ defined by Eq. (3.30).

$$\begin{aligned} x(t) &= X(0) + X(1)t + X(2)t^2 + \cdots X(K)t^K \\ y(t) &= Y(0) + Y(1)t + Y(2)t^2 + \cdots Y(K)t^K \end{aligned} \tag{3.29}$$

$$\begin{aligned} X(k) &= \left[P_{sv}, P_m, V_s, E'_d, E'_q, E''_d, E''_q, \Delta, S_m, E_{fd}, V_f, V_r, V_{ts}\right] \\ Y(k) &= [I_d, I_q, P_e, I_x, I_y, V_d, V_q, V_x, V_y, S_e, V_t] \end{aligned} \tag{3.30}$$

At each simulation step, $\{X(0), Y(0)\}$ are known and $\{X(k), Y(k), k = 1, 2, \ldots, K\}$ are to be solved. Algorithm 3.1 is designed to calculate the $(k+1)$th-order coefficients from the coefficients of orders $0, 1, \ldots, k$. The algorithm is executed $(K - 1)$ times to obtain all the coefficients starting from $\{X(0), Y(0)\}$. Since the operations of differential transformation are purely linear, the calculation process is explicit.

Toy Example: The differential transformation-based simulation scheme is illustrated below using a single-machine infinite-bus system given in Eq. (3.31) with parameters $H = 3$ seconds, $D = 3$ p.u., $\omega_s = 2\pi \times 60$ rad/s, $P_{max} = 1.7$ p.u., $P_m = 0.44$ p.u., and the initial state $\delta(0) = 0.26$ rad and $\omega(0) = 0.002$ p. u.

$$\begin{cases} \dot{\delta} = \omega_s \omega \\ \dot{\omega} = \dfrac{1}{2H}(P_m - P_{max} \sin \delta - D\omega) \end{cases} \tag{3.31}$$

To obtain the trajectories of rotor angle and rotor speed, the first step is to apply the differential transformation to Eq. (3.31). Similar to the steps described in Section 3.2.1, its differential transformation is given in Eqs. (3.32) and (3.33):

Algorithm 3.1 Solve Coefficients for the ODE Model
Input: $X(0:k)$, $Y(0:k)$
Output: $X(0:k+1)$, $Y(0:k+1)$

1. Calculate $P_{sv}(k+1)$ from Eq. (3.8).
2. Calculate $P_m(k+1)$ from Eq. (3.10).
3. Calculate $E'_d(k+1), E'_q(k+1), E''_d(k+1), E''_q(k+1), \Delta(k+1), S_m(k+1)$ from Eq. (3.17).
4. Calculate $I_d(k)$, $I_q(k)$, $P_e(k)$ from Eq. (3.18).
5. Calculate $V_x(k)$, $V_y(k)$ from Eq. (3.19).
6. Calculate $\Phi(k)$, $\Psi(k)$ from Eq. (3.3).
7. Calculate $V_d(k)$, $V_q(k)$, $I_x(k)$, $I_y(k)$ from Eq. (3.20).
8. Calculate $S_e(k)$, $V_t(k)$ from Eqs. (3.25)–(3.27).
9. Calculate $E_{fd}(k+1)$, $V_r(k+1)$, $V_f(k+1)$, $V_{ts}(k+1)$ from Eq. (3.24).
10. Calculate $V_s(k+1)$ from Eq. (3.12).

$$(k+1)\Delta(k+1) = \omega_s W(k) \tag{3.32}$$

$$2H(k+1)W(k+1) = P_m\varrho(k) - P_{max}\Phi(k) - DW(k) \tag{3.33}$$

With the values of parameters plugged in, the recursive formula to calculate the coefficients becomes

$$\begin{aligned} \Delta(k+1) &= \frac{377}{k+1}W(k) \\ W(k+1) &= \frac{1}{k+1}[0.073\varrho(k) - 0.283\Phi(k) - 0.5W(k)] \end{aligned} \tag{3.34}$$

The coefficients $\Delta(0)$, $W(0)$, $\Phi(0)$, and $\Psi(0)$, are obtained in Eq. (3.35) using Proposition 3.1(a).

$$\begin{aligned} \Delta(0) &= \delta(0) = 0.26 \\ W(0) &= w(0) = 2 \times 10^{-3} \\ \Phi(0) &= \sin(\delta(0)) = 0.2571 \\ \Psi(0) &= \cos(\delta(0)) = 0.9664 \end{aligned} \tag{3.35}$$

The values of $\Delta(k)$, $W(k)$ for arbitrary k can be calculated starting from Eq. (3.35). For example, the coefficients with $k = 1$ are calculated in Eq. (3.36) with $\varrho(0) = 1$.

$$\begin{aligned} \Delta(1) &= 377W(0) = 0.754 \\ W(1) &= 0.073 - 0.283\Phi(0) - 0.5W(0) = -7.6 \times 10^{-4} \\ \Phi(1) &= \Psi(0)\Delta(1) = 0.7287 \\ \Psi(1) &= -\Phi(0)\Delta(1) = -0.1938 \end{aligned} \tag{3.36}$$

Such a recursive process continues until k reaches a desired order K. The rotor angle and rotor speed expressions for $K = 3$ are approximated by

$$\begin{aligned} \delta(t) &= \Delta(0) + \Delta(1)t + \Delta(2)t^2 + \Delta(3)t^3 \\ w(t) &= W(0) + W(1)t + W(2)t^2 + W(3)t^3 \end{aligned} \tag{3.37}$$

3.2.3 Case Study

In this section, the differential transformation-based simulation scheme is tested on the IEEE 10-machine 39-bus system and the Polish 327-machine 2383-bus system [9]. The accuracy of the differential transformation (DT) approach is validated by various disturbances. The numerical stability, accuracy, and time performance are compared with five commonly used numerical methods [10, 11]: the modified Euler method (ME), fourth-order Runge–Kutta method (RK4), Gill's version of RK method (RKG), Trapezoidal method, and Gear method. In Sections 3.2.3.1, 3.2.3.2, and 3.2.3.3, the benchmark result is given by the RK4 method with a very small time step of 0.3 ms. Errors of each method are considered as its differences from the benchmark result. Simulations are performed in MATLAB R2017a on a computer with an i5-7200U CPU.

3.2.3.1 Scanning Contingencies

The purpose of this test is to demonstrate the differential transformation method's capability and accuracy in detecting various types of contingencies. For the 39-bus system, three stability scenarios, i.e. one stable case, one marginally stable case, and one unstable case, are simulated for 20 seconds. The stable case has a three-phase fault at bus 3 that starts at $t = 1$ seconds and is cleared after five cycles by tripping lines 3–4. The marginal stable and unstable cases are created by clearing the fault after 12.935 cycles and 12.940 cycles, respectively. Figures 3.1–3.3 show the trajectories of rotor angles from both the differential transformation and benchmark RK4 methods, as well as the rotor angle errors of the differential transformation method. Machine 1 at bus 30 is selected as the reference. The results of the differential transformation method with $K = 8$ and the

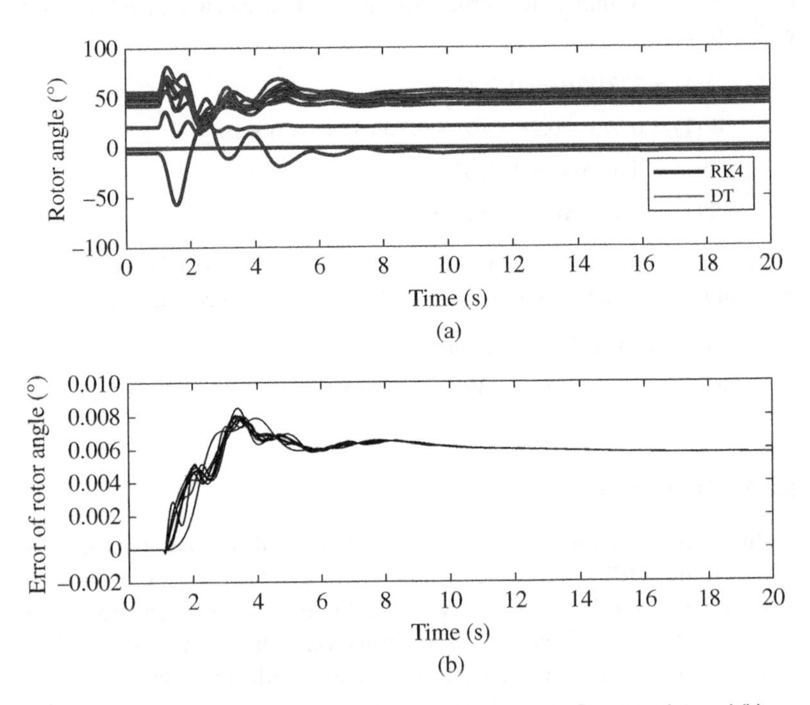

Figure 3.1 39-bus system under a stable contingency. (a) Rotor angles and (b) error of rotor angles.

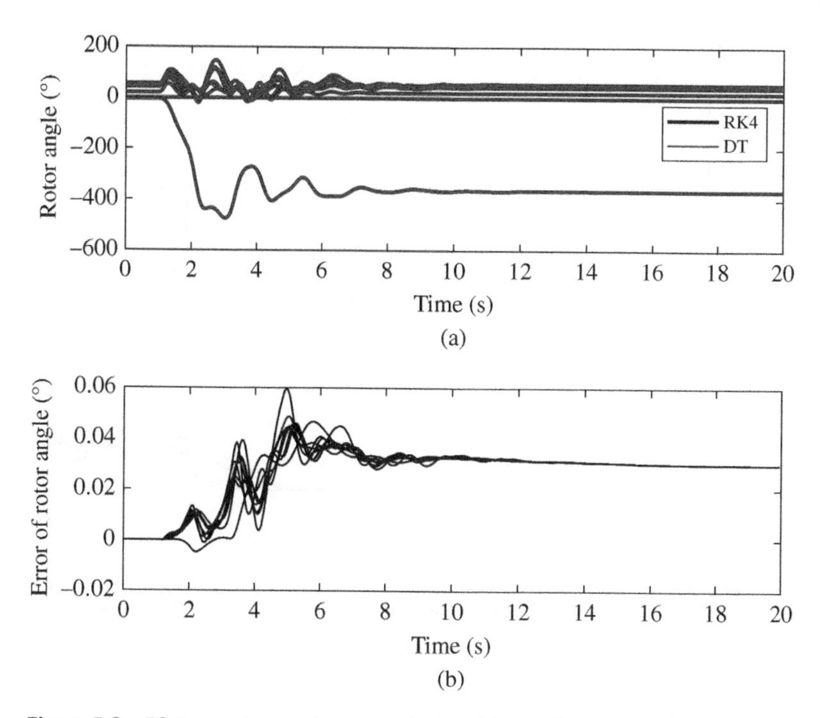

Figure 3.2 39-bus system under a marginal stable contingency. (a) Rotor angles and (b) errors of rotor angles.

time step length of 0.05 seconds accurately match the benchmark results for all three scenarios. The maximum errors of all state variables (including rotor angles, rotor speeds, transient and sub-transient voltages in d-axis and q-axis, field voltages, and all other variables) over the entire 20-second simulation period are 1.48×10^{-4}, 1.00×10^{-3}, and 1.30×10^{-3} p.u. for the three scenarios, respectively.

The same tests are also conducted on the Polish 2383-bus system, and the differential transformation method can still accurately assess stability. Figures 3.4–3.6 show the trajectories of rotor angles from both the differential transformation and RK4 methods, as well as the rotor angle errors of the differential transformation method using the time step length of 0.016 seconds. Machine 1 at bus 10 is selected as the reference. The maximum errors of all state variables over the entire 20-second simulation period are 5.2×10^{-3}, 3.9×10^{-3}, and 3.8×10^{-3} p.u. for the three scenarios, respectively.

To demonstrate the reliable performance of the differential transformation method across various scenarios, the impacts of various factors on the

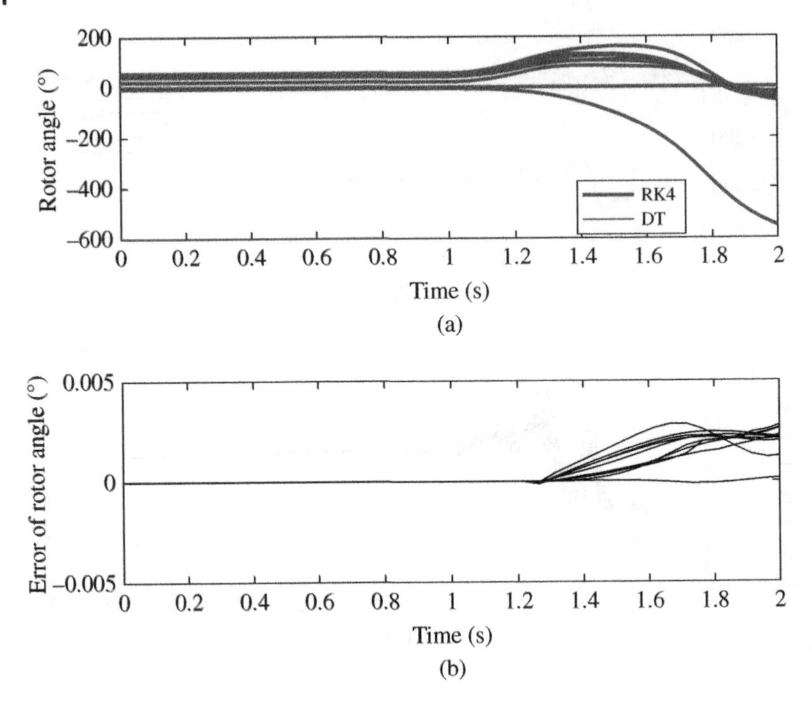

Figure 3.3 39-bus system under marginally unstable contingency. (a) Rotor angles and (b) errors of rotor angles.

simulation of the differential transformation method are analyzed, including fault types, fault locations, the modeling of PSS, and the modeling of saliency. Apply these two types of faults to all buses and lines of the 39-bus system: (i) temporary three-phase short-circuit fault lasting for five cycles on each bus (total, 39 faults); (ii) permanent three-phase short-circuit line fault near each end of every line that is tripped by opening the line after five cycles (total, $46 \times 2 = 92$ faults). Figures 3.7 and 3.8 show that the maximum errors are within 1.1×10^{-3} p.u. for any state variables of the system in all cases. Besides, an 830 MW generator trip on bus 38 is simulated, and the maximum error is 2.26×10^{-4} p.u.

Also, the impact of the PSS is simulated in many cases. Figure 3.9 shows the maximum errors are within 3.0×10^{-4} p.u. and are not affected much by the modeling of PSS. Meanwhile, the impact of saliency is being studied. Figure 3.10 shows the maximum errors are within 2.6×10^{-4} p.u., and the modeling of saliency does not introduce additional errors.

3.2.3.2 Numerical Stability

To study error propagation and numerical stability, three scenarios are designed for the 39-bus system with three different time step lengths: 0.050,

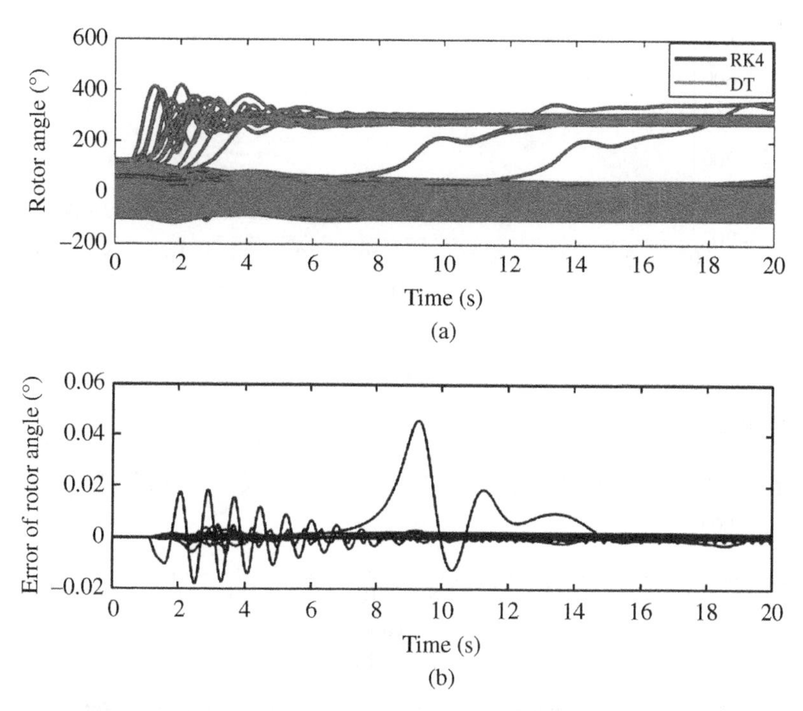

Figure 3.4 2383-bus system under a stable contingency. (a) Rotor angles and (b) errors of rotor angles.

0.067, and 0.100 seconds. The simulated disturbance is a three-phase fault at bus 3, cleared after five cycles by tripping the branch 3–4. Figure 3.11a–c shows the error propagation in the three scenarios where all methods start from the same initial state at $t = 0$ seconds. To take a detailed look at the transient dynamics right after the fault, the figure focuses on the first one second. The logarithmic vertical axis is the maximum error of all state variables at the end of each time step. The errors do not propagate much for the four methods when the time step length is 0.050 seconds. Meanwhile, Figure 3.11b shows that the error of the ME method propagates when the time step length is increased to 0.067 seconds and approaches 10^2 p.u. at $t = 1.0$ seconds, indicating divergence of the ME method. Figure 3.11c shows that the errors of ME, RK4, and RKG methods increase significantly along time steps, indicating the tendency toward numerical instability.

Table 3.1 summarizes whether the four methods tend to be numerically stable or unstable under three scenarios. All methods work well when the time step length is 0.050 seconds. The ME method has a numerical instability issue when the time step length is 0.067 seconds. Only the differential transformation method is numerically stable in all three scenarios.

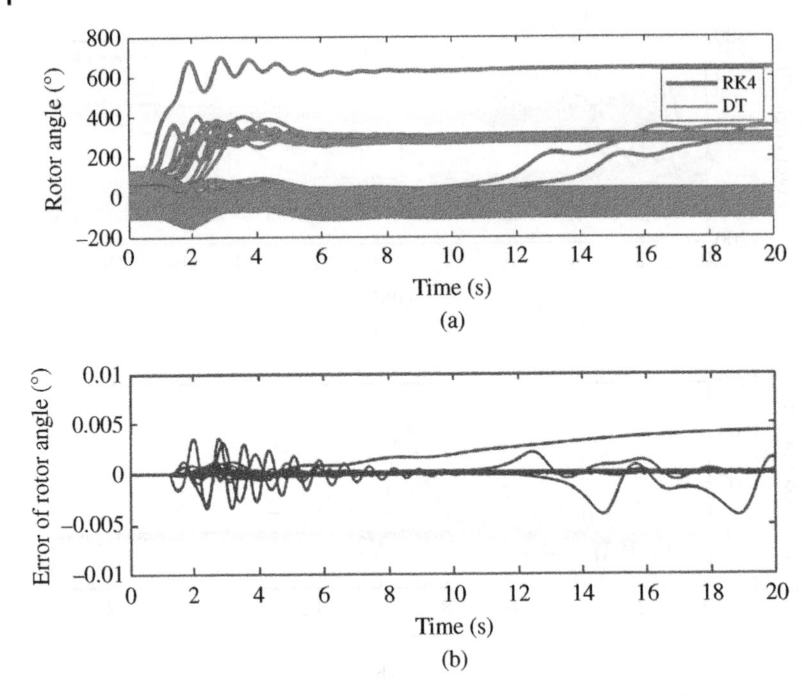

Figure 3.5 2383-bus system under a marginally stable contingency. (a) Rotor angles and (b) errors of rotor angles.

Table 3.1 Comparison of numerical stability under three scenarios.

Time step length (s)	ME	RK4	RKG	DT
0.050	√	√	√	√
0.067	×	√	√	√
0.100	×	×	×	√

Table 3.2 summarizes the maximum time step lengths of the four methods to maintain numerical stability for both the 39-bus system and the 2383-bus system by extensive simulations with gradually increased time step lengths until numerical instability occurs. The differential transformation approach can maintain numerical stability using much larger time step lengths than the ME, RK4, and RKG methods. But it will diverge when the time step length is larger than 0.125 seconds for the 39-bus system and 0.017 seconds for the 2383-bus system. By contrast, the implicit methods such as the

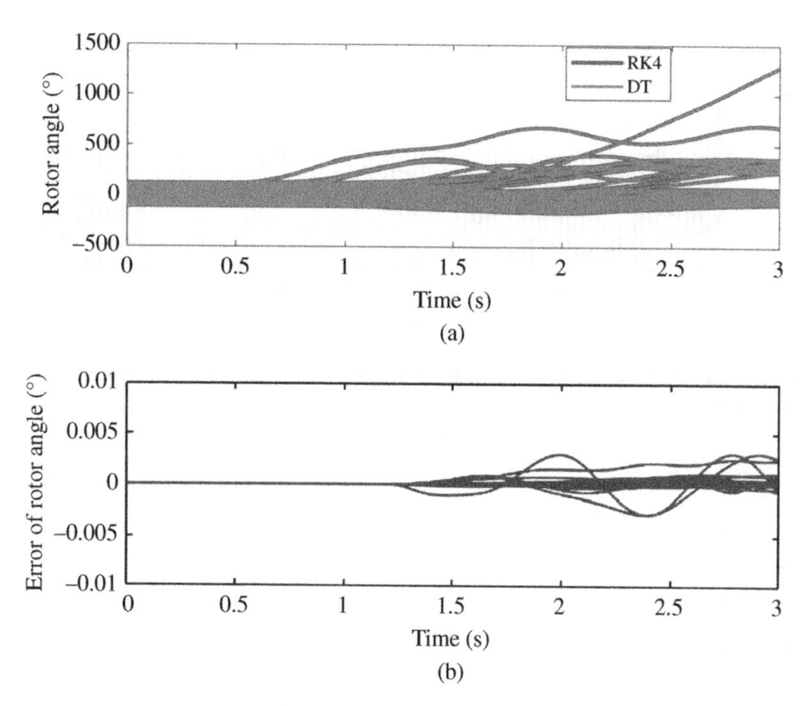

Figure 3.6 2383-bus system under a marginally unstable contingency. (a) Rotors and (b) errors of rotor angles.

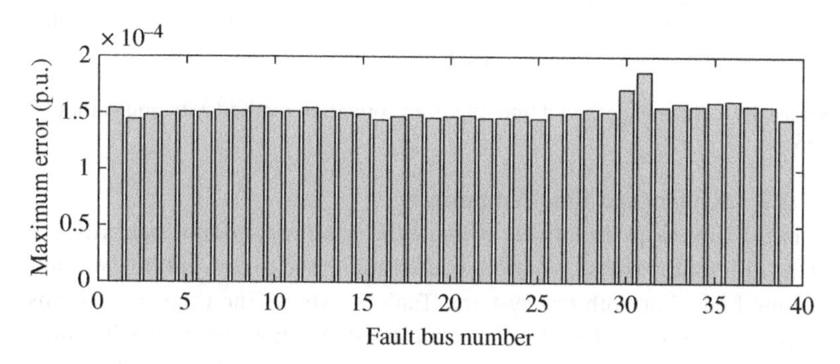

Figure 3.7 Accuracy of the differential transformation method when faults occur at all buses.

Trapezoidal method and the Gear method are very stable, and there is no need to limit their time step lengths from the numerical stability perspective. These results indicate the numerical stability of the differential transformation approach is better than the ME, RK4, and RKG methods but weaker than the Trapezoidal method and the Gear method.

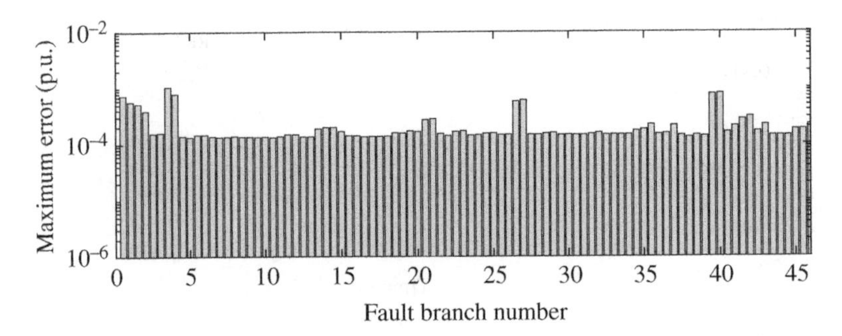

Figure 3.8 Accuracy of the differential transformation method when faults occur at all branches.

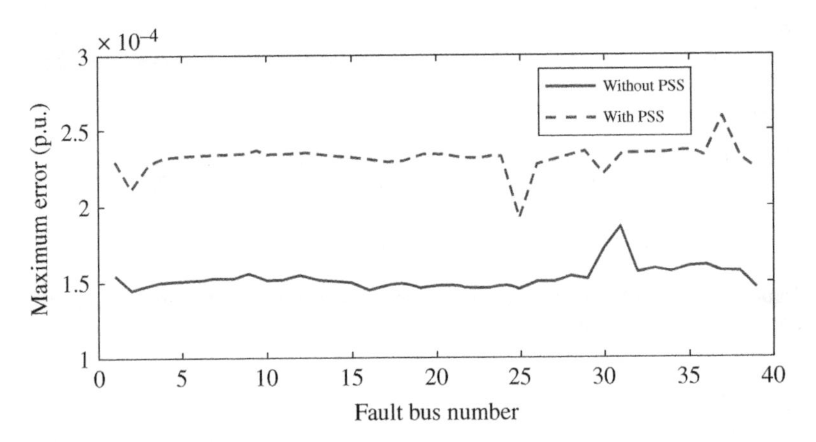

Figure 3.9 Accuracy of the differential transformation method with and without PSS.

3.2.3.3 Accuracy and Time Performance

This test aims to quantitatively evaluate the accuracy and time performance of the differential transformation method compared to conventional numerical methods. For both test systems, Table 3.3 shows the time step lengths of the ME, RK4, RKG, and differential transformation methods with three desired error tolerances, i.e. the maximum error of all state variables over the entire simulation period is in the magnitude of 10^{-2}, 10^{-3}, and 10^{-4} p.u., respectively. For the Trapezoidal and Gear methods, the average time step lengths are shown since they are implemented by MATLAB solvers ode23t and ode15s, respectively, with variable time step lengths.

For the 39-bus system, the time step lengths of the differential transformation can be increased compared with the ME method to 7.5 times, 23.3 times, and 50.0 times while still meeting the requirements of three error tolerances,

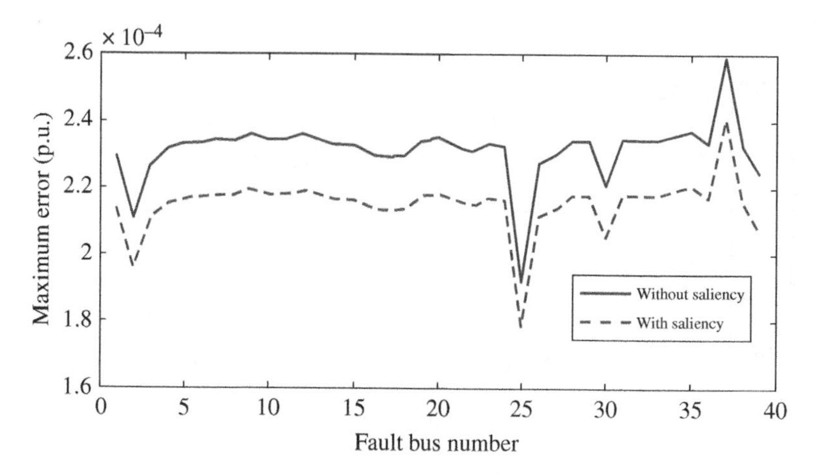

Figure 3.10 Accuracy of the differential transformation method with and without saliency.

Table 3.2 Comparison of maximum time step length to maintain the numerical stability (Unit: s).

Test systems	ME	RK4	RKG	DT
39-bus system	0.059	0.077	0.077	0.125
2383-bus system	0.007	0.010	0.010	0.017

respectively. Compared with both the RK4 and RKG methods, it is increased by 2.0 times, 2.7 times, and 4.5 times. Also, the differential transformation method increases the time step length to 4.5 times and 1.7 times compared to the Trapezoidal and Gear methods, respectively, under the same error tolerance. Similar results are also observed on the 2383-bus system. These results show that the differential transformation approach can prolong the time step lengths with the other methods for the same level of error tolerance.

For both test systems, Table 3.4 shows the computation times of the six methods using the time step lengths in Table 3.3. For the 39-bus system, the time costs of the differential transformation method are reduced by 33.3%, 66.4%, and 86.8% compared to the ME method under three error tolerances, respectively. It is reduced by 15.1%, 17.8%, and 43.2% compared with both the RK4 and RKG methods. Also, the differential transformation reduces the time cost by 61.1% and 11.9% compared to the Trapezoidal and Gear methods, respectively, under the same error tolerances. Similar results are

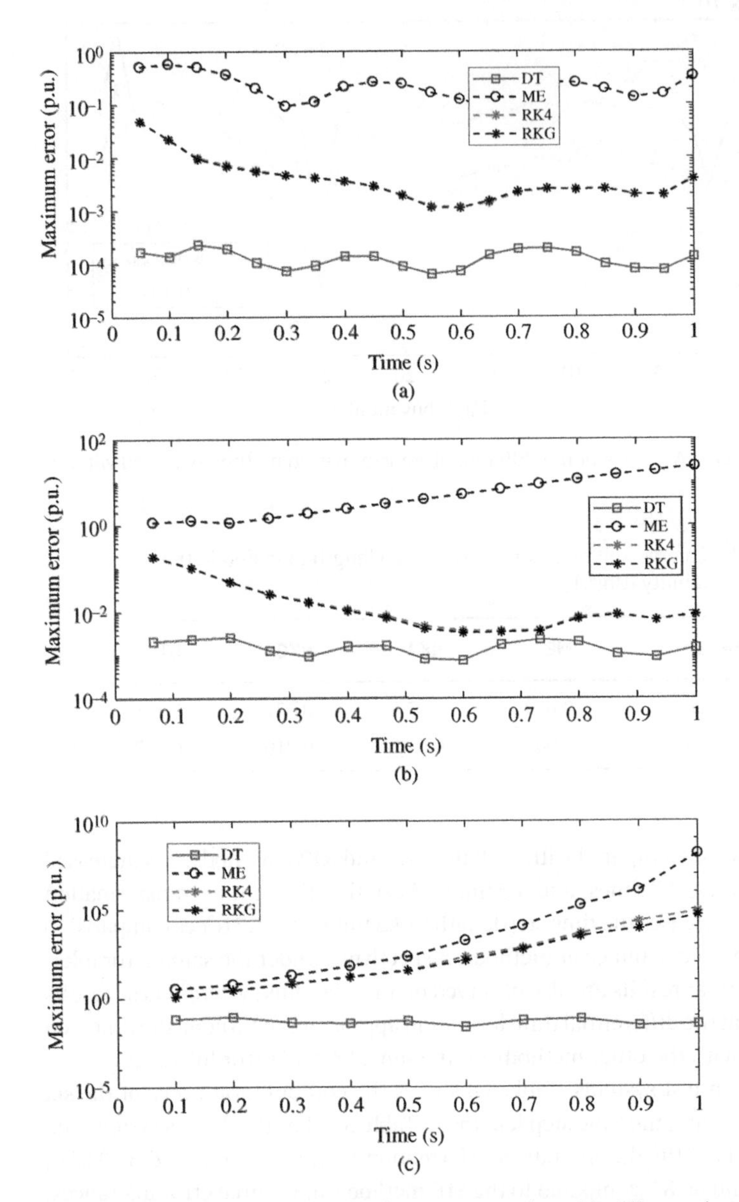

Figure 3.11 Error propagation for the three scenarios with different time steps. (a) Time step = 0.050 seconds, (b) time step = 0.067 seconds, and (c) time step = 0.100 seconds.

Table 3.3 Comparison of time step length under different error tolerances (Unit: s).

Test systems	Error (p.u.)	ME	RK4	RKG	DT	Trapezoidal	Gear
39-bus system	10^{-2}	0.012	0.045	0.045	0.090	0.020	—
	10^{-3}	0.003	0.026	0.026	0.070	—	0.041
	10^{-4}	0.001	0.011	0.011	0.050	—	—
2383-bus system	10^{-2}	0.003	0.010	0.010	0.017	0.004	—
	10^{-3}	0.001	0.007	0.007	0.016	—	0.009
	10^{-4}	0.0005	0.004	0.004	0.012	—	—

Table 3.4 Comparison of computation time under different error tolerances (Unit: s).

Test systems	Error (p.u.)	ME	RK4	RKG	DT	Trapezoidal	Gear
39-bus system	10^{-2}	0.42	0.33	0.33	0.28	0.72	—
	10^{-3}	1.10	0.45	0.46	0.37	—	0.42
	10^{-4}	3.71	0.86	0.87	0.49	—	—
2383-bus system	10^{-2}	4.67	3.33	3.29	3.02	1320	—
	10^{-3}	14.	4.04	4.06	3.67	—	376
	10^{-4}	42.67	8.34	8.29	5.10	—	—

also observed on the 2383-bus system. These results show the differential transformation approach can reduce computation time compared with the other methods for the same level of error tolerance.

3.3 Solving the Differential-Algebraic Equation Model

This section focuses on the use of the differential transformation method for solving DAEs in power system modeling. Solving DAEs requires additional techniques compared to solving ODEs due to the presence of the algebraic nonlinear power flow equation. In this section, we will first introduce the basic ideas of the differential transformation method for DAEs. Then, we will describe the derivation process for how to handle the nonlinear power flow equation using the method. After that, we will present the solution algorithm for solving the DAEs. Finally, we will provide case study results to illustrate the effectiveness of the method.

3.3.1 Basic Idea

Consider the power system DAE model in the state-space representation given in Eq. (3.38), where $\boldsymbol{x}$ is the state vector, $\mathbf{v}$ is the vector of bus voltages, $\boldsymbol{f}$ represents a vector field determined by differential equations on dynamic devices such as synchronous generators and associated controllers, $\boldsymbol{i}$ is the vector-valued function on current injections from all generators and load buses, and $\boldsymbol{Y_{bus}}$ is the network admittance matrix.

$$\begin{aligned} \dot{\boldsymbol{x}} &= \boldsymbol{f}(\boldsymbol{x}, \boldsymbol{v}) \\ \boldsymbol{Y_{bus}}\boldsymbol{v} &= \boldsymbol{i}(\boldsymbol{x}, \boldsymbol{v}) \end{aligned} \tag{3.38}$$

In the presented differential transformation-based solution method, the solution of both state variables and the nonstate variables, bus voltages, is approximated by Kth-order power series in time in Eq. (3.39). The major task is to solve power series coefficients of orders from 0 to K. The two steps to obtain these coefficients are conceptually shown below and then elaborated in Sections 3.3.2 and 3.3.3, respectively.

$$\begin{aligned} \boldsymbol{x} &= \sum_{0}^{K} \boldsymbol{X}(k)t^k \\ \boldsymbol{v} &= \sum_{0}^{K} \boldsymbol{V}(k)t^k \end{aligned} \tag{3.39}$$

The differential transformations of the DAE model (3.38) will be derived in Section 3.3.2 and have the general form in Eq. (3.40). Compared with the original DAE model (3.38), each variable or function $\boldsymbol{x}$, $\boldsymbol{v}$, $\boldsymbol{f}$, and $\boldsymbol{i}$ is transformed to their power series coefficients $\boldsymbol{X}(k)$, $\boldsymbol{V}(k)$, $\boldsymbol{F}(k)$, and $\boldsymbol{I}(k)$ (denoted by their corresponding capital letters), coupled by a new set of equations in (3.40). It can be observed that the left-hand side of Eq. (3.40) only contains the $(k+1)$th-order coefficients of state variables and kth-order coefficients of bus voltages, respectively, while the right-hand side couples the 0th- to kth-order coefficients of both state variables and bus voltages by nonlinear functions $\boldsymbol{F}$ and $\boldsymbol{I}$.

$$(k+1)\boldsymbol{X}(k+1) = \boldsymbol{F}(k) = \boldsymbol{F}(\boldsymbol{X}(l), \boldsymbol{V}(l)), l = 0 \cdots k \tag{3.40a}$$

$$\boldsymbol{Y_{bus}}\boldsymbol{V}(k) = \boldsymbol{I}(k) = \boldsymbol{I}(\boldsymbol{X}(l), \boldsymbol{V}(l)), l = 0 \cdots k \tag{3.40b}$$

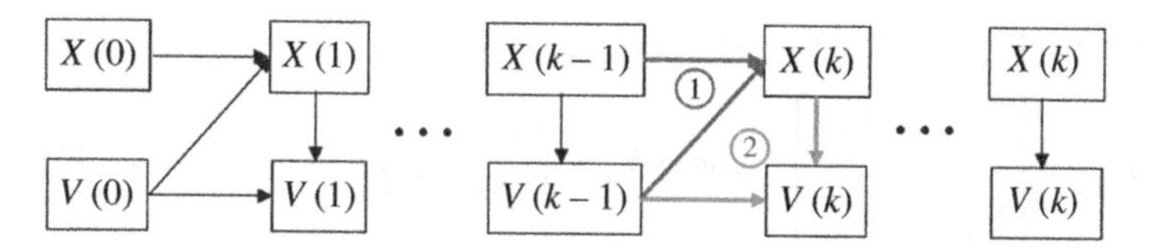

Figure 3.12 Recursive process to solve power series coefficients.

Then, the main task is to solve power series coefficients $\boldsymbol{X}(k)$ and $\boldsymbol{V}(k)$ $(k \geq 1)$ from the $(k-1)$th-order coefficients, as indicated by two circled numbers in Figure 3.12. Thus, any-order coefficients are solvable from $\boldsymbol{X}(0)$ and $\boldsymbol{V}(0)$.

Rewrite Eq. (3.40a) as Eq. (3.41) by replacing k with $k-1$. Note that $\boldsymbol{X}(k)$ only appears on the left-hand side, and the right-hand side only contains coefficients up to order $k-1$. Therefore, $\boldsymbol{X}(k)$ can be explicitly solved from calculated lower order coefficients.

$$\boldsymbol{X}(k) = \frac{1}{k}\boldsymbol{F}(\boldsymbol{X}(l), \boldsymbol{V}(l)), l = 0 \cdots k-1 \tag{3.41}$$

In contrast, from (3.40b), solving $\boldsymbol{V}(k)$ is not straightforward since it appears on both the left-hand side, and right-hand side and the vector-valued function $\boldsymbol{I}(\cdot)$ is nonlinear. Later in Section 3.3.3, we will show that the coefficients of current injection $\boldsymbol{I}(k)$ satisfy a formally linear equation (3.42) about $\boldsymbol{V}(k)$.

$$\begin{gathered} \boldsymbol{I}(k) = \boldsymbol{A}\boldsymbol{V}(k) + \boldsymbol{B} \\ \begin{cases} \boldsymbol{A} = \boldsymbol{A}(\boldsymbol{X}(l_1), \boldsymbol{V}(l_2)) \\ \boldsymbol{B} = \boldsymbol{B}(\boldsymbol{X}(l_1), \boldsymbol{V}(l_2)) \end{cases}, \begin{cases} l_1 = 0 \cdots k \\ l_2 = 0 \cdots k-1 \end{cases} \end{gathered} \tag{3.42}$$

Note that matrices $\boldsymbol{A}$ and $\boldsymbol{B}$ still contain nonlinear functions that only involve the $(k-1)$th and lower order coefficients on bus voltages, so they do not affect the solvability of $\boldsymbol{V}(k)$. Finally, $\boldsymbol{V}(k)$ is explicitly solved in Eq. (3.43) after substituting Eq. (3.42) into Eq. (3.40b). The detailed derivation of matrices $\boldsymbol{A}$ and $\boldsymbol{B}$ is presented in Section 3.3.3.

$$\boldsymbol{V}(k) = (\boldsymbol{Y}_{\boldsymbol{bus}} - \boldsymbol{A})^{-1}\boldsymbol{B} \tag{3.43}$$

For complex variables and parameters in Eqs. (3.38)–(3.43), such as current injection vector $\boldsymbol{i}$, bus voltage vector $\boldsymbol{v}$, differential transformations $\boldsymbol{I}(k)$ and $\boldsymbol{V}(k)$, and admittance matrix $\boldsymbol{Y}_{bus}$, their real and imaginary parts are separate as follows, where N is the number of buses.

$$\begin{aligned} \boldsymbol{i} &= [i_{x,1}, i_{y,1} \cdots i_{x,N}, i_{y,N}]^T \\ \boldsymbol{v} &= [v_{x,1}, v_{y,1} \cdots v_{x,N}, v_{y,N}]^T \\ \boldsymbol{I}(k) &= [I_{x,1}(k), I_{y,1}(k) \cdots I_{x,N}(k), I_{y,N}(k)]^T \end{aligned}$$

$$\boldsymbol{V}(k) = [V_{x,1}(k), V_{y,1}(k) \cdots V_{x,N}(k), V_{y,N}(k)]^T$$

$$\boldsymbol{Y_{bus}} = \begin{bmatrix} \boldsymbol{Y}_{11} & \dots & \boldsymbol{Y}_{1N} \\ \vdots & \ddots & \vdots \\ \boldsymbol{Y}_{N1} & \dots & \boldsymbol{Y}_{NN} \end{bmatrix}, \text{ where } \boldsymbol{Y}_{ij} = \begin{pmatrix} G_{ij} & B_{ij} \\ -B_{ij} & G_{ij} \end{pmatrix} \tag{3.44}$$

Remark: There are two important observations: (i) from Eq. (3.42) that current injections and bus voltages, which are coupled by nonlinear network equations in Eq. (3.38), turn out to have linear relationships in terms of their coefficients after differential transformation; (ii) coefficients on bus voltages can be explicitly solved by Eq. (3.43) and then used to calculate bus voltages by Eq. (3.39) in a straightforward manner, which is different from using a conventional power flow solver to calculate bus voltages by numerical iterations. The presented differential transformation-based method for solving DAEs differentiates itself from the traditional solution schemes, which rely on iterative numerical methods such as the family of Newton–Raphson (NR) methods.

3.3.2 Derivation Process

3.3.2.1 Current Injection of Generators

Consider the detailed sixth-order synchronous generator model. The current injection using the *d–q* coordinate system is given in Eq. (3.45). The coordination transformation between *d–q* and *x–y* coordinate systems is given in Eq. (3.46). Variables i_d and i_q are the *d*-axis and *q*-axis stator currents; e''_d and e''_q are *d*-axis and *q*-axis sub-transient voltages; v_d and v_q are the *d*-axis and *q*-axis terminal voltages; δ is the rotor angle. Parameters x''_d, x''_q, and r_a are the *d*-axis and *q*-axis sub-transient reactance and internal resistance, respectively.

$$\begin{bmatrix} i_d \\ i_q \end{bmatrix} = \boldsymbol{y_a} \left(\begin{bmatrix} e''_d \\ e''_q \end{bmatrix} - \begin{bmatrix} v_d \\ v_q \end{bmatrix} \right), \text{ where } \boldsymbol{y_a} = \begin{bmatrix} r_a & -x''_q \\ x''_d & r_a \end{bmatrix}^{-1} \tag{3.45}$$

$$\begin{bmatrix} i_x \\ i_y \end{bmatrix} = \boldsymbol{r} \begin{bmatrix} i_d \\ i_q \end{bmatrix}, \begin{bmatrix} v_x \\ v_y \end{bmatrix} = \boldsymbol{r} \begin{bmatrix} v_d \\ v_q \end{bmatrix}, \text{ where } \boldsymbol{r} = \begin{bmatrix} \sin\delta & \cos\delta \\ -\cos\delta & \sin\delta \end{bmatrix} \tag{3.46}$$

The current injection under the *x–y* axis is given in Eq. (3.47) by combining Eqs. (3.45) and (3.46).

$$\begin{bmatrix} i_x \\ i_y \end{bmatrix} = \tau \begin{bmatrix} e''_d \\ e''_q \end{bmatrix} - \lambda \begin{bmatrix} v_x \\ v_y \end{bmatrix}, \text{ where } \begin{cases} \tau = \boldsymbol{r}\boldsymbol{y_a} \\ \lambda = \tau \boldsymbol{r^T} \end{cases} \tag{3.47}$$

The differential transformation of Eq. (3.47) is given in Eq. (3.48).

$$\begin{bmatrix} I_x(k) \\ I_y(k) \end{bmatrix} = \Gamma(k) \otimes \begin{bmatrix} E''_d(k) \\ E''_q(k) \end{bmatrix} - \Lambda(k) \otimes \begin{bmatrix} V_x(k) \\ V_y(k) \end{bmatrix} \tag{3.48}$$

3.3.2.2 Current Injection of Loads

Consider the ZIP load model in Eq. (3.49) where p and q are the active and reactive power loads, respectively; v_t is the bus voltage magnitude defined in Eq. (3.50) and u equals its square; p_0, q_0, and v_{t0} are the steady-state active power, reactive power, and bus voltage; a_p and a_q are the percentages of constant impedance load; b_p and b_q are the percentages of constant current load; and c_p and c_q are the percentages of constant power load. There are $a_p + b_p + c_p = 1$ and $a_q + b_q + c_q = 1$.

$$\begin{cases} p = p_0\left(a_p\left(\dfrac{v_t}{v_{t0}}\right)^2 + b_p\left(\dfrac{v_t}{v_{t0}}\right) + c_p\right) \\ q = q_0\left(a_q\left(\dfrac{v_t}{v_{t0}}\right)^2 + b_q\left(\dfrac{v_t}{v_{t0}}\right) + c_q\right) \end{cases} \tag{3.49}$$

$$v_t = \sqrt{u}, \quad u = v_x^2 + v_y^2 \tag{3.50}$$

The current injected into the network can be calculated from the active and reactive power injections and is written in matrix forms in Eq. (3.51), where $\boldsymbol{\beta}_a$, $\boldsymbol{\beta}_b$, and $\boldsymbol{\beta}_c$ are constant matrices.

$$\begin{aligned} \begin{bmatrix} i_x \\ i_y \end{bmatrix} &= \begin{bmatrix} i_x \\ i_y \end{bmatrix}_{\boldsymbol{z}} + \begin{bmatrix} i_x \\ i_y \end{bmatrix}_{\boldsymbol{i}} + \begin{bmatrix} i_x \\ i_y \end{bmatrix}_{\boldsymbol{p}} \\ &\triangleq \frac{1}{v_{t0}^2}\boldsymbol{\beta}_a \begin{bmatrix} v_x \\ v_y \end{bmatrix} + \frac{1}{v_{t0}}\frac{1}{v_t}\boldsymbol{\beta}_b \begin{bmatrix} v_x \\ v_y \end{bmatrix} + \frac{1}{u}\boldsymbol{\beta}_c \begin{bmatrix} v_x \\ v_y \end{bmatrix} \\ \boldsymbol{\beta}_a &= \begin{bmatrix} p_0 a_p & q_0 a_q \\ -q_0 a_q & p_0 a_p \end{bmatrix}, \boldsymbol{\beta}_b = \begin{bmatrix} p_0 b_p & q_0 b_q \\ -q_0 b_q & p_0 b_p \end{bmatrix}, \boldsymbol{\beta}_c = \begin{bmatrix} p_0 c_p & q_0 c_q \\ -q_0 c_q & p_0 c_p \end{bmatrix} \end{aligned} \tag{3.51}$$

Differential transformations of u and v_t are given in Eqs. (3.52) and (3.53). Then, differential transformations of the right-hand-side terms in Eq. (3.51) can be obtained, as explained in the following.

$$U(k) = V_x(k) \otimes V_x(k) + V_y(k) \otimes V_y(k) \tag{3.52}$$

$$V_t(k) = \frac{1}{2V_t(0)}U(k) - \frac{1}{2V_t(0)}\sum_{m=1}^{k-1} V_t(m)V_t(k-m) \tag{3.53}$$

Differential transformations of the three terms on the right-hand side of Eq. (3.51), i.e. the current injection of constant impedance load, constant current load, and constant power load, are given in Eqs. (3.54)–(3.56), respectively.

$$\begin{bmatrix} i_x \\ i_y \end{bmatrix}_{\boldsymbol{z}} = \frac{1}{v_{t0}^2}\boldsymbol{\beta}_a \begin{bmatrix} v_x \\ v_y \end{bmatrix} \rightarrow \begin{bmatrix} I_x(k) \\ I_y(k) \end{bmatrix}_{\boldsymbol{z}} = \frac{1}{v_{t0}^2}\boldsymbol{\beta}_a \begin{bmatrix} V_x(k) \\ V_y(k) \end{bmatrix} \tag{3.54}$$

$$\begin{bmatrix} i_x \\ i_y \end{bmatrix}_i = \frac{1}{v_{t0}} \frac{1}{v_t} \beta_b \begin{bmatrix} v_x \\ v_y \end{bmatrix} \rightarrow$$

$$\begin{bmatrix} I_x(k) \\ I_y(k) \end{bmatrix}_i = \frac{1}{v_{t0}} \frac{1}{V_t(0)} \left(\beta_b \begin{bmatrix} V_x(k) \\ V_y(k) \end{bmatrix} - \sum_{m=0}^{k-1} V_t(k-m) \begin{bmatrix} i_x(m) \\ i_y(m) \end{bmatrix}_i \right) \tag{3.55}$$

$$\begin{bmatrix} i_x \\ i_y \end{bmatrix}_p = \frac{1}{u} \beta_c \begin{bmatrix} v_x \\ v_y \end{bmatrix} \rightarrow$$

$$\begin{bmatrix} I_x(k) \\ I_y(k) \end{bmatrix}_p = \frac{1}{U(0)} \left(\beta_c \begin{bmatrix} V_x(k) \\ V_y(k) \end{bmatrix} - \sum_{m=0}^{k-1} U(k-m) \begin{bmatrix} I_x(m) \\ I_y(m) \end{bmatrix}_p \right) \tag{3.56}$$

Finally, the differential transformation of the current injection equation (3.51) is given in Eq. (3.57) by summing the differential transformations of all three terms.

$$\begin{bmatrix} I_x(k) \\ I_y(k) \end{bmatrix} = \begin{bmatrix} I_x(k) \\ I_y(k) \end{bmatrix}_z + \begin{bmatrix} I_x(k) \\ I_y(k) \end{bmatrix}_i + \begin{bmatrix} I_x(k) \\ I_y(k) \end{bmatrix}_p \tag{3.57}$$

3.3.2.3 Transmission Network Equation

The network equation is in Eq. (3.58), which couples the current injections of all generators and loads. Its differential transformation is given in Eq. (3.59).

$$\boldsymbol{i} = \boldsymbol{Y_{bus}} \boldsymbol{v_{bus}} \tag{3.58}$$

$$\boldsymbol{I}(k) = \boldsymbol{Y_{bus}} \boldsymbol{V}(k) \tag{3.59}$$

3.3.3 Solution Algorithm

Proposition 3.7 The transformed current injections in Eqs. (3.48) and (3.57) for generators and loads, respectively, satisfy Eqs. (3.60) and (3.61), which are formally linear.

The proofs of Proposition 3.7 and the detailed expressions of matrices $\boldsymbol{A_g}$, $\boldsymbol{B_g}$, $\boldsymbol{A_l}$, and $\boldsymbol{B_l}$ could be found in [3]. Current injections for all buses in Eq. (3.42) are obtained from this proposition with $\boldsymbol{A}$ and $\boldsymbol{B}$ in Eq. (3.62).

$$\begin{bmatrix} I_x(k) \\ I_y(k) \end{bmatrix} = \boldsymbol{A_g} \begin{bmatrix} V_x(k) \\ V_y(k) \end{bmatrix} + \boldsymbol{B_g} \tag{3.60}$$

$$\begin{bmatrix} I_x(k) \\ I_y(k) \end{bmatrix} = \boldsymbol{A_l} \begin{bmatrix} V_x(k) \\ V_y(k) \end{bmatrix} + \boldsymbol{B_l} \tag{3.61}$$

$$\boldsymbol{A} = diag(\boldsymbol{A}_1, \boldsymbol{A}_2 \cdots \boldsymbol{A}_N), \boldsymbol{A}_n = \begin{cases} \boldsymbol{A}_g, \text{for generator buses} \\ -\boldsymbol{A}_l, \text{for load buses} \\ \boldsymbol{A}_g - \boldsymbol{A}_l, \text{for buses with both} \\ \text{generators and loads} \end{cases}$$

$$\boldsymbol{B} = \left[\boldsymbol{B}_1{}^T, \boldsymbol{B}_2{}^T \cdots \boldsymbol{B}_N{}^T\right]^T, \boldsymbol{B}_n = \begin{cases} \boldsymbol{B}_g, \text{for generator buses} \\ -\boldsymbol{B}_l, \text{for load buses} \\ \boldsymbol{B}_g - \boldsymbol{B}_l, \text{for buses with both} \\ \quad \text{generators and loads} \end{cases} \tag{3.62}$$

Following the basic idea in Figure 3.12, Algorithm 3.2 is further designed to solve power series coefficients of both state variables and bus voltages up to any desired order. Note that all the coefficients are explicitly calculated with no iteration.

Algorithm 3.2 Solve Coefficients for the DAE Model
Input: initial values of state variables and bus voltages $\boldsymbol{x}_0$, $\boldsymbol{v}_0$
Output: any order coefficients $\boldsymbol{X}(k)$, $\boldsymbol{V}(k)$, $k = 0 \cdots K$
Steps:
Initialization: $\boldsymbol{X}(0) = \boldsymbol{x}_0$, $\boldsymbol{V}(0) = \boldsymbol{v}_0$
1. Calculate the matrix $\boldsymbol{A}$
    1.1 Calculate the matrix $\boldsymbol{A}_g$ for generators
    1.2 Calculate the matrix $\boldsymbol{A}_l$ for loads
2. Calculate the matrix $(\boldsymbol{Y}_{\boldsymbol{bus}} - \boldsymbol{A})$ and solve $(\boldsymbol{Y}_{\boldsymbol{bus}} - \boldsymbol{A})^{-1}$
for $k = 1:K$
3. Solve $\boldsymbol{X}(k) = \frac{1}{k}\boldsymbol{F}(\boldsymbol{X}(l), \boldsymbol{V}(l)), l = 0 \cdots k-1$
4. Calculate the matrix $\boldsymbol{B}$
    4.1 Calculate the matrix $\boldsymbol{B}_g$ for generators
    4.2 Calculate the matrix $\boldsymbol{B}_l$ for loads
5. Solve $\boldsymbol{V}(k) = (\boldsymbol{Y}_{\boldsymbol{bus}} - \boldsymbol{A})^{-1}\boldsymbol{B}$
end

3.3.4 Case Study

To validate the accuracy, time performance, and robustness of the differential transformation-based method for solving practical high-dimensional nonlinear DAEs, the 327-machine, 2383-bus Polish system [9] with detailed models on generators, exciters, governors, turbines, and ZIP loads is used. In the ZIP load model, the percentages of each component are 20%, 30%, and 50%, respectively.

Two widely used solution approaches are implemented for comparison [11]: (i) **TRAP-NR** method, where the differential equations are algebraized by implicit trapezoidal method (TRAP) first and then solved *simultaneously* with the network equations by NR method. (ii) **ME-NR** method using a partitioned scheme where the differential and network equations are *alternatively* solved by explicit ME and NR methods, respectively. The time step length of both the TRAP-NR method and ME-NR method is 1×10^{-3} seconds, while the differential transformation-based method prolongs the time step length to 10 times and still achieves better accuracy.

For a fair comparison, the benchmark result is given by the TRAP-NR method using a small enough time step length of 1×10^{-4} seconds, and errors of the differential transformation-based method, the TRAP-NR method, and the ME-NR method are calculated by their differences from the benchmark result. Simulations are conducted in MATLAB R2018b on a personal computer with i7-6700U CPU.

3.3.4.1 Accuracy and Time Performance

Both stable and unstable scenarios are simulated for the Polish system to validate the accuracy and time performance of the differential transformation-based method. Respectively, for two scenarios, Figures 3.13 and 3.14 show the transient responses of rotor angles, rotor speeds, and bus voltages simulated by the differential transformation-based method. Machine 1 is selected as the reference to calculate relative rotor angles. The maximum errors of rotor angles, rotor speeds, and bus voltages compared with the benchmark results are 3.02×10^{-5} degree, 4.27×10^{-7} Hz, and 3.33×10^{-7} p.u. for stable scenario and 2.02×10^{-5} degree, 3.00×10^{-7} Hz, and 2.78×10^{-7} p.u. for unstable scenario, respectively. It shows the differential transformation-based method can accurately simulate both stable and unstable contingencies in the transient stability simulation.

Table 3.5 gives the maximum errors of all state variables (including rotor angles, rotor speeds, transient and sub-transient voltages, and field voltages) and bus voltages, respectively, as well as the computation time per one-second simulation. The errors of the state variables and the bus voltages using the differential transformation-based method are, respectively, two orders of magnitude lower and one order of magnitude lower than those using the TRAP-NR method and the ME-NR method. Also, the computation speed of the differential transformation-based method is around 10 times faster than the other two methods. These results show the differential transformation-based method is more efficient and accurate.

Since a large computation burden with transient stability simulation lies in solving linear equations, both sparse matrix and LU

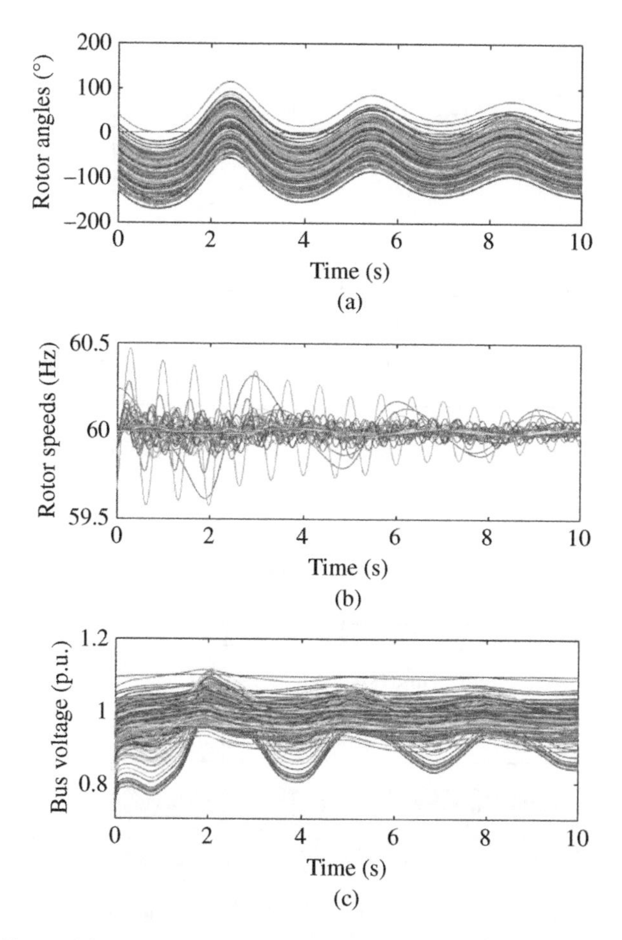

Figure 3.13 Trajectories of the stable scenario for the 2383-bus system. (a) Rotor angles, (b) rotor speeds, and (c) bus voltages.

factorization techniques are implemented in this article for the differential transformation-based method, TRAP-NR method, and ME-NR method. Table 3.6 compares the total number N_{LU} of times of LU factorization with three methods in a one-second simulation. It is calculated by $N_{LU} = n_{LU} \times M$, where n_{LU} is the number of times LU factorization occurs within each time step, and M is the total number of time steps. Within each time step, both the TRAP-NR and the ME-NR methods need to perform LU factorization for each iteration unless a so-called very dishonest NR method is applied, but the differential transformation-based method only needs to perform LU factorization once. Also, the differential transformation-based method only takes 1/10 of time steps of the other two methods. Therefore, the differential transformation-based method can significantly reduce the number of times LU factorization occurs in a simulation.

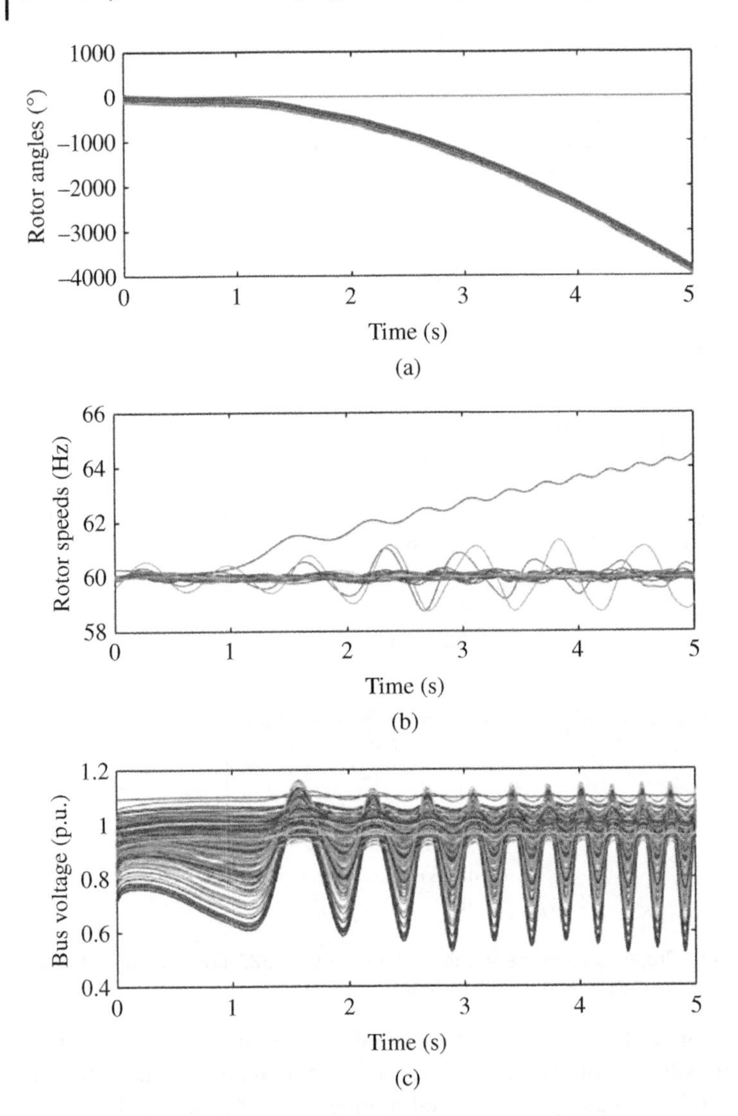

Figure 3.14 Trajectories of the unstable scenario for the 2383-bus system. (a) Rotor angles, (b) rotor speeds, and (c) bus voltages.

3.3.4.2 Robustness

The robustness of the differential transformation-based method is validated in three sets of cases: (i) stable and unstable scenarios; (ii) disturbances with different severities; and (iii) different percentages of constant power load.

By comparing the results of stable and unstable scenarios in Table 3.5, it is observed that the TRAP-NR method and ME-NR method are less accurate

Table 3.5 Comparison of accuracy and time performance.

Scenarios	Methods	Error of state variables (p.u.)	Error of bus voltages (p.u.)	Computation time (s)
Stable	DT	2.69×10^{-6}	3.33×10^{-7}	18.76
	TRAP-NR	1.30×10^{-4}	1.10×10^{-6}	176.43
	ME-NR	2.63×10^{-4}	2.26×10^{-6}	191.40
Unstable	DT	1.89×10^{-6}	2.78×10^{-7}	18.85
	TRAP-NR	1.41×10^{-4}	1.61×10^{-6}	182.76
	ME-NR	2.79×10^{-4}	2.93×10^{-6}	196.02

Table 3.6 Comparison of number of LU factorization.

Methods	n_{LU}	M	$N_{LU} = n_{LU} \times M$
DT	1	100	100
TRAP-NR	3.004	1000	3004
ME-NR	2.060	1000	2006

and slower in simulating the unstable scenario than the stable scenario, but the differential transformation-based method performs almost the same in both scenarios. This is because the system states change significantly in the unstable scenario, and the NR method takes more iterations to converge. At each time step, the TRAP-NR method takes an average of 3.004 iterations in the stable scenario and 3.118 iterations in the unstable scenario. For ME-NR method, it takes 2.060 and 2.132 iterations, respectively. In contrast, the differential transformation-based method does not require iterations in the solving process, thus having better robustness in unstable scenarios.

Figure 3.15 gives the time performance and the average number of iterations of the NR method under different disturbances with increasing severities using the three methods. It shows the computation time of the differential transformation-based method is almost the same for different disturbances, but both the TRAP-NR method and ME-NR method take a longer time when simulating larger disturbances due to the increased number of iterations.

The time performance and the average number of iterations of the NR method under different percentages of constant power load are shown in Figure 3.16. The higher percentage of constant power load brings stronger

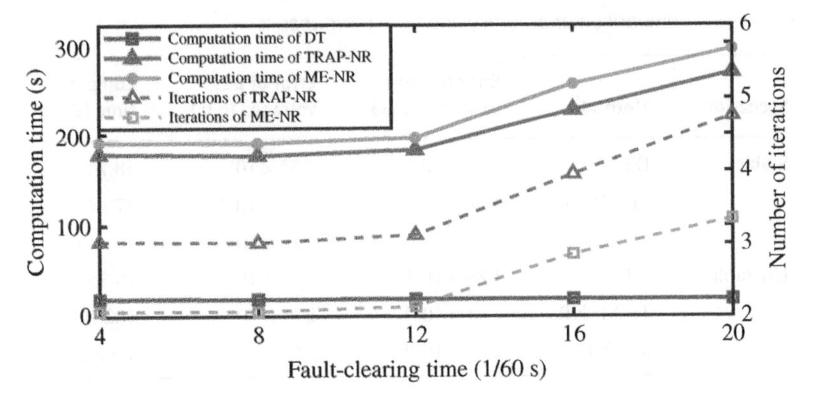

Figure 3.15 Robustness against different disturbances.

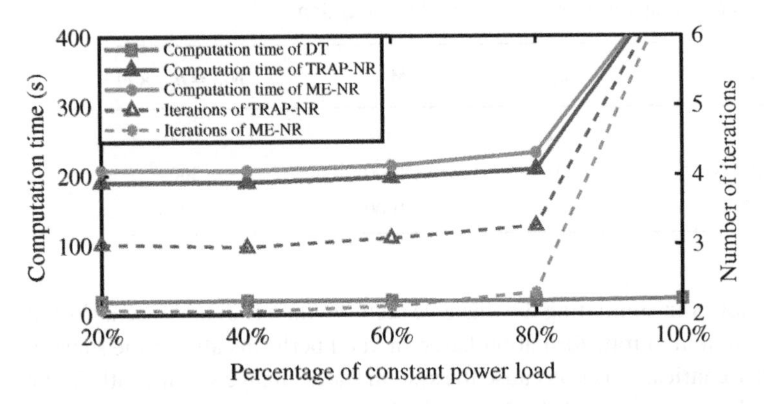

Figure 3.16 Robustness against different percentages of constant power load.

nonlinearity to the DAEs, thus making the NR method more difficult or failing to converge. In Figure 3.16, the computation time of both the TRAP-NR method and ME-NR method increases significantly with the higher percentage of the constant power load. But the differential transformation-based method does not need iterations, and its computation time is not affected much, showing it is more robust to handle the strong nonlinearity caused by constant power load.

3.4 Broader Applications

In addition to its use in solving ODEs and DAEs presented in Sections 3.2 and 3.3, the differential transformation method has also found applications in addressing a wide range of challenging power system problems.

Examples of three selected applications are briefly described below. Readers are referred to the published studies of Liu et al. [3, 7, 8] for more details.

In the first application, the differential transformation method was used to convert the nonlinear algebraic current injection equations into formally linear equations, enabling a noniterative algorithm for solving power system nonlinear DAE models in transient stability assessment. This idea was applied to nonlinear AC power flow equations to design a noniterative dynamized power flow algorithm for static voltage stability assessment [3]. The algorithm was more efficient than the conventional continuation power flow approach for tracing solution curves.

In the second application, the arbitrary high-order approximation of differential equations provided by the differential transformation method was used to predict the power system frequency response in a faster-than-real-time manner [7]. An adaptive switching control strategy for wind turbine generators was designed based on the predicted frequency response. This control strategy allows wind turbine generators to maintain maximum power point tracking mode most of the time and provides frequency support only when necessary, overcoming the conservativeness of the conventional deadband-based switching control strategy and reducing the profit loss of wind farms by providing frequency support.

In the third application, the flexibility of the differential transformation method in balancing accuracy and time performance was utilized to design a variable-order, variable-step differential transformation method [8]. This method has found further applications in high-performance computing environments when combined with the Parareal, a temporal parallelization approach. The developed adaptive Parareal method utilizing the variable-order variable-step differential transformation method as its coarse solver was found to be more efficient and robust than the conventional Parareal method equipped with a numerical integration solver.

3.5 Conclusions and Future Directions

In this chapter, we present a differential transformation-based approach for power system simulation. We summarized the transformation rules of various widely used generic nonlinear functions in the power system model and detailed the key algorithms and test results for solving power system ODE models and DAE models. In addition, we discussed some broader applications of the differential transformation method, including the static voltage stability assessment problem, the frequency control of wind turbine generators, and the temporal parallelization of high-performance power system simulations.

Moving forward, there are several directions worth exploring. First, we suggest exploring broader applications of the differential transformation method beyond what has been presented in the literature. The differential transformation method has the potential to design efficient high-order noniterative algorithms for many other power system problems that are modeled as nonlinear ODEs, DAEs, or algebraic equations. For example, the differential transformation method can be utilized for designing high-order noniterative algorithms for many optimization problems, either in the long-term and middle-term planning problems over several years or the scheduling problems in several hours or 15 minutes, which often involve a lot of computational burdens in solving nonlinear algebraic equations.

Second, practical issues still need to be addressed to make the differential transformation-based algorithms a viable commercial software solution to benefit broader audiences in the power industry and beyond. For example, the differential transformation method needs to be further improved to deal with nonsmooth models such as power system models with control limits or other discrete events. Addressing these practical issues will ensure that the differential transformation-based algorithms can be used effectively to solve complex real-world power system problems.

References

1 Liu, Y., Sun, K., Yao, R., and Wang, B. (2019). Power system time domain simulation using a differential transformation method. *IEEE Transactions on Power Systems* 34 (5): 3739–3748.

2 Liu, Y. and Sun, K. (2020). Solving power system differential algebraic equations using differential transformation. *IEEE Transactions on Power Systems* 35 (3): 2289–2299.

3 Liu, Y., Sun, K., and Dong, J. (2020). A dynamized power flow method based on differential transformation. *IEEE Access* 8: 182441–182450.

4 Ev Pukhov, G.G. (1982). Differential transforms and circuit theory. *International Journal of Circuit Theory and Applications* 10 (3): 265–276.

5 Hassan, I.A.H. (2007). Application to differential transformation method for solving systems of differential equations. *Applied Mathematical Modelling* 32 (12): 2552–2559.

6 Yu, R. and Gu, W. (2022). Non-iterative calculation of quasi-dynamic energy flow in the heat and electricity integrated energy systems. *IEEE*

Transactions on Power Systems https://doi.org/10.1109/TPWRS.2022.3210167.

7 Liu, Y., Zhang, Y., Sun, K., and Zhao, X. (2022). *Adaptive Switching Control of Wind Turbine Generators for Necessary Frequency Response*. Denver: IEEE Power & Energy Society General Meeting.

8 Liu, Y., Park, B., Sun, K. et al. (2023). Parallel-in-time power system simulation using a differential transformation based adaptive Parareal method. *IEEE Open Access Journal of Power and Energy* 10: 61–72.

9 Zimmerman, R.D., Murillo-Sánchez, C.E., and Thomas, R.J. (2010). MATPOWER: steady-state operations, planning, and analysis tools for power systems research and education. *IEEE Transactions on Power Systems* 26 (1): 12–19.

10 Kundur, P.S. and Malik, O.P. (2022). *Power System Stability and Control*. McGraw-Hill Education.

11 Milano, F. (2010). *Power System Modelling and Scripting*. Springer Science & Business Media.

4

Accelerated Power System Simulation Using Analytic Continuation Techniques

Chengxi Liu

School of Electrical Engineering and Automation, Wuhan University, Wuhan, China

Power system dynamic simulation is of great importance for the power system operation, which evaluates the dynamic response of a power system under an operating condition following disturbances or contingencies. Nowadays, because of the large scales of expanding power networks with increasing numbers of buses, the integration of power electronic devices with fast dynamics and the penetration of renewable energy with more uncertainties, the time-domain simulation for a large-scale power system may be time-consuming, and not yet full-fledged for real time or even faster than real-time applications.

In Chapters 2 and 3, power series about time t is obtained as the basic semi-analytical solutions (SASs) for accelerating power system simulation. The essential difference between Taylor Series Expansion in Chapter 2, and Differential Transformation Method in Chapter 3 are that the former involves an offline stage to derive the symbolic expression of each state variables, and the latter does not contain an offline stage but online derives the expression only about time t. Nevertheless, both methods in Chapters 2 and 3 find the SAS in the form of power series, which is then truncated to a finite series up to a specified order. Examples have verified in Chapters 2 and 3 that both methods have better performance in simulation speed and potential for parallel simulation.

It can be expected that the computation burden of these power series increases significantly with the number of orders. Moreover, these power series solutions have limited convergence regions. In the application of power system simulation, the accuracy can be ensured by the short period

Power System Simulation Using Semi-Analytical Methods, First Edition. Edited by Kai Sun.

Published 2024 by John Wiley & Sons, Inc.

after initial state, but the error is increased rapidly over the time window. Therefore, the simulation speed is still restricted by the short time interval, which also constrains the performance in parallel simulation.

This chapter describes a technique of analytic continuation to prolong the time interval of the power series form SAS and still can guarantee the accuracy. Section 4.1 introduces the definition of analytic continuation, exemplifies the regions of convergence before and after analytic continuation using several simple functions, then introduces two methods, i.e. matrix method, and Viskovatov method to transfer the power series to the forms of Padé approximation and continued fractions, respectively. Section 4.2 finds the SASs using Padé approximants to prolong the time interval. Section 4.3 finds the SASs using continued fractions which can be deemed as a special but more applicable form of Padé approximants, then the adaptive time interval is also achieved based on the priori error bound of continued fractions.

As the accelerator of power series form SAS, these analytic continued techniques are optional, but the acceleration performance is verified in the examples with detailed dynamic models, i.e. the IEEE 3-generator 9-bus power system, the IEEE 10-generator 39-bus New England power system, the 327-generator 2383-bus Polish system.

4.1 Introduction to Analytic Continuation

Essentially, analytic continuation is a technique to extend the domain of definition of a given analytic function. Originally, the analytic function should be in the complex domain, which can also be generalized to other domains. This technique often succeeds in extending further values of a function, for instance, in an extended region where an infinite series can present, however, it is initially defined becomes divergent with finite series.

Suppose f is an analytic function defined on a non-empty open subset U of the complex plane $\mathbf{C}$. If V is a larger subset of $\mathbf{C}$, i.e. containing U, and F is function defined on V such that

$$F(z) = f(z), \qquad \forall z \in U \tag{4.1}$$

then F is called an analytical continuation of f.

Analytic continuation is unique in the **original domain** in the following sense: If V is the connected domain of two analytic functions F_1 and F_2, such that U is contained in V for all z in U, i.e. $U \subset V$

$$F_1(z) = F_2(z) = f(z), \qquad \forall z \in U \tag{4.2}$$

then

$$F_1 = F_2 \tag{4.3}$$

on all of V. This is because $F_1 - F_2$ is an analytic function that vanishes on the open, connected domain U of f and hence must vanish on its entire domain. This follows directly from the identity theorem for holomorphic functions (en.wikipedia.org).

Firstly, let us consider a simple series $f_1(s)$ given by

$$f_1(s) = 1 + s + s^2 + s^3 + \dots \tag{4.4}$$

Only for $|s| < 1, f_1(s)$ can be represented equivalently as:

$$f_1(s) = 1 + s + s^2 + s^3 + \cdots = \frac{1}{1-s}, \quad \forall |s| < 1 \tag{4.5}$$

It is obvious that the series $f_1(s)$ converges only for values of s that satisfy $|s| < 1$.

Based on the aforementioned definition of analytic continuation, to extend the domain of convergence of the original function, analytic continuation is exploited to construct a different function that

(1) Coincides with the original function within its radius of convergence, i.e. U;
(2) More importantly, extends the domain of convergence of the original function, e.g. V, s.t. $U \subset V$.

Then, consider another function $f_2(s)$ given by an integral function

$$f_2(s) = \int_0^{\infty} e^{-(1-x)x} \cdot dx \tag{4.6}$$

For the complex domain that the real part of s is less than 1, i.e. $\text{Re}(s) < 1$, $f_2(s)$ can be explicitly evaluated as:

$$\begin{aligned} f_2(s) &= \int_0^{\infty} e^{-(1-x)x} \cdot dx = \lim_{A\to\infty} \int_0^{A} e^{-(1-x)x} \cdot dx \\ &= \lim_{A\to\infty} \frac{1 - e^{-(1-s)A}}{1-s} = \frac{1}{1-s}, \\ &\forall\, \text{Re}(s) < 1 \end{aligned} \tag{4.7}$$

Firstly, function $f_2(s)$ is equal to the original series accurately when $|s| < 1$. Additionally, $f_2(s)$ has a larger domain of convergence compared with the original function. Therefore, the integral function $f_2(s)$ is the analytic continuation based on the series function $f_1(s)$.

It should be noted that the analytic continuation of an original function to **extended domain** is not unique. For example, consider the function $f_3(s)$

defined below:

$$f_3(s) = \frac{1}{1-s} \tag{4.8}$$

The fractional function $f_3(s)$ coincides with the original functions $f_1(s)$ and $f_2(s)$ within their respective radius of convergence. In addition, $f_3(s)$ is valid for all values of $s\epsilon \boldsymbol{C}$, except at $s = 1$. The fractional function $f_3(s)$ is defined as the maximal analytic continuation of the original function $f_1(s)$, which has the maximum domain of convergence among all the analytic continuation functions of $f_1(s)$.

The regions of convergence of the functions $f_1(s)$, $f_2(s)$, and $f_3(s)$ are shown in Figure 4.1.

Intuitively, for the dynamic simulation based on SAS, if the maximal analytic continuation extends the original series function $f(t)$ on the time domain to an extended convergence domain, the simulation step size can be greatly prolonged to some extent. Although, different approaches can be adopted to achieve the analytic continuation of an original function, according to Stahl's Padé convergence theory [1, 2], "For an analytic function f with finite singularities, the close to diagonal sequence of diagonal Padé approximants converge in minimal logarithmic capacity to the original function in the extremal domain."

It means that the following Padé approximants have the largest domain of convergence when the orders in nominator L is close to the orders in denominator M, i.e. diagonal Padé approximants.

$$\text{Padé}[L/M] = \frac{a_0 + a_1 s + a_2 s^2 + \cdots + a_L s^L}{b_0 + b_1 s + b_2 s^2 + \cdots + b_M s^M} = \frac{a(s)}{b(s)} \tag{4.9}$$

Therefore, it is suggested that Padé approximants can be exploited to evaluate the maximal analytic continuation of the original series function,

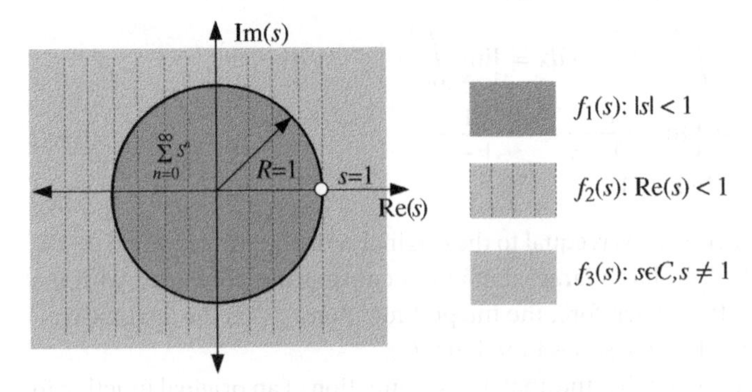

Figure 4.1 Domains of Convergence of $f_1(s)$, $f_2(s)$, and $f_3(s)$.

favorably in the Padé fractional function like (4.9), whose orders in numerator and denominators are close.

Usually, three main approaches can be used for the calculation of Padé approximants: Subramanian [3] and the first two are introduced in Sections 4.1.1 and 4.1.2:

(1) Direct method (or matrix method)
(2) Continued fractions (i.e. Viskovatov method)
(3) Epsilon algorithm

4.1.1 Direct Method (or Matrix Method)

The direct method was developed by Henri Padé in 1890, so the Padé approximant is named after him [4]. The core idea of this method is to represent the original function in the form of a rational fraction composed of two power series [5].

Consider the power series representation of an analytic function $c(s)$:

$$c(s) = c_0 + c_1 s + c_2 s^2 + c_3 s^3 + \cdots = \sum_{n=0}^{\infty} c[n] s^n \tag{4.10}$$

The notation $c[n]$ or c_n represents the coefficient of the power series of degree n. For a power series given by (4.10), the Padé approximant is a rational fraction of the following form:

$$\begin{aligned} c(s) &= c_0 + c_1 s + c_2 s^2 + \cdots + c_{L+M} s^{L+M} + O(s^{L+M+1}) \\ &= \text{Padé}[L/M] = \frac{a_0 + a_1 s + a_2 s^2 + \cdots + a_L s^L}{b_0 + b_1 s + b_2 s^2 + \cdots + b_M s^M} = \frac{a(s)}{b(s)} \end{aligned} \tag{4.11}$$

where $O(s^{L+M+1})$ indicates the truncation error of Padé approximants at order of $[L/M]$.

In (4.11), there are $L+M+1$ known coefficients in the power series, i.e. c_0 to c_{L+M}. However, there are $L+M+2$ unknowns in the Padé approximant, i.e. a_0 to a_L and b_0 to b_M. Therefore, one of the coefficients can be chosen as a free variable to scale the entire equation. For simplification, let us fix the first order of denominator, e.g. $b_0 = 1$. After multiplying (4.11) by $b(s)$ on both sides and equating the coefficients from s^{L+1} to s^{L+M}, a system of M linear equations is obtained:

$$\begin{bmatrix} c_{L-M+1} & c_{L-M+2} & \cdots & c_L \\ c_{L-M+2} & c_{L-M+3} & \cdots & c_{L+1} \\ \vdots & \vdots & \ddots & \vdots \\ c_L & c_{L+1} & \cdots & c_{L+M-1} \end{bmatrix} \begin{bmatrix} b_M \\ b_{M-1} \\ \vdots \\ b_1 \end{bmatrix} = - \begin{bmatrix} c_{L+1} \\ c_{L+2} \\ \vdots \\ c_{L+M} \end{bmatrix} \tag{4.12}$$

Table 4.1 A portion of Padé table for the exponential function exp(x).

L \ M	0	1	2	3
0	$\frac{1}{1}$	$\frac{1}{1-x}$	$\frac{1}{1-x+\frac{1}{2}x^2}$	$\frac{1}{1-x+\frac{1}{2}x^2+\frac{1}{6}x^3}$
1	$\frac{1+x}{1}$	$\frac{1+\frac{1}{2}x}{1-\frac{1}{2}x}$	$\frac{1+\frac{1}{3}x}{1-\frac{2}{3}x+\frac{1}{6}x^2}$	$\frac{1+\frac{1}{4}x}{1-\frac{3}{4}x+\frac{1}{4}x^2-\frac{1}{24}x^3}$
2	$\frac{1+x+\frac{1}{2}x^2}{1}$	$\frac{1+\frac{2}{3}x+\frac{1}{6}x^2}{1-\frac{1}{3}x}$	$\frac{1+\frac{1}{2}x+\frac{1}{12}x^2}{1-\frac{1}{2}x+\frac{1}{12}x^2}$	$\frac{1+\frac{1}{5}x+\frac{1}{20}x^2}{1-\frac{3}{5}x+\frac{3}{20}x^2-\frac{1}{60}x^3}$
3	$\frac{1+x+\frac{1}{2}x^2+\frac{1}{6}x^3}{1}$	$\frac{1+\frac{3}{4}x+\frac{1}{4}x^2+\frac{1}{24}x^3}{1-\frac{1}{4}x}$	$\frac{1+\frac{3}{5}x+\frac{3}{20}x^2+\frac{1}{60}x^3}{1-\frac{2}{5}x+\frac{1}{20}x^2}$	$\frac{1+\frac{1}{2}x+\frac{1}{10}x^2+\frac{1}{120}x^3}{1-\frac{1}{2}x+\frac{3}{10}x^2-\frac{1}{120}x^3}$

With traditional LU factorization computation techniques, (4.12) can be solved to obtain the denominator polynomial coefficients, i.e. b_1 to b_M.

Likewise, by equating the coefficients of power series of s on both sides from s^0 to s^L in (4.11), the numerator polynomial coefficients, i.e. a_0 to a_L, can also be calculated, shown as the linear equations (4.13).

$$a_l = \sum_{k=0}^{l} c_k b_{l-k}, \quad l = 0, 1, \ldots, L \tag{4.13}$$

Based on the Stahl's Padé convergence theory, if a Padé approximant $[L/M]\frac{a(s)}{b(s)}$ is diagonal/near-diagonal, this Padé approximant is the maximal analytic continuation of the original function. A diagonal Padé approximant is a rational approximant whose numerator and denominator polynomial degrees are exactly equal (i.e. $L = M$). Similarly, a near-diagonal Padé approximant is a rational approximant where the difference between the degree of the numerator and denominator polynomial is 1 (i.e. $|L-M| = 1$).

Here is an example of Padé table for the exponential function e^x. It can be noted that the second column for denominator's order $M = 0$ coincides with Taylor Series at $x = 0$, also called Maclaurin Series.

Only the diagonal Padé approximants (diagonal cells in Table 4.1) or the near-diagonal Padé approximants (near diagonal cells closest to the diagonal cells in Table 4.1) have the maximal analytical continuation domain of the original function e^x, converge faster than the Maclaurin Series.

4.1.2 Continued Fractions (i.e. Viskovatov Method)

There is an intimate connection between regular continued fractions and Padé tables with normal approximants along the main diagonal and two steps should be followed for the Viskovatov method to find the easiest continued fraction:

(1) Taking the original power series and convert it to a continued fraction.
(2) Converting the continued fraction to a rational function.

The first step can be achieved by recursively inverting partial series, which requires that all the inverses of the partial series exist [6]. Taking the Taylor Series representation of an analytic function $c(s)$ in (4.10) as an example, the procedure of Viskovatov method is explained:

$$\begin{aligned} c(s) &= c[0] + c[1]s + c[2]s^2 + \cdots + c[n]s^n + \cdots \\ &= c[0] + s(c[1] + c[2]s + \cdots + c[n]s^{n-1} + \cdots) \\ &= c[0] + \cfrac{s}{\cfrac{1}{c[1] + c[2]s + \cdots + c[n]s^{n-1} + \cdots}} \\ &= c[0] + \frac{s}{c^{(1)}(s)} \end{aligned} \tag{4.14}$$

In (4.14), $c^{(1)}(s)$ is given by:

$$c^{(1)}(s) = \frac{1}{c[1] + c[2]s + \cdots + c[n]s^{n-1} + \cdots} \tag{4.15}$$

Applying the technique described above recursively, Eq. (4.14) becomes:

$$c(s) = c[0] + \cfrac{s}{c^{(1)}[0] + \cfrac{s}{c^{(2)}[0] + \cfrac{s}{c^{(3)}[0] + \ldots}}} \tag{4.16}$$

Equation (4.16) is the analytic continuation of power series (4.10) in the form of continued fraction. Proved by the principle of mathematical induction [5] and the iterative re-expansion of the continued fraction [7], the continued fraction is equivalent to either a diagonal or near-diagonal Padé approximants in turns, depending on the truncated orders of continued fraction. Hence, the maximal analytic continuation of the original function is obtained as well.

4.2 Finding Semi-Analytical Solutions Using Padé Approximants

Padé approximants can be used for power system dynamic simulations, so as to extend its domain of convergence to enlarge the step size of SAS [8]. In Section 4.2, A two-stage scheme is introduced to ensure the performance of the acceleration method.

In the offline derivation stage, this method first derives an SAS in the form of an explicit Padé function of symbolic variables, such as simulation time, initial states, and all other parameters of system operating conditions. In the online simulation stage, the dynamic simulation is implemented by only substituting actual values to symbolic expressions at each time interval.

Although the accuracy of an SAS can only be guaranteed for a limited time interval, due to its gradually increased error, the entire simulation period can be made of many consecutive time intervals whose lengths are adaptively adjusted according to a prespecified error tolerance. Meanwhile, the use of Padé approximants can effectively prolong each time interval window, in order to accelerate the simulation speed.

In Section 4.2, an SAS using Padé approximants extends the convergence region. Compared with the conventional SAS in the form of power series in Chapters 3 and 4, the SAS in the form of Padé approximants with the same order has better performance of extended time interval and improved simulation speed. The advantage of this method for contingency simulations is validated on the IEEE 3-generator 9-bus system and IEEE 10-generator 39-bus system with detailed 6-order generator models.

4.2.1 Semi-Analytical Solution Using Padé Approximants

4.2.1.1 Offline Solving Differential Equations Using Power Series Expansion

The differential equation (DE) of power system can be expressed by (4.17), where state vector $\mathbf{x}$ consists of $K \times 1$ state variables and f is a vector of nonlinear functions.

$$\begin{cases} \dot{\mathbf{x}} = \mathbf{f}(\mathbf{x}) \\ \mathbf{x}(\mathbf{t_0}) = \mathbf{x_0} \end{cases}$$
$$\mathbf{x}(t) = [x_1(t), x_2(t), \cdots x_K(t)]^T \tag{4.17}$$

To obtain the SAS of the DE, the ith state variable $x_i(t)$ is firstly given in the form of a power series up to prespecified order N, as defined in (4.18).

$$x_i(t) = a_{i0} + a_{i1}t + a_{i2}t^2 + \cdots + a_{in}t^n \cdots + a_{iN}t^N + O(t^{N+1}) \approx \sum_{n=0}^{N} a_{in}t^n \tag{4.18}$$

Assume (4.18) to be the approximation of the solution to (4.17) with initial condition x_{i0} at $t = 0$, where a_{i0} to a_{iN} are the unknowns to be determined.

The derivatives of $\mathbf{x(t)}$ is easy to obtain as (4.19).

$$\dot{\mathbf{x}}(\mathbf{t}) = \begin{pmatrix} \dot{x}_1(t) \\ \dot{x}_2(t) \\ \vdots \\ \dot{x}_K(t) \end{pmatrix} = \begin{pmatrix} a_{11} + 2a_{12}t + \cdots + na_{1n}t^{n-1} + \cdots + Na_{1N}t^{N-1} \\ a_{21} + 2a_{22}t + \cdots + na_{2n}t^{n-1} + \cdots + Na_{2N}t^{N-1} \\ \vdots \\ a_{K1} + 2a_{K2}t + \cdots + na_{Kn}t^{n-1} + \cdots + Na_{KN}t^{N-1} \end{pmatrix} \tag{4.19}$$

The nonlinear function f can also be transformed to the series form as defined in (4.20).

$$\mathbf{f}(\mathbf{x}) = \mathbf{f}\begin{pmatrix} x_1(t) \\ x_2(t) \\ \vdots \\ x_K(t) \end{pmatrix} = \mathbf{f}\begin{pmatrix} a_{10} + a_{11}t + \cdots + a_{1n}t^n + \cdots + a_{1N}t^N \\ a_{20} + a_{21}t + \cdots + a_{2n}t^n + \cdots + a_{2N}t^N \\ \vdots \\ a_{K0} + a_{K1}t + \cdots + a_{Kn}t^n + \cdots + a_{KN}t^N \end{pmatrix}$$
$$\approx \begin{pmatrix} f_{10} + f_{11}t + \cdots + f_{1f}t^f + \cdots + f_{1F}t^F \\ f_{20} + f_{21}t + \cdots + f_{2f}t^f + \cdots + f_{2F}t^F \\ \vdots \\ f_{K0} + f_{K1}t + \cdots + f_{Kf}t^f + \cdots + f_{KF}t^F \end{pmatrix} \tag{4.20}$$

There are several methods to obtain the expansion of $\mathbf{f(x)}$ in (4.20), for example, Taylor series expansion expressed in (4.21), where $F = M-1$.

$$f_i[\mathbf{x}(t)] = f_i(t_0) + \frac{f_i'(t_0)}{1!}(t - t_0) + \frac{f_i''(t_0)}{2!}(t - t_0)^2$$
$$+ \cdots + \frac{f_i^{(F)}(t_0)}{F!}(t - t_0)^F \tag{4.21}$$

or the Adomian Decomposition Method (4.22), which will be introduced in Chapter 6,

$$\begin{cases} f_i[\mathbf{x}(t)] = \sum_{n=0}^{\infty} A_{in}(\mathbf{x}_0, \mathbf{x}_1, \cdots, \mathbf{x}_n), i = 1 \cdots K \\ A_{in} = \frac{1}{n!} \sum_{n=0}^{\infty} \left[\frac{\partial^n}{\partial \lambda^n} f_i \left(\sum_{i=0}^{n} \mathbf{x}_i \lambda^i \right) \right] \Bigg|_{\lambda=0} \end{cases} \tag{4.22}$$

where λ is the grouping factor [9, 10].

Equating both sides in (4.20) and (4.21), one can obtain the symbolic expression of a_{in} ($i = 1$ to K and $n = 1$ to N) in (4.23).

$$\begin{cases} f_{i,0} - a_{i,1} = 0 \\ f_{i,1} - 2a_{i,2} = 0 \\ f_{i,n-1} - n \cdot a_{i,n} = 0 \\ \vdots \\ f_{i,N-1} - N \cdot a_{i,N} = 0 \end{cases} \quad \text{for } i = 1\text{-}K \tag{4.23}$$

4.2.1.2 Offline Transforming Power Series Expansion to the Padé Approximants

Padé approximants with the specified fraction order $[L/M]$ are used to equate the power series function of $x_i(t)$ in the time domain, as defined in (4.24).

$$\begin{aligned} x_i(t) &= a_{i0} + a_{i1}t + a_{i2}t^2 + \cdots + a_{iN}t^N \\ &= \frac{Num_L(t)}{Den_M(t)} = a_{i0} + \frac{n_0 + n_1 t + n_2 t^2 + \cdots + n_L t^L}{d_0 + d_1 t + d_2 t^2 + \cdots + d_M t^M} \end{aligned} \tag{4.24}$$

According to Stahl's Padé convergence theorem, the orders in numerator and denominator in (4.24) preferably have the same or almost the same order, as expressed in (4.25).

$$L = M \quad \text{or} \quad |L - M| = 1 \tag{4.25}$$

The Viskovatov's method is used to obtain the coefficients of Padé approximants [11], in which the initial values are the parameters of power series in (4.26).

$$\begin{cases} c_{0,0} = 1, \quad c_{0,n} = 0 \\ c_{1,0} = a_{i1}, \quad c_{1,1} = a_{i2}, c_{1,n} = a_{in}, \end{cases} \quad \text{for } n = 1, \cdots, N \tag{4.26}$$

Then, the inductive process is carried out to calculate the matrix of Padé approximants, as defined in (4.27).

$$c_{j,n} = \frac{c_{j-2,n+1}}{c_{j-2,1}} - \frac{c_{j-1,n+1}}{c_{j-1,1}}, \quad \text{for } j = 2, 3, \cdots, 2N \tag{4.27}$$

Finally, the numerator and denominator of Padé approximants can be calculated inductively by (4.28) and (4.29).

$$\begin{aligned} &Num_0(t) = a_{i1}, Den_0(t) = 1 \\ &Num_1(t) = a_{i1} \cdot c_{2,0} + t, Den_1(t) = c_{2,0} \end{aligned} \tag{4.28}$$

$$\frac{Num_j(t)}{Den_j(t)} = \frac{Num_{j-1}(t) + c_{j+1,0}t \cdot Num_{j-2}(t)}{Den_{j-1}(t) + c_{j+1,0}t \cdot Den_{j-2}(t)} \text{ for } j \leq 2N - 1 \tag{4.29}$$

In (4.29), $c_{j+1,0}$ is the Viskovatov coefficients, j is the order or Padé approximants to adjust the performance of the trajectories. Larger j leads to a higher accuracy close to the initial point t_0, but may worsen the performance with a larger time window, vice versa. A relatively good performance of overall convergence is usually achieved when $j \approx N$.

4.2.1.3 Online Evaluating SAS Within a Time Window

For online dynamic simulation, e.g. a time-domain simulation for contingency scenarios, the SAS is evaluated by replacing the symbolic values by the values of the initial/current states. In order to accelerate the process of online evaluation, the symbolic values are composed of three different groups.

Group A: the simulation time t.
Group B: initial/current values of state variables $\boldsymbol{x}_0$ and other variables for system conditions $\boldsymbol{y}$.
Group C: admittance matrix $\mathbf{Y}$.

To create the continuous curves of an SAS, the symbolic variable of time in *Group A* is substituted by the simulation time t in (4.29). For every time interval T_s, variables in *Group B* are substituted by the ending values in the previous interval, while *Group C* is only updated at the time point of network topology change, such as fault starting and fault clearing. Figure 4.2 illustrates the process of such online stage. The simulation is divided into pre-fault, fault-on, and post-fault periods by red vertical lines. The black solid line is the SAS curve by plugging in the simulation t.

The simulation time interval T_s, either adaptive or fixed, is vital to the simulation speed because most of symbolic variables are assigned by the updated values in every T_s. The operations of assignment are the major computation burden for online stage in this approach.

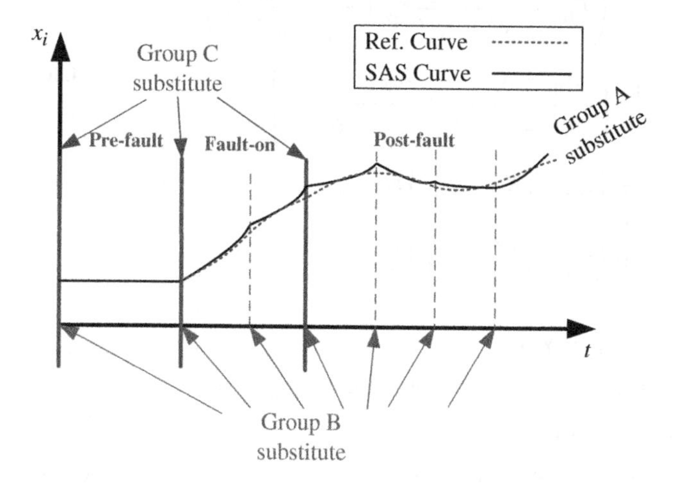

Figure 4.2 Demonstration of the process of online stage.

If the fixed time interval is adopted, T_s can be conservatively chosen to be a small value that ensures the maximum error of state variables over the whole simulation period is less than the prespecified error tolerance ε. If the adaptive time interval is adopted, the error is estimated, then the time interval is adaptively changing according to the estimated error.

4.2.2 Padé Approximants of Power System Differential Equations

For a K-generator N-bus power system, the DE of the kth generator is modeled by a sixth-order d–q axis model. In (4.30), δ_k, $\Delta\omega_k$, e'_{qk}, e'_{dk}, e''_{qk}, e''_{dk} are the state variables of rotor angle, relative rotor speed, q- and d-axis transient voltage, q-axis and d-axis sub-transient voltage, respectively. P_{mk}, P_{ek}, H_k, D_k, T'_{d0k}, T'_{q0k}, T''_{d0k}, T''_{q0k}, X'_{dk}, X'_{qk}, X''_{dk}, X''_{qk}, X_{dk}, X_{qk} are mechanical and electrical powers, the parameters of inertia, damping, d- and q-axis transient and sub-transient time constants and reactance and synchronous reactance, respectively.

All K generators are coupled by a nonlinear algebraic equation (AE) defined in (4.31), in which P_{ek}, e_{dk}, e_{qk}, i_{dk}, i_{qk} are the electrical power, d- and q-axis voltage, and current. $\mathbf{Y}_k^*$ is the kth row of $K \times K$ admittance matrix reduced from original $N \times N$ network admittance matrix [12].

$$\begin{cases} \dot{\delta}_k = \Delta\omega_k \\ \Delta\dot{\omega}_k = \dfrac{\omega_R}{2H_k}\left(P_{mk} - P_{ek} - D_k\dfrac{\Delta\omega_k}{\omega_R}\right) \\ \dot{e}'_{qk} = \dfrac{1}{T'_{d0k}}\left(E_{fdk} - \dfrac{X_{dk} - X''_{dk}}{X'_{dk} - X''_{dk}}e'_{qk} + \dfrac{X_{dk} - X'_{dk}}{X'_{dk} - X''_{dk}}e''_{qk}\right) \\ \dot{e}'_{dk} = \dfrac{1}{T'_{q0k}}\left(-\dfrac{X_{qk} - X''_{qk}}{X'_{qk} - X''_{qk}}e'_{dk} + \dfrac{X_{qk} - X'_{qk}}{X'_{qk} - X''_{qk}}e''_{dk}\right) \\ \dot{e}''_{qk} = \dfrac{1}{T''_{d0k}}\left(e'_{qk} - \left(X'_{dk} - X''_{dk}\right)i_d - e''_{qk}\right) \\ \dot{e}''_{dk} = \dfrac{1}{T''_{q0k}}\left(e'_{dk} + \left(X'_{qk} - X''_{qk}\right)i_q - e''_{qk}\right) \end{cases} \tag{4.30}$$

$$\begin{aligned} &I_{tk} = i_{Rk} + ji_{Ik} \equiv \mathbf{Y}_\mathbf{k}^*\mathbf{E} \\ &i_{qk} = i_{Ik}\sin\delta_k + i_{Rk}\cos\delta_k, i_{qk} = i_{Rk}\sin\delta_k - i_{Ik}\cos\delta_k \\ &P_{ek} = e_{qk}i_{qk} + e_{dk}i_{dk} \\ &e_{qk} = e'_{qk} - X'_{dk}i_{dk}, e_{dk} = e'_{dk} + X'_{qk}i_{qk} \end{aligned} \tag{4.31}$$

In order to illustrate the process of power series expansion and Padé approximants of a real power system, a single-machine infinite-bus (SMIB) system is modeled by (4.30) and (4.31). After the calculation of power series expansion introduced in Section 4.2.1, the trajectory of the state variables can be expressed as (4.32), where the parameters of power series can be expressed by all the symbolic variables. For the sake of simplicity, only the first-order parameters are shown as (4.33).

$$\begin{cases} \delta(t) = \delta_0(t_0) + \sum_{n=1}^{N} \delta_n (t-t_0)^n \\ \omega(t) = \omega_0(t_0) + \sum_{n=1}^{N} \omega_n (t-t_0)^n \\ e_q^{/}(t) = e_{q0}^{/}(t_0) + \sum_{n=1}^{N} e_{qn}^{/} (t-t_0)^n \end{cases} \quad \begin{cases} e_d^{/}(t) = e_{d0}^{/}(t_0) + \sum_{n=1}^{N} e_{dn}^{/} (t-t_0)^n \\ e_q^{//}(t) = e_{q0}^{//}(t_0) + \sum_{n=1}^{N} e_{qn}^{//} (t-t_0)^n \\ e_d^{//}(t) = e_{d0}^{//}(t_0) + \sum_{n=1}^{N} e_{dn}^{//} (t-t_0)^n \end{cases} \tag{4.32}$$

$$\begin{cases} \delta_1 = \omega_0 \\ \omega_1 = -\left(Y_R\omega_R\left(e_{q0}^{//2} + e_{d0}^{//2}\right) + D\omega_0 - P_m\omega_R\right)/2H \\ e_{q1}^{/} = \left(E_{fd} + e_{q0}^{//}\left(X_d - X_q^{/}\right)/\left(X_d^{/} - X_q^{//}\right)\right. \\ \qquad \left. - e_{q0}^{/}\left(X_d - X_d^{//}\right)/\left(X_d^{/} - X_d^{//}\right)\right)/T_{d0}^{/} \\ e_{d1}^{/} = -\left(e_{d0}^{/}\left(X_q - X_q^{//}\right) - e_{d0}^{//}\left(X_q - X_q^{/}\right)\right)/\left[T_{d0}^{/}\left(X_q^{/} - X_q^{//}\right)\right] \\ e_{q1}^{//} = \left(e_{q0}^{/} - e_{q0}^{//} + e_{q0}^{//}Y_I\left(X_d^{/} - X_d^{//}\right) + e_{d0}^{//}Y_R\left(X_d^{//} - X_d^{/}\right)\right)/T_{d0}^{//} \\ e_{d1}^{//} = \left(e_{d0}^{/} - e_{d0}^{//} + e_{d0}^{//}Y_I\left(X_q^{/} - X_q^{//}\right) + e_{q0}^{//}Y_R\left(X_q^{/} - X_q^{//}\right)\right)/T_{q0}^{//} \end{cases} \tag{4.33}$$

$\delta_0(t_0)$, $\omega_0(t_0)$, $e'_{q0}(t_0)$, $e'_{d0}(t_0)$, $e''_{q0}(t_0)$, $e''_{d0}(t_0)$ in (4.32) represent the initial values of the state variables in each time interval, which can be replaced by the ending values of previous interval. Each higher order parameters, e.g. a_n, can be expressed as a function of previous lower order parameters a_0 to a_{n-1}. For each state variable, the power series is truncated at a specified order N. Higher order N induces higher computation burden.

After plugging the current values of all symbolic variables into the power series, Padé approximation is implemented to increase the convergence region, as introduced in Section 4.2.1. Taking $N = 3$ for example, the Viskovatov coefficients ($c_{j+1,0}$ in (4.29)) of angle $\delta(t)$ are calculated as (4.34).

$$
\begin{aligned}
c_{2,0} &= \delta_1, \\
c_{3,0} &= -\delta_2/2\delta_1 \\
c_{4,0} &= \left(3\delta_2^2 - 4\delta_1\delta_3\right)/6\delta_1\delta_2 \\
c_{5,0} &= 8\delta_3^2\delta_1/\left(12\delta_1\delta_2\delta_3 - 9\delta_2^3\right) \\
c_{6,0} &= 32\delta_2\delta_3/\left(3\delta_2^2 - 4\delta_1\delta_3\right)
\end{aligned} \tag{4.34}
$$

The highest orders of Padé's numerator and denominator are derived from N order power series and subjected to Stahl's theory $[N/2, N/2]$ (if N is even) or $[(N-1)/2+1, (N-1)/2]$ (if N is odd). However, the order of Padé approximants is recommended as a lower value in order to find the optimal balance between local precision and overall convergence.

4.2.3 Examples

4.2.3.1 Case A. Test on the IEEE 9-Bus Power System

The IEEE 3-generator 9-bus test system is chosen to demonstrate the proposed Padé approximants-based approach for power system dynamic simulation (icseg.iti.illinois.edu). Bus 9 is subjected to a 3-phase fault and Line 6–9 is tripped after 5 cycles. The post-fault curves of state variables between Padé approximants, Taylor series, and the fourth-order Runge–Kutta (RK4) method are compared. The numerical integration method RK4 is carried out with very short time interval, so it is accurate enough for being the reference.

Example 4.1 ***Adaptive Time Interval with Specified Error Limit***

In this example, the adaptive time interval is chosen. The time interval is adjusted according to the specified error limit, i.e. 0.001 rad for angle δ and 0.001 p.u. for other state variables. Taking the first time interval of post-fault for illustration, Figure 4.3 compares the curves of different SASs. RK4 in a black dashed line is the reference. The 6th-order Taylor series is shown in red and the 3rd-order to 6th-order Padé approximants are shown in other colors. It can be observed that the Padé approximants-based SAS has better approximation performance and longer time-step for the given error tolerance. Figure 4.4 compares the adaptive time interval of the 8th-order Padé series and 8th-order Taylor series over the whole five-seconds transient simulation. In overall, the Padé approximants have relatively larger time intervals than that of the Taylor series.

Example 4.2 ***Fixed Time Interval***

In this simulation, the fixed time interval is chosen for the SASs based on the Taylor series and Padé approximants, respectively. The 5th-order Taylor series and 5th-order Padé approximants are compared, where the

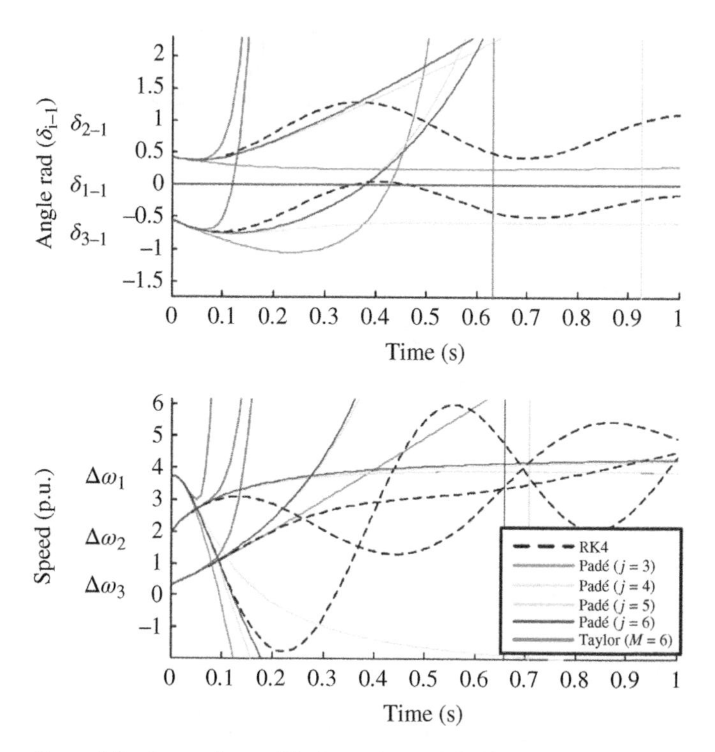

Figure 4.3 Comparison of Taylor series and Padé approximants at the first step of post-fault response (adaptive time step).

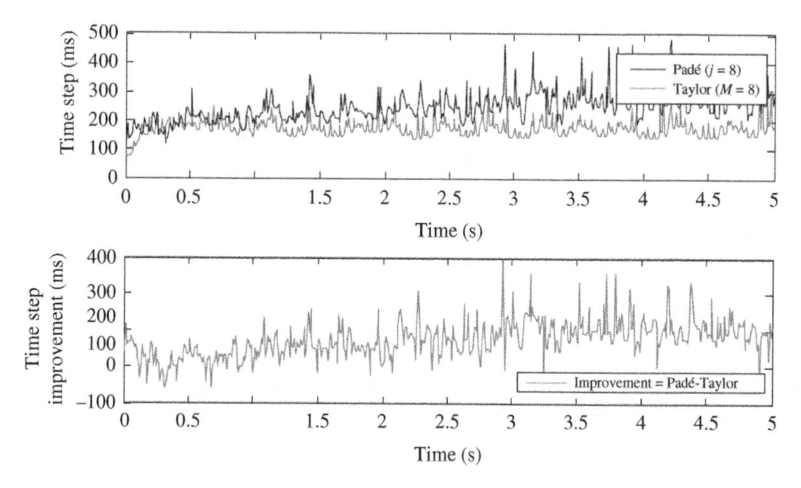

Figure 4.4 Comparison of time intervals between Taylor series and Padé approximants (adaptive time interval).

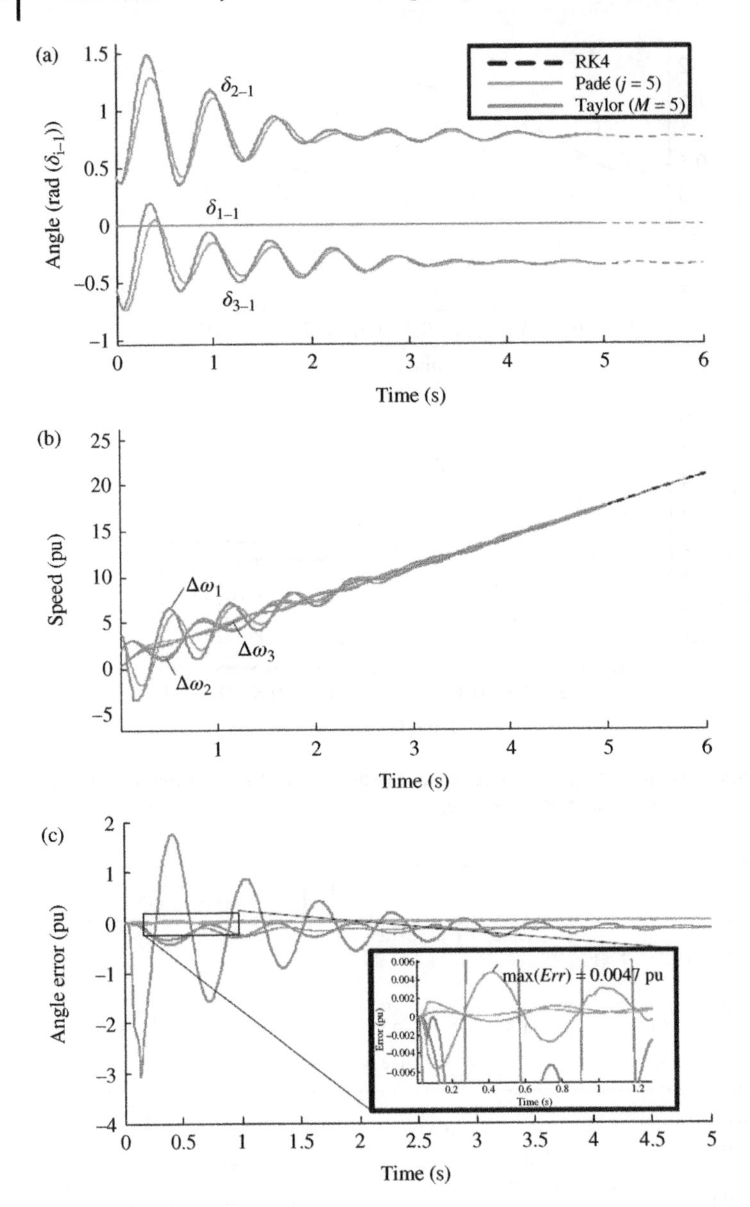

Figure 4.5 Comparison of curves between 5th-order Taylor series and 5th-order Padé approximants (fixed time interval).

time interval is chosen as 4 cycles, i.e. 66.68 ms. The post-fault responses of state variable angle δ and speed $\Delta\omega$ are show in Figure 4.5a and b, respectively. The green curve of Padé-based SAS basically overlaps with the reference by RK4, but the red curves of Taylor-based SAS have noticeable derivation from the RK4. The error of angle is shown in Figure 4.5c, where the maximum error of Padé-based SAS is 0.047 p.u.

4.2.3.2 Case B. Test on the IEEE 39-Bus Power System

The Padé approximants approach for power system dynamic simulation is also demonstrated on the IEEE 10-generator, 39-bus system as shown in Figure 4.6 [13]. If the time interval is chosen as 2 cycles (33.34 ms), the

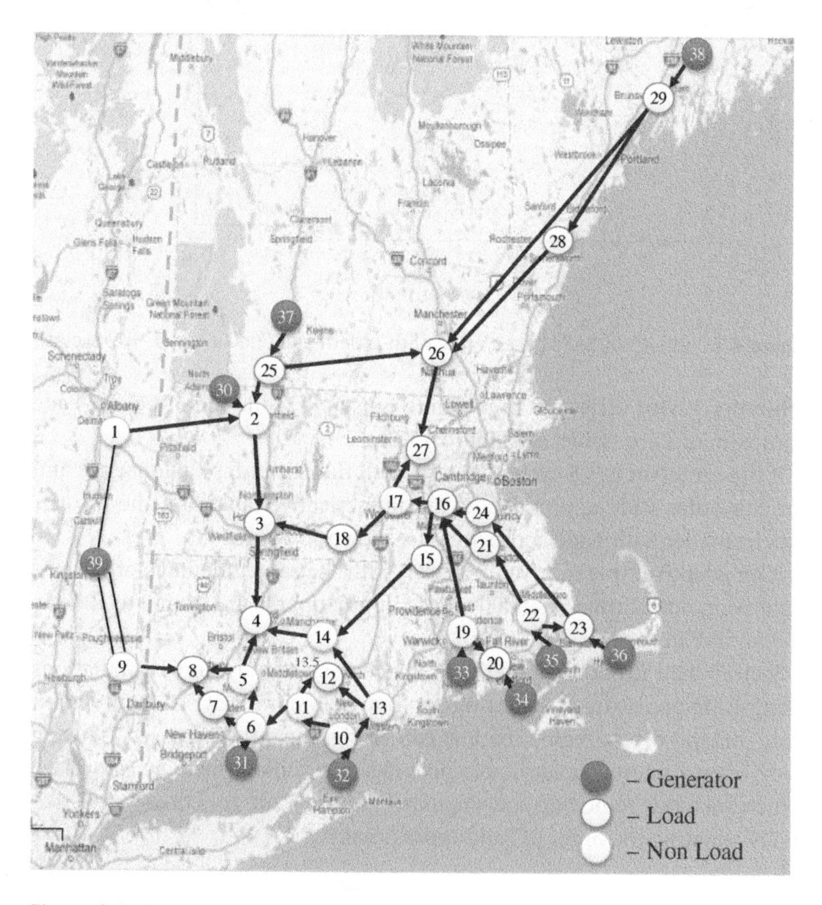

Figure 4.6 Map of IEEE 10-generator, 39-bus New England system.

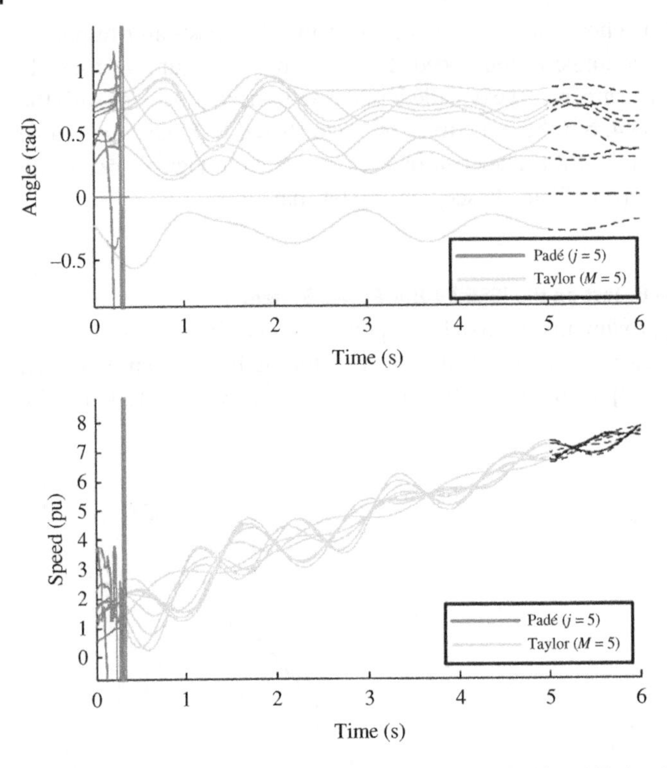

Figure 4.7 Curves of SASs based on the 5th-order Padé and Taylor ($T_s = 2$ cycles).

5th-order Taylor series are numerically unstable while the 5th-order Padé approximants are stable, see Figure 4.7.

If time interval is 1.5 cycles (25 ms), both the 5th-order Taylor series and the 5th-order Padé approximants are numerically stable, but the error is much smaller with Padé approximants, shown in Figure 4.8.

If an adaptive time interval is chosen with the same tolerance of Case A, improvements of time interval by Padé approximants are from 0 to 100 ms, shown in Figure 4.9.

Section 4.2 introduces a method for power system dynamic simulations using Padé approximants to solve the power system DEs. The simulation time interval of SAS-based method can be significantly increased by using Padé approximants. Readers are encouraged to achieve further researches, which includes removing the singularity of the Padé functions, and optimally adapting the order of Padé approximants.

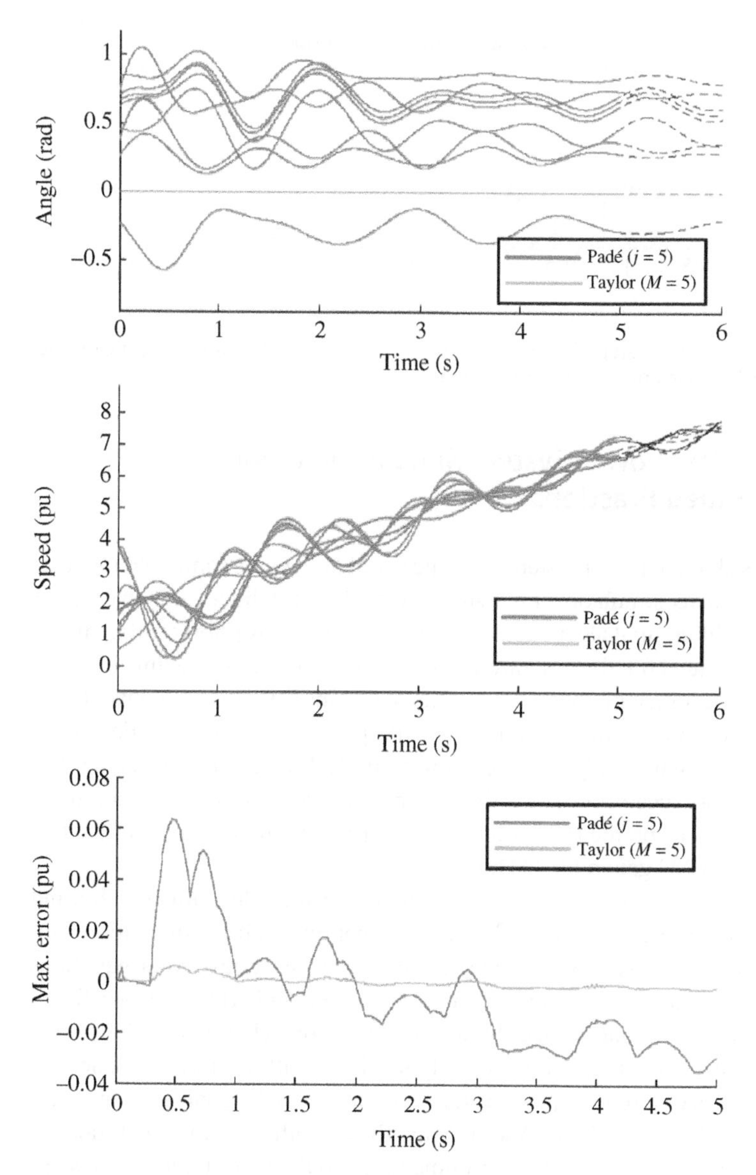

Figure 4.8 Curves of SASs based on the 5th-order Padé & Taylor ($T_s = 1.5$ cycles).

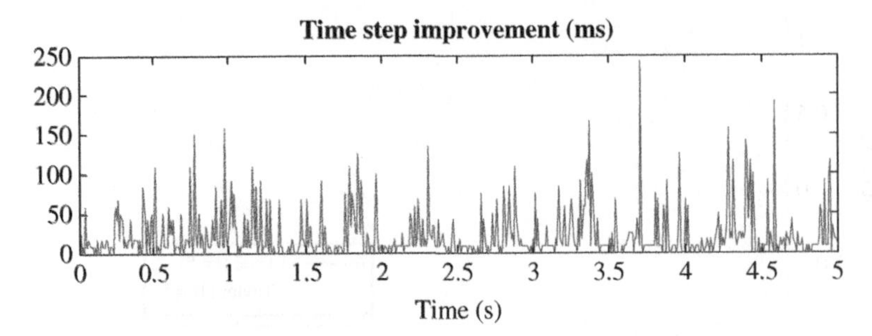

Figure 4.9 Time step of IEEE 10-generator, 39-bus New England system with the 5th-order Padé and the 5th-order Taylor.

4.3 Fast Power System Simulation Using Continued Fractions

In this chapter, power system dynamic simulation using continued fractions to extend its simulation time step is introduced, which can solve power system differential algebraic equations (DAEs) [14]. A priori error bound of the continued fractions is introduced to adaptively adjust the time intervals for the accuracy of analytical solutions. Compared with the conventional numerical integration methods, the proposed continued fraction-based method is semi-analytical and semi-numerical, and it has advantages in simulation speed and suitability for parallel computing. The method is demonstrated on the IEEE 9-bus system, IEEE 39-bus system and tested on the Polish 327-generator 2383-bus system.

Numerical integration methods, such as Forward Euler, Runge–Kutta, and Trapezoidal methods are widely used in commercialized simulation software to solve power system DAEs comprising ordinary differential equations (ODEs) on dynamic devices such as generators and AEs on power flows. These conventional methods basically separate the whole simulation period into many short integration steps, typically in milliseconds, and calculate the DEs as difference equations for every time step. Undoubtably, the computation burden is high. Additionally, these methods do not naturally fit well in parallel computing environment due to their short integration steps and time-sequential mechanism [15–17], unless additional techniques, such as network decomposition methods [18–20] and Parareal in time method [21–23], are applied.

Many SAS methods can be utilized as an alternative simulation approach, which is an approximate but explicit solution of power system ODEs. Using the SAS, online simulation can be speeded up due to the following reasons.

First, a major portion of the computation burden can be transformed into an offline stage and thus online simulation becomes just an evaluation of a SAS, i.e. substituting actual values for symbolic variables of the SAS for sequential time intervals to make up the simulation period. Second, the SAS is a nonlinear expression derived from the power system ODE model, so its time interval of accuracy can be much longer than the time step of traditional numerical integration methods. Third, the evaluation of a SAS can be easily parallelized due to its analytical form in, e.g. power series.

An SAS in the form of a power series in time for simulation of fault-on trajectories has been introduced [24] and an enhanced method to solve more general DAEs models of large-scale power systems has also been investigated, which derives SASs only for ODEs and adopts the numerical approach to solve the network algebraic equations [25]. However, an SAS in the form of truncated power series has a limited radius of convergence, close to which the error of the solution increases rapidly. Therefore, an SAS has to be evaluated for relatively short time intervals to ensure its convergence, which restricts the simulation speed.

This chapter also introduces a similar scheme that solves ODEs and AEs of a large power system respectively by an SAS and a numerical approach. It proposes a new SAS in the form of a continued fraction (for short, CF-SAS). Compared with the power series-based SAS (for short, PS-SAS) [25], the proposed CF-SAS has an extended radius of convergence and hence faster online simulation speed. With longer time intervals for SAS evaluation, network AEs are solved less often, so it provides more room for parallel computing techniques to speed up simulation. A priori error bound of the CF-SAS is also introduced, which has the ability to adaptively adjust the time intervals.

4.3.1 The Proposed Two-Stage Simulation Scheme

The complete power system model is composed of a set of first-order ODEs on dynamic elements, such as synchronous machines and controllers, which are coupled through the power network, and a set of algebraic equations on the power flows and other static components.

$$\begin{cases} \dot{\mathbf{x}} = \mathbf{f}(\mathbf{x}, \mathbf{y}) \\ \mathbf{0} = \mathbf{g}(\mathbf{x}, \mathbf{y}) \end{cases} \tag{4.35}$$

Figure 4.10 shows the flow chart of using a CF-SAS for power system dynamic simulation. Similar to Section 4.2, a double-stage scheme is proposed, i.e. the offline stage and the online stage. The offline stage obtains the SAS, i.e. an explicit function regarding all symbolic variables and time.

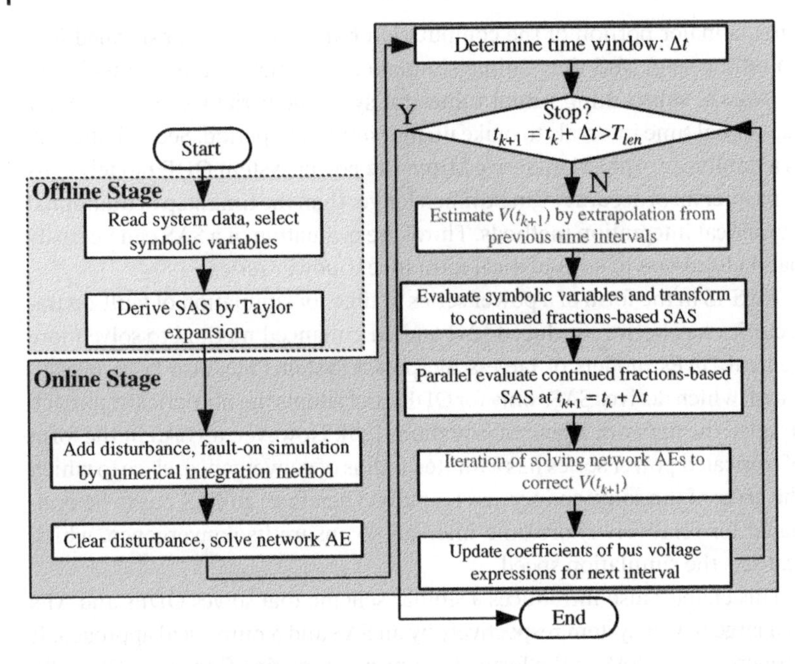

Figure 4.10 Flow chart of the proposed CF-SAS approach.

The online stage only evaluates the obtained SAS by plugging the symbolic variables with actual values over sequential time intervals. The accuracy of the SAS can be ensured by adjusting the length of time intervals during the online stage. More specifically, for every time interval, the initial values of state variables take their final values of the previous interval.

Different from the SAS approach that derives SASs only for ODEs and adopts the numerical approach, the SAS evaluation process first substitutes values for symbolic variables except for time to obtain the coefficient values of the power series, which is then transformed into a continued fraction about time, i.e. the CF-SAS. Determine the length of time interval Δt based on the radius of convergence of the continued fraction and finally evaluate the CF-SAS to get the trajectories of solutions.

4.3.1.1 Solving Power System DAEs Using a Partitioned Dynamic Bus Method

Take an N-bus K-generator L-load power system for example. The sets of buses, generators, and loads are denoted by $\mathcal{N}$, $\mathcal{K}$, and $\mathcal{L}$ respectively. The terminal voltage V_k and terminal injected current I_k of the kth generator are calculated by the nonlinear algebraic equation (4.36), in which v_{dk}, v_{qk}, i_{dk}, and i_{qk} are d-axis and q-axis voltages and currents, respectively. p_{ek}

is the generator's electrical power in its armature.

$$\begin{cases} v_{dk} + jv_{qk} = V_k e^{-j(\delta_k - \pi/2)} \\ i_{dk} = \left(\left(e''_{dk} - v_{dk} \right) R_{ak} + \left(e''_{qk} - v_{qk} \right) X''_{qk} \right) / \left(R_{ak}^2 + X''_{dk} X''_{qk} \right) \\ i_{qk} = \left(-\left(e''_{dk} - v_{dk} \right) X''_{dk} + \left(e''_{qk} - v_{qk} \right) R_{ak} \right) / \left(R_{ak}^2 + X''_{dk} X''_{qk} \right), \\ \qquad \forall k \in \mathcal{K} \\ I_k = (i_{dk} + j i_{qk}) e^{j(\delta_k - \pi/2)} \\ p_{ek} = v_{dk} i_{dk} + v_{qk} i_{qk} \end{cases} \tag{4.36}$$

The ZIP static load model can be considered for all the L loads in the system, defined by (4.37), where I_{Il} and I_{Pl} represent the currents from constant-current load and constant-power load at the lth load bus. The constant-impedance load is merged into the network admittance matrix in $\mathbf{Y}_{\mathcal{N}}$ in (4.38). V_{0l}, P_{0l}, and Q_{0l} are the initial bus voltage, active and reactive loads at the lth load bus at the initialization of the simulation; p_{Il}, q_{Il}, p_{Pl}, and q_{Pl} are the percentages of constant-current and constant-power components of the active and reactive loads at the lth load bus. Note that $p_{Zl} + p_{Il} + p_{Pl} = 1$ and $q_{Zl} + q_{Il} + q_{Pl} = 1$ for active and reactive loads, where p_{Zl} and q_{Zl} represent the percentages of constant-impedance load.

$$I_{Il} + I_{Pl} = \frac{p_{Il} P_{0l} \dfrac{|V_l|}{|V_{0l}|} - j q_{Il} Q_{0l} \dfrac{|V_l|}{|V_{0l}|} + p_{Pl} P_{0l} - j q_{Pl} Q_{0l}}{V_l^*}, \forall l \in \mathcal{L} \tag{4.37}$$

All the K generators and L loads are coupled by the network algebraic equations (4.38), in which $\mathbf{I}_{\mathcal{N}}$ and $\mathbf{V}_{\mathcal{N}}$ are vectors about bus injection currents and bus voltages.

$$\mathbf{Y}_{\mathcal{N}} \mathbf{V}_{\mathcal{N}} = \mathbf{I}_{\mathcal{N}} \tag{4.38}$$

Note that in (4.38), at $t_{k+1} = t_k + \Delta t$, injections $\mathbf{I}_{\mathcal{N}}(t_{k+1})$ of all generators and loads depend on terminal voltages $\mathbf{V}_{\mathcal{N}}(t_{k+1})$ at the same time instant, which are solved as follows. The partitioned method is applied in this chapter to extend the SAS from ODEs to general power system DAEs. The basic idea is to introduce a dynamic bus whose voltage is expressed by a function extrapolated from the voltages of previous time intervals, e.g. $\mathbf{V}_{\mathcal{N}}(t_{k+1}) = \mathbf{h}(\mathbf{V}_{\mathcal{N}}(t_k), \mathbf{V}_{\mathcal{N}}(t_{k-1}), \ldots)$. For ODEs, state variables in $\mathbf{x}(t_{k+1})$ of the SAS are evaluated using the function of terminal voltages $\mathbf{V}_{\mathcal{N}}(t_{k+1})$. While for network algebraic equations, the current injections in $\mathbf{I}_{\mathcal{N}}(t_{k+1})$ are calculated in an iterative process to correct the terminal voltages in $\mathbf{V}_{\mathcal{N}}(t_{k+1})$. Therefore, seen from each dynamic element, the impacts from all

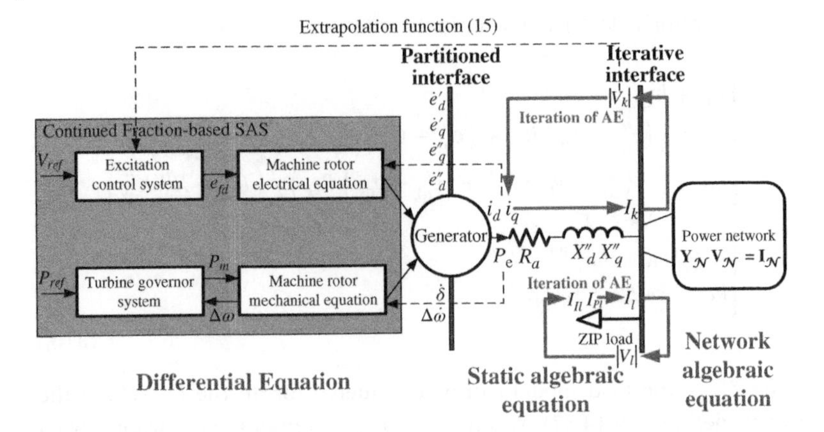

Figure 4.11 Partitioned method for solving power system Differential-algebraic equations.

other elements in the network are reflected by the dynamics of its terminal bus voltage. Then, for each time interval, the analytical expression about each dynamic element is independent from the others.

In Eq. (4.39), $V_i(t)$ is in the polar form of time–power series for voltages at bus i. $V_{\mathrm{m}i}(t)$ and $V_{\mathrm{a}i}(t)$ are the magnitude and angle of the time–power series for voltages at bus i. Both can be truncated up to the N_vth order, where V_{min} and V_{ain} are respectively the nth order coefficients of voltage magnitude and the angle at bus i, which are calculated by the linear least square regression and updated for the next time interval.

$$V_i(t) = V_{\mathrm{m}i}(t)e^{jV_{\mathrm{a}i}(t)} \approx \left(\sum_{n=0}^{N_v} V_{\mathrm{m}in}t^n\right) e^{j\left(\sum_{n=0}^{N_v} V_{\mathrm{a}in}t^n\right)}, \forall i \in \mathcal{N} \tag{4.39}$$

Figure 4.11 shows the procedure of the partitioned method for solving power system DAEs using SAS. Please note that, for each interval, the SAS is evaluated for each iteration of solving algebraic equations, shown by the green arrow lines, which is to correct the terminal voltage at each node.

4.3.2 Continued Fractions-Based Semi-Analytical Solutions

4.3.2.1 Online Evaluation of SAS Over a Time Interval

The simulation can be divided into pre-fault, fault-on, and post-fault periods by red vertical lines. The length of Δt is significant to the simulation speed because most of the symbolic variables are assigned by updating the values of the previous time interval and the network algebraic equations are solved at the end of each time interval. Additionally, to accelerate the simulation speed, the operations of assignment can be computed in a paralleled mechanism, but the process of solving network algebraic equations cannot be parallelized. Therefore, the process of solving network algebraic equations becomes the major computation burden in the online stage. To realize real-time simulation, larger time intervals are desirable to reduce the times of solving network algebraic equations.

The simulation time interval Δt can be either fixed or adaptive. If the fixed time interval is chosen, time interval Δt can be conservatively specified to be a small value, which ensures that the maximum error of state variables over the whole simulation period is always less than a prespecified error tolerance ε. If the adaptive time interval is chosen, the error can be estimated along with the simulation. Then, the time interval Δt can be adaptively changing on the basis of the estimated error $\hat{\varepsilon}$.

4.3.2.2 Transformation from Power Series to Continued Fractions

A continued fraction is an expression obtained through a recursive process of addition and division. It also calls attention that algebraic continued fractions commonly have a more extended radius of convergence than that of the corresponding power series [26]. In this scheme, the Stieltjes fractions with a specified fraction order $[M]$ in the form of (4.40) are adopted to express the function of SAS, where $\mathbf{K}$ represents the operator of continued fraction and $c_{m,0}$ represents the coefficient of the mth order.

$$S(t) = a_0 + \mathbf{K}_{m=1}^{M}\left(\frac{c_{m,0}t}{1}\right) = a_0 + \cfrac{c_{1,0}t}{1+\cfrac{c_{2,0}t}{1+\cfrac{\ddots}{1+c_{M,0}t}}} \tag{4.40}$$

As mentioned in Section 4.3.1, the state variable $x_i(t)$ is evaluated from the PS-SAS, defined by (4.41), where a_i are the coefficients of the truncated power series up to the Nth order.

$$x(t) = a_0 + a_1t + a_2t^2 + \cdots a_nt^n + \cdots + a_Nt^N \tag{4.41}$$

Viskovatov's method is then adopted to obtain the coefficients of an equivalent continued fraction [11]. To equate the power series function in (4.41), the second row of the continued fraction matrix expressed by (4.42) are filled with power series coefficients.

$$
\begin{array}{|c|ccccc}
1 & 0 & 0 & 0 & 0 & \cdots \\
c_{1,0} & c_{1,1} & c_{1,2} & \cdots & c_{1,n} & \cdots \\
c_{2,0} & c_{2,1} & c_{2,2} & \cdots & c_{2,n} & \cdots \\
\vdots & \vdots & \vdots & \vdots & \vdots & \cdots \\
c_{m,0} & c_{m,1} & c_{m,2} & \cdots & c_{m,n} & \cdots \\
\vdots & \vdots & \vdots & \vdots & \vdots & \ddots
\end{array}
\tag{4.42}
$$

$$
\begin{cases} c_{0,0} = 1, \;\; c_{0,n} = 0 \\ c_{1,0} = a_1, \;\; c_{1,1} = a_2, \cdots, c_{1,n} = a_n, \end{cases} \qquad \forall n \geq 1 \tag{4.43}
$$

Then, the coefficients in the continued fraction matrix (4.42) are calculated term by term by an inductive process, defined as (4.44).

$$
c_{m,n} = \frac{c_{m-2,n+1}}{c_{m-2,1}} - \frac{c_{m-1,n+1}}{c_{m-1,1}}, \qquad \forall m \geq 2 \text{ and } n \geq 0 \tag{4.44}
$$

Finally, the coefficients of a continued fraction in (4.40) take the values in the first column of the matrix, i.e. $c_{1,0}$–$c_{m,0}$ in (4.44).

For demonstration, a SMIB system without governor and exciter is adopted to show the deriving process of CF-SAS. The SMIB system is modeled by Eqs. (4.30) and (4.36). As introduced by Section 4.3.2.1, the curves of state variables at the time t_0 can be approximated by a PS-SAS, expressed as (4.45), where all the parameters of PS-SAS, i.e. δ_n, ω_n, e'_{qn}, e'_{dn}, e''_{qn}, and e''_{dn}, are expressed by symbolic variables.

$$
\begin{cases}
\delta(t) = \delta_0(t_0) + \sum\limits_{n=1}^{N} \delta_n (t-t_0)^n \\
\omega(t) = \omega_0(t_0) + \sum\limits_{n=1}^{N} \omega_n (t-t_0)^n \\
e'_q(t) = e'_{q0}(t_0) + \sum\limits_{n=1}^{N} e'_{qn} (t-t_0)^n
\end{cases}
\quad
\begin{cases}
e'_d(t) = e'_{d0}(t_0) + \sum\limits_{n=1}^{N} e'_{dn} (t-t_0)^n \\
e''_q(t) = e''_{q0}(t_0) + \sum\limits_{n=1}^{N} e''_{qn} (t-t_0)^n \\
e''_d(t) = e''_{d0}(t_0) + \sum\limits_{n=1}^{N} e''_{dn} (t-t_0)^n
\end{cases}
\tag{4.45}
$$

where $\delta_0(t_0)$, $\omega_0(t_0)$, $e'_{q0}(t_0)$, $e'_{d0}(t_0)$, $e''_{q0}(t_0)$, and $e''_{d0}(t_0)$ represent the initial values of the state variables in each time interval. Moreover, all the power series of state variables are truncated at a specified order N. Higher order N means a heavier computation burden in both offline stage and online stage, but higher accuracy for each time interval. Each higher order parameter, e.g. δ_N for order N, can be expressed as an analytical function of lower order parameters, i.e. δ_n, ω_n, e'_{qn}, e'_{dn}, e''_{qn}, and e''_{dn} for the 1st order to the $(N-1)$th order.

For the purpose of demonstration, only the 1st order parameter of (4.45) is analytically expressed as (4.46).

$$\begin{cases} \delta_1 = \omega_0 \\ \omega_1 = -\left(Y_{\mathrm{R}}\omega_{\mathrm{R}} \left(e''^2_{\mathrm{q}0} + e''^2_{\mathrm{q}0} \right) + D\omega_0 - P_{\mathrm{m}}\omega_{\mathrm{R}} \right) / 2H \\ e'_{\mathrm{q}1} = \left(E_{\mathrm{fd}} + e''_{\mathrm{q}0} \left(X_{\mathrm{d}} - X'_{\mathrm{q}} \right) / \left(X'_{\mathrm{d}} - X''_{\mathrm{q}} \right) \right. \\ \qquad \left. - e'_{\mathrm{q}0} \left(X_{\mathrm{d}} - X''_{\mathrm{d}} \right) / \left(X'_{\mathrm{d}} - X''_{\mathrm{d}} \right) \right) / T'_{\mathrm{d}0} \\ e'_{\mathrm{d}1} = -\left(e'_{\mathrm{d}0} \left(X_{\mathrm{q}} - X''_{\mathrm{q}} \right) - e''_{\mathrm{d}0} \left(X_{\mathrm{q}} - X'_{\mathrm{q}} \right) \right) / \left[T'_{\mathrm{d}0} \left(X'_{\mathrm{q}} - X'' \right) \right] \\ e''_{\mathrm{q}1} = \left(e'_{\mathrm{q}0} - e''_{\mathrm{q}0} + e''_{\mathrm{q}0} Y_{\mathrm{I}} \left(X'_{\mathrm{d}} - X''_{\mathrm{d}} \right) + e''_{\mathrm{d}0} Y_{\mathrm{R}} \left(X''_{\mathrm{d}} - X' \right) \right) / T''_{\mathrm{d}0} \\ e''_{\mathrm{d}1} = \left(e'_{\mathrm{d}0} - e''_{\mathrm{d}0} + e''_{\mathrm{q}0} Y_{\mathrm{I}} \left(X'_{\mathrm{q}} - X''_{\mathrm{q}} \right) + e''_{\mathrm{q}0} Y_{\mathrm{R}} \left(X'_{\mathrm{q}} - X''_{\mathrm{q}} \right) \right) / T''_{\mathrm{q}0} \end{cases} \tag{4.46}$$

After all symbolic variables of each power series being replaced by their values, it is transformed into a continued fraction to increase the radius of convergence.

For a given N-order PS-SAS, the specified order M of the CF-SAS is flexible, but it is recommended to equal N for a good balance between overall convergence and local accuracy.

4.3.3 Adaptive Time Interval Based on Priori Error Bound of Continued Fractions

Given a set of credible initial values, the CF-SAS can maintain accuracy for a time interval. Therefore, in the online stage, the entire simulation period should be assembled by consecutive time intervals whose lengths are fixed or adaptively adjusted according to the comparison between the estimation of error and a prespecified error tolerance. A priori upper error bound of the continued fraction which only depends upon its coefficients is used to conservatively estimate the errors of analytical solutions in the dynamic simulations with limited computation burden.

4.3.3.1 Priori Error Bound of Continued Fractions

The continued fractions with a specified fraction order $[m]$ in the form of (4.40) can be expressed as $x_m(t; w_m)$:

$$x_m(t; w_m) = x_0 + \cfrac{c_{1,0}t}{1 + \cfrac{c_{2,0}t}{1 + \cfrac{\ddots}{1 + \frac{c_{m,0}t}{w_m}}}} \tag{4.47}$$

in which w_m is called the mth truncated approximant, and $c_{1,0}$ to $c_{m,0}$ are the coefficients of continued fraction. The objective is to find the maximum ρ, to ensure the estimated error of continued fraction is less than a given error tolerance ε_{th} for $t < \rho$. The parabola theorem for error bound of complex-valued continued fractions has been investigated [27, 28]. This chapter introduces the error bound estimation of a real-valued continued fraction.

Theorem 4.1 A continued fraction is convergent if and only if there exists a maximum radius ρ, and for any given error tolerance ε_{th}, the inequality (4.48) is satisfied if $t < \rho$.

$$|f(t) - f_m(t)| \le 2c_{1,0}\rho \prod_{k=2}^{m} \frac{\sqrt{1 + 4c_{k,0}\rho} - 1}{\sqrt{1 + 4c_{k,0}\rho} + 1} = \hat{\varepsilon}(\rho), \qquad \text{for } m \ge 2 \tag{4.48}$$

where $\{f_m(t)\} = \{x_m(t; w_m)\}$ is the sequence of the continued fraction. Therefore, $\hat{\varepsilon}(\rho)$ is the error upper bound of the continued fraction $\{f_m(t)\}$ and ρ is the maximum time interval with respect to a prespecified error tolerance ε_{th}.

The proof of the error-bound estimation of continued fractions regarding the lengths of the time interval is given as follows.

Proof: Set the error bound of the mth approximant $|f(\rho) - f_m(\rho)| \le 1/d_{m+1}$, where convergence radius d_m with variant m belongs to a sequence about convergence radius ($d_m > 0$), which monotonically increases with m and set as the quadratic function regarding $|w_m|$. The numerator and denominator multiply $d_1 - d_m$ at the same time, so (4.49) is obtained.

$$|f(\rho) - f_m(\rho)| \le \frac{1}{d_1} \times \frac{d_1}{d_2} \times \cdots \times \frac{d_{m-1}}{d_m} \times \frac{d_m}{d_{m+1}} \tag{4.49}$$

From (4.47), it can be derived that $w_m = 1 + (c_{m+1,0} \times \rho)/w_{m+1}$, so

$$\frac{d_{m+1}}{d_m} = \frac{w_m + \tau_{m+1}|w_m|^2}{w_m - \sigma_m} \tag{4.50}$$

where σ_m and τ_m are parameters of d_m regarding w_m.

Define two new real variables ξ and λ, subjecting to $\xi := w_m/\sigma_m$ and $\lambda := \sigma_m \tau_{m+1} = 1/4c_{m,0} \times \rho$. Then,

$$\frac{d_{m+1}}{d_m} = \frac{w_m + \tau_{m+1}|w_m|^2}{w_m - \sigma_m} \ge \frac{\xi + \lambda\xi^2}{\xi - 1} := \varphi(\xi) \tag{4.51}$$

with $\tau_{m+1} = (1 - \sigma_m)/c_{m,0}\rho$.

The function on the right-hand side of (4.51) is

$$\varphi(\xi) = \frac{\xi + \lambda\xi^2}{\xi - 1} \tag{4.52}$$

whose maximum occurs at

$$\xi^* = 1 + \sqrt{1 + \lambda^{-1}} > 2 \tag{4.53}$$

So that the maximum value of $\varphi(\xi)$ from (4.51) is

$$\varphi(\xi^*) = \frac{\sqrt{1+\lambda^{-1}} - 1}{\sqrt{1+\lambda^{-1}} + 1} = \frac{\sqrt{1+4c_{k,0}\rho} - 1}{\sqrt{1+4c_{k,0}\rho} + 1} \tag{4.54}$$

Choose $\sigma_m = 1/2$ and note that $d_1 = 1/(2c_{1,0}\rho)$. Then, (4.48) is proven by combining (4.54) regarding all orders from $1/d_1$ to d_m/d_{m+1} in (4.49). ■

4.3.3.2 Adaptive Time Interval for Analytical Solution-Based Dynamic Simulations

Figure 4.12 shows the process of exterminating the current time interval Δt_k based on the priori error bound of CF-SAS. For each interval, an initial radius of convergence ρ is given based on the last time interval Δt_{k-1}, and a multiplication factor β is given to limit the increasing rate between the current interval and the previous interval, as defined by (4.55).

$$\rho = \beta \cdot \Delta t_{k-1} \; (\beta > 1) \tag{4.55}$$

Then the prior error bound for each state variable is calculated by (4.48) and the maximum error bound among all state variables is found by (4.56), where N_s is the number of state variables.

$$\hat{\varepsilon}_{\max}(\rho) = \max\{\hat{\varepsilon}_1(\rho), \hat{\varepsilon}_2(\rho) \cdots, \hat{\varepsilon}_{Ns}(\rho)\} \tag{4.56}$$

If the maximum error bound is larger than a prespecified error tolerance ε_{th}, multiply the time interval by a shrinking factor α ($\alpha < 1$) until the error tolerance ε_{th} is satisfied.

The first time interval of simulation is set to be very small, i.e. $\Delta t_0 = 0.1$ ms. The following intervals increase to β ($\beta > 1$) times of their previous intervals. From the perspective of the whole time sequence, this is a "bottom-up" scheme, since the length of each time interval grows based on its previous one at a limited speed defined by β.

Although the selection of the "error-tolerance" depends on the power system model, it is usually appropriate to set a smaller error tolerance for the state variables with faster dynamics. In this way, the time interval for fast dynamics is constrained to ensure that the simulated dynamic responses are accurate enough. For example, for the simulations of small-scale and mid-scale power systems with transient and sub-transient models, the orders of magnitude for error tolerance are recommended to be 10^{-2} degrees for rotor angles, 10^{-3} p.u. for mechanical powers and 10^{-4} p.u. for transient and sub-transient voltages.

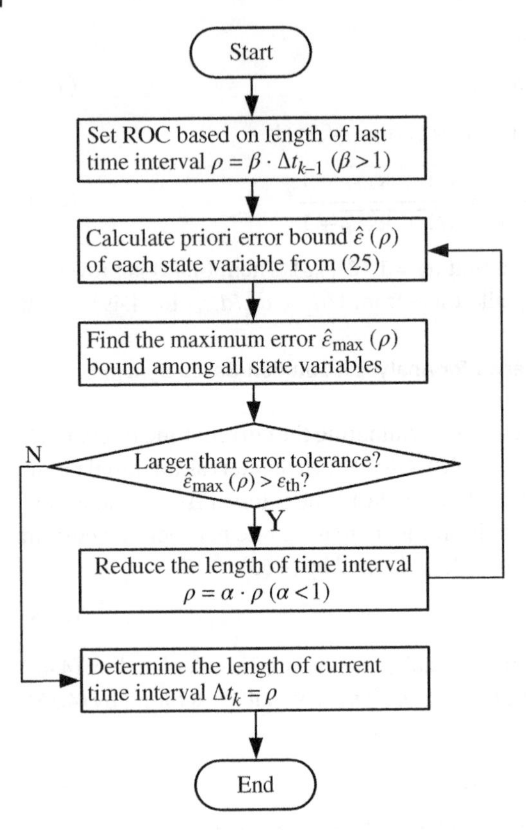

Figure 4.12 Flow chart of determination of length of time interval.

4.3.4 Examples

The approach introduced in this chapter is implemented in MATLAB and tested using three examples on a desktop computer with Intel Core i7-6700 CPU at 3.40 GHz with 16.00 GB RAM. The constant impedance load is included in the network admittance matrix, so the power system dynamic model is represented by a set of nonlinear ODEs. In order to clearly show the superiority of continued fractions over existing SAS methods, a detailed 6-order generator model without a governor or exciter is adopted in the simulations.

The 327-generator 2383-bus Polish system with a detailed 6-order generator model integrated with a 1-order governor and a 1-order exciter [29] is used to demonstrate the CF-SAS approach for power system DAEs. The ZIP (constant impedance (Z), constant current (I), and constant power (P)) load model is also considered for all load buses. The dynamic bus method is used to decouple the dynamic impacts from different elements.

The Polish power system with 2383 buses and 327 generators provided by MATPOWER [29] is adopted to test the performance of a 10-seconds dynamic simulation. All generators use the 6-order detailed model (4.30) equipped with a 1-order exciter model (4.57) and a 1-order governor model (4.58).

$$\dot{e}_{\mathrm{fd}k} = (K_k(V_{\mathrm{ref}k} - |V_k|) - e_{\mathrm{fd}k})/T_{ek}, \quad \forall k \in \boldsymbol{\mathcal{K}} \tag{4.57}$$

$$\dot{p}_{\mathrm{m}k} = (P_{\mathrm{ref}k} - p_{\mathrm{m}k} - \Delta\omega_k/R_k)/T_{gk}, \quad \forall k \in \boldsymbol{\mathcal{K}} \tag{4.58}$$

All loads are represented by the ZIP load model (4.37) with 20% constant impedance, 30% constant current, and 50% constant power components, i.e. $p_{\mathrm{Z}l} = q_{\mathrm{Z}l} = 20\%$, $p_{\mathrm{I}l} = q_{\mathrm{I}l} = 30\%$ and $p_{\mathrm{P}l} = q_{\mathrm{P}l} = 50\%$. The winter peak of the year 1999–2000 is the pre-fault operating condition. A 3-phase short circuit occurs at Bus 10 and is cleared after 4 cycles (66.67 ms) without tripping any line [30]. The Forward Euler numerical iteration method with a fixed time-step of 0.2 ms is selected as the reference for accuracy. The SAS adopts the adaptive time interval with priori error bound estimation. The error tolerance of each interval is selected as 10^{-2} degrees, 10^{-3} and 10^{-4} p.u. for all rotor angles, voltages, and mechanical powers respectively. All simulations use the 1st order polar form time–power series to calculate terminal voltages $V_i(t)$, i.e. $N_v = 1$ in Eq. (4.39).

Figure 4.13a–d respectively show the result of time-domain simulation of generators' rotor angle, speed, q-axis, and d-axis transient voltages for the 3rd order CF-SAS (i.e. $M = 3$) and the Forwsard Euler method. Five generators at Bus 10, Bus 29, Bus 294, Bus 346, and Bus 347, the most adjacent to the contingency, are plotted to examine the result. It can be noticed that the simulation results overlap with the reference (the dotted lines).

Figure 4.14a and b respectively show the length of time intervals and maximum errors of all voltages using the 3-order, 5-order, and 6-order CF-SAS. (i.e. $M = 3, 5, 6$). Using higher-order CF-SAS can prolong the time intervals to speed up the simulation, but it introduces slightly larger errors.

It can be also noticed that the time of solving algebraic equations using the Forward Euler method takes almost 90% of the whole time cost, i.e. 52.60 seconds out of 58.03 seconds, which can be reduced to 3.680, 1.824, and 1.454 seconds if using the 3-order, 5-order, and 6-order CF-SASs. The reason is that long time intervals reduce the total number of times to solve network algebraic equations.

The simulation speed is compared in Table 4.2 between the traditional Forward Euler method and the CF-SAS method with different orders. The Forward Euler method selects a fixed time step of 1 ms for a faster simulation and the SAS method uses executable C code, offline compiled by MATLAB

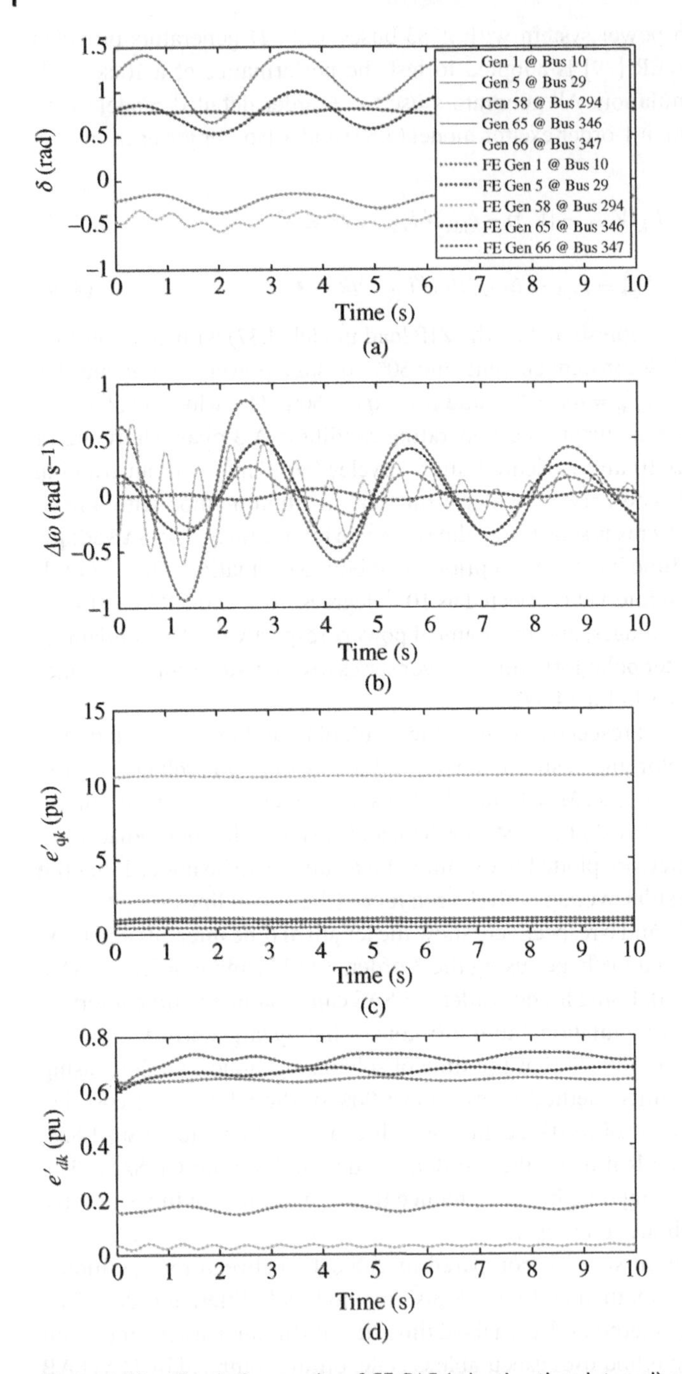

Figure 4.13 Simulation results of CF-SAS (adaptive time interval) and the reference (the Forward Euler method with 0.2 ms fixed time-step). (a) Generator root angle, (b) generator root speed, (c) generator q-axis transient voltage. (d) generator d-axis transient voltage.

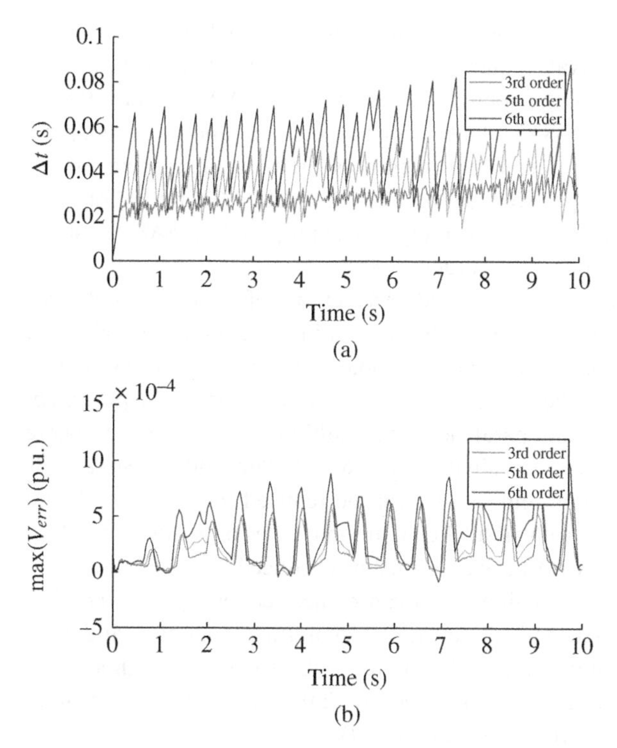

Figure 4.14 (a) Time interval and (b) maximum error of voltage using the 3-, 5- and 6-order CF-SAS (adaptive time interval).

Coder™. It takes 58.03 seconds to complete the 10 seconds simulation using the Forward Euler method. While the 3-, 5-, and 6-order CF-SAS take only 28.14, 25.16, and 44.29 seconds, respectively.

The 5-order CF-SAS takes less simulation time than the 3-order CF-SAS because a higher-order CF-SAS has a larger radius of convergence and thus has larger time intervals. See the 3rd row of Table 4.1. The 3-order CF-SAS needs 817 intervals to complete the 10 seconds simulation, while the 5-order CF-SAS only needs 294 intervals. However, the 6-order CF-SAS takes more simulation time than the 5-order CF-SAS. It is because the number of intervals for the 6-order CF-SAS does not reduce significantly, i.e. from 294 to 223, but the time cost of SAS evaluation takes more time due to the more complex continued fractions. In *Example A*, the 5-order CF-SAS is the optimal.

It can be also noticed that the time of solving algebraic equations using the Forward Euler method takes almost 90% of the whole the time cost, i.e. 52.60 seconds out of 58.03 seconds, which can be reduced to 3.680, 1.824, and

1.454 seconds, if using the 3-order, 5-order, and 6-order CF-SAS. The reason is that long time intervals reduce the frequency of solving network AEs.

Table 4.3 compares the total time cost between PS-SAS and CF-SAS. Unlike CF-SAS, the PS-SAS adapts error-rate upper bound for an adaptive time interval [25] The error-rate tolerances limit the increasing rate of errors, which are respectively set as 2 °/s, 0.01 and 0.001 p.u./s for rotor angles, mechanical powers, and voltages. The proposed CF-SAS can speed up 2.4–3.8 times over the same order PS-SAS.

As mentioned before, the proposed SAS methods offer more potential for parallel computations to speed up the simulations, because the evaluation process of SAS can be easily paralleled. Table 4.4 shows the time cost reduction if adopting a parallel mechanism for SAS evaluation. If the process of SAS evaluation is ideally paralleled by a multiprocessor CPU, e.g. 4-core CPU. the time cost of SAS evaluation is 15.698, 18.062, and 36.489 seconds for the 3-order, 5-order, and 6-order can be reduced to 16.38, 11.62, and 16.92 seconds, respectively. If a CPU with more cores or a GPU is used for parallel simulations, the time can be furtherly reduced.

In summary, the simulation speed can be accelerated by 2.3 times (i.e. 58.03 seconds/25.16 seconds) if no parallel computing is adopted, and if the computation is ideally paralleled in a 4-core CPU of a PC, the method introduced in this chapter can accelerate the simulation speed by almost 5 times (i.e. 58.03 seconds/11.61 seconds) (Table 4.4).

In *Examples B* and *C*, the inertias of all 327 generators in the Polish power system are increased by 2 times and 5 times, respectively. Figure 4.15 shows the time intervals of 4-order CF-SAS for *Example A (original H)*, *Example B* (2× *original H*), and *Example C* (5× *original H*). It can be noticed that, for

Table 4.2 Comparison of simulation between CF-SAS and forward Euler.

	FE	3-order CF-SAS	5-order CF-SAS	6-order CF-SAS
Time step	Fixed	Adaptive	Adaptive	Adaptive
No. of interval	10 000	817	294	223
Avg. time step (ms)	1	12.24	30.01	44.84
Time cost of algebraic equation (s)	52.60	3.680	1.824	1.454
Time cost of SAS evaluation (s)	–	15.698	18.062	36.489
Total time cost (s)	58.03	28.14	25.16	44.29

Table 4.3 Total time cost between CF-SAS and PS-SAS.

	3-order	4-order	5-order	6-order
CF-SAS(s)	28.14	27.48	25.16	44.29
PS-SAS(s)	66.85	74.26	90.14	166.13

Table 4.4 Time cost using ideal parallel computation (compared with 58.03 seconds for Forward-Euler).

	3-order CF-SAS	5-order CF-SAS	6-order CF-SAS
Without parallel computation (s)	28.14	25.16	44.29
With parallel computation of 4 cores (s)	16.38	11.61	16.92

the same power system with different dynamic parameters, the bigger the inertia, the longer time intervals.

Section 4.3 introduces a fast dynamic simulation approach for general power system DAEs. The SASs of power system DEs are developed offline and a partitioned dynamic bus method is adopted to generalize the SAS approach to DAEs, suitable for large-scale power systems. During online stage, CF-SAS is proposed to extend the radius of convergence of the SASs. Compared with the conventional PS-SAS, the proposed CF-SAS has a better computational performance by extending the time interval and improving

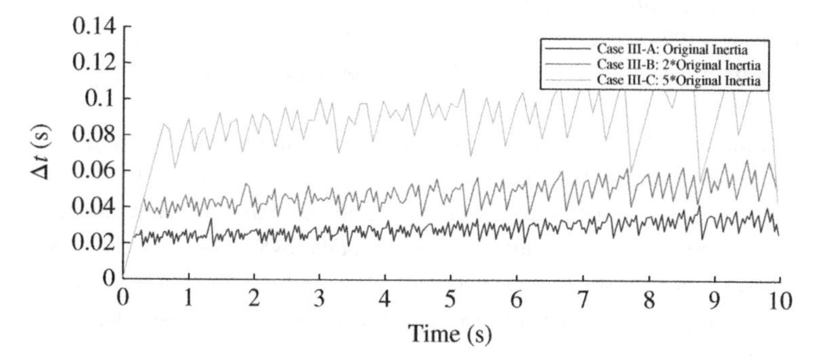

Figure 4.15 Time interval using the 4-order CF-SAS with different inertia parameters (Example A: 1×H, Example B: 2×H, Example C: 5×H).

the online simulation speed. A priori error bound of the continued fractions has also been used to efficiently estimate the maximum error of all the state variables. An adaptive time interval based on prior error bound is also proposed to guarantee the accuracy of the analytical solution. Compared with the conventional numerical integration methods, the CF-SAS introduced in Section 4.3 has advantages in simulation speed and suitability for parallel computation.

4.4 Conclusions

This chapter applies analytic continuation on the time domain to extend the region of convergence of power series-form semi-analytic solutions. Two typical analytic continuation techniques are introduced, i.e. Padé Approximation and Continued Fractions. Better computational performance can be achieved by extending the time intervals and improving the online simulation speed. An adaptive time interval simulation based on prior error bound of continued fractions is also proposed to guarantee the accuracy of the analytical solution. These analytic continuation techniques can be employed for both ODE and DAE of power system simulations. Compared with the conventional methods, these techniques have advantages in simulation speed, longer time interval, and suitability for parallel simulation.

References

1 Stahl, H. (1989). On the convergence of generalized Padé approximants. *Constructive Approximation* 5: 221–240.
2 Stahl, H. (1997). The convergence of Padé approximants to functions with branch points. *Journal of Approximation Theory* 91 (2): 139–204.
3 Subramanian, M. K. (2014). Application of holomorphic embedding to the power-flow problem, Master thesis. Arizona State University.
4 Padé, H. (1892). Sur la représentation approchée d'une fonction par des fractions rationnelles. *Annales scientifiques de l'ÉNS* 9: 3–93. (in French).
5 Baker, G. and Graves-Morris, P. (1996). *Padé approximants Series: Encyclopaedia of Mathematics and its applications*. New York: Cambridge University Press.
6 Baker, G.A. (1965). The theory and application of the Padé approximant method. *Advances in Theoretical Physics* 1: 1–58.

7 Small, G. (2010). *Expansions and Asymptotics for Statistics*, 85–88. Chapman and Hall/CRC Publications.
8 Liu, C., Wang, B., and Sun, K. (2017). Fast power system simulation using semi-analytical solutions based on Padé approximants. In: *Proceedings of IEEE PES General Meeting*, Chicago, IL.
9 Adomian, G. (1994). *Solving Frontier Problem of Physics: The Decomposition Method.* Dordrecht: Kluwer Academic.
10 Duan, N. and Sun, K. (2015). Application of the Adomian decomposition method for semi-analytic solutions of power system differential algebraic equations. In: *Proceedings of Powertech*, Eindhoven, Netherlands.
11 Cuyt, A., Petersen, V.B. et al. (2008). *Handbook of Continued Fraction for Special Functions.* Springer.
12 Wang, B. and Sun, K. (2015). Power system differential-algebraic equations, arXiv prepoint arXiv:1512.05185.
13 Hu, F., Sun, K. et al. (2016). Measurement-based real-time voltage stability monitoring for load areas. *IEEE Transactions on Power Systems* 31 (4): 2787–2798.
14 Liu, C., Wang, B., and Sun, K. (2018). Fast power system dynamic simulation using continued fractions. *IEEE Access* 6: 62687–62698.
15 Hollman, J.A. and Marti, J.R. (2003). Real time network simulation with PC-cluster. *IEEE Transactions on Power Systems* 18 (2): 563–569.
16 Mitra, J. (2014). A Lyapunov function based remedial action screening tool using real-time data. In: *Proceedings of IEEE PES General Meeting*, National Harbor, MD.
17 Palmer, B., Perkins, W., Chen, Y. et al. (2014). GridPACK™: a framework for developing power grid simulations on high performance computing platforms. In: *Proceedings of the 4th International Workshop on Domain-Specific Languages and High-Level Frameworks for High Performance Computing*, New Orleans, LA, 68–77.
18 Zadkhast, S. et al. (2015). A multi-decomposition approach for accelerated time-domain simulation of transient stability problems. *IEEE Transactions on Power Systems* 30 (5): 2301–2311.
19 Aristidou, P. et al. (2016). Power system dynamic simulations using a parallel two-level Schur-complement decomposition. *IEEE Transactions on Power Systems* 31 (5): 3984–3995.
20 Tomim, M.A. et al. (2010). MATE network tearing techniques for multiprocessor solution of large power system networks. In: *Proceedings of IEEE PES General Meeting*, Minneapolis, MN, (25–29 Jul. 2010).
21 Duan, N. and Sun, K. (2017). Power system simulation using the multi-stage Adomian decomposition method. *IEEE Transactions on Power Systems* 32 (1): 430–441.

22 Gurrala, G., Dimitrovski, A. et al. (2015). Parareal in time for fast power system dynamic simulations. *IEEE Transactions on Power Systems* 99: 1–11.

23 Duan, N., Dimitrovski, A., Simunovic, S. et al. (2016). Applying reduced generator models in the coarse solver of Parareal in time parallel power system simulation. In: *Proceedings of IEEE PES ISGT Europe Conference*, Ljubljana, Slovenia.

24 Abreut, E., Wang, B., and Sun, K. (2017). Semi-analytical fault-on trajectory simulation and its application in direct methods. In: *Proceedings of IEEE PES General Meeting*, Chicago, IL.

25 Wang, B., Duan, N., and Sun, K. (2019). A time-power series based semi-analytical approach for power system simulation. *IEEE Transactions on Power Systems* 34 (2): 841–851.

26 Van Vleck, E.B. (1901). On the convergence of continued fractions with complex elements. *Transactions of the American Mathematical Society* 2 (3): 215–233.

27 Gragg, W.B. et al. (1983). Two constructive results in continued fractions. *SIAM Journal on Numerical Analysis* 20 (6): 1187–1197.

28 Christoffel, E.B. et al. (1981). *The Influence of His Work on Mathematics and the Physical Sciences*, 203–211. Basel: Birhäuser.

29 Zimmerman, R.D. et al. (2011). Matpower: steady-state operations, planning and analysis tools for power systems research and education. *IEEE Transactions on Power Systems* 26 (1): 12–19.

30 Gurrala, G., Dinesha, D.L., Dimitrovski, A. et al. (2017). Large multi-machine power system simulations using multi-stage Adomian decomposition. *IEEE Transactions on Power Systems* 32 (5): 3594–3606.

5

Power System Simulation Using Multistage Adomian Decomposition Methods

Nan Duan

Transmission Planning, Midcontinent Independent System Operator, Inc., Carmel, IN, USA

5.1 Introduction to Adomian Decomposition Method

The true solution of a system of nonlinear differential equations may be approached by summating infinite terms of some series expansion. A semi-analytical solution can be defined as the sum of a finite number of terms that is accurate over a time window. Such a series expansion can be derived using the Adomian Decomposition Method (ADM) [1]. The method applies the sum of infinite Adomian polynomials to approach any nonlinear expression. Compared to other decomposition methods like Taylor series expansion, the ADM is able to keep the nonlinearity of the system model. Such a semi-analytical approach also suggests a viable, alternative paradigm for fast stochastic simulation. For example, early works by Adomian in the 1970s utilized the ADM to solve nonlinear stochastic differential equations [2] by embedding explicitly stochastic processes into the terms of a semi-analytical solution.

5.1.1 Solving Deterministic Differential Equations

Consider a nonlinear dynamic system with M state variables modeled by nonlinear DE (5.1)

$$\dot{\mathbf{x}}(t) = \mathbf{f}(\mathbf{x}(t)) \tag{5.1}$$

$$\mathbf{x}(t) = [x_1(t)\ x_2(t)\ \cdots\ x_M(t)]^T$$

$$\mathbf{f}(\cdot) = \left[f_1(\cdot)\ f_2(\cdot)\ \cdots\ f_M(\cdot)\right]^T$$

Power System Simulation Using Semi-Analytical Methods, First Edition. Edited by Kai Sun.

Published 2024 by John Wiley & Sons, Inc.

To solve $\mathbf{x}(t)$, the first step of the ADM is to apply Laplace transform $\mathscr{L}[\cdot]$ to transform Eq. (5.1) into algebraic equations about complex frequency s [3, 4], and then solve $\mathscr{L}$ [$\mathbf{x}$] to obtain Eq. (5.2).

$$\mathscr{L}[\mathbf{x}] = \frac{\mathbf{x}(0)}{s} + \frac{\mathscr{L}[\mathbf{f}(\mathbf{x})]}{s} \tag{5.2}$$

Assume that $\mathbf{x}(t)$ can be decomposed as Eq. (5.3). Then, use Eq. (5.4) to decompose each $f_i(\cdot)$, i.e. $\mathbf{f}(\cdot)$'s ith element, as a sum of infinite Adomian polynomials given by Eq. (5.5), where λ is called a grouping factor [5].

$$\mathbf{x}(t) = \sum_{n=0}^{\infty} \mathbf{x}_n(t) \tag{5.3}$$

$$f_i(\mathbf{x}) = \sum_{n=0}^{\infty} A_{i,n}(\mathbf{x}_0, \mathbf{x}_1, \dots, \mathbf{x}_n), \quad i = 1 \cdots M \tag{5.4}$$

$$A_{i,n} = \frac{1}{n!}\left[\frac{\partial^n}{\partial \lambda^n} f_i\left(\sum_{i=0}^{n} \mathbf{x}_i \lambda^i\right)\right]\Bigg|_{\lambda=0} \tag{5.5}$$

Matching the terms of $\mathbf{x}(t)$ and $\mathbf{f}(\cdot)$ with the same index [6], we can easily derive recursive formulas (5.6) and (5.7) for $\mathscr{L}$ [$\mathbf{x}_n$] ($n \geq 0$), where $\mathbf{A}_n = [A_{1,n}, \dots, A_{M,n}]^T$.

$$\mathscr{L}[\mathbf{x}_0] = \mathbf{x}(0)/s \tag{5.6}$$

$$\mathscr{L}[\mathbf{x}_{n+1}] = \mathscr{L}[\mathbf{A}_n]/s \quad n \geq 0 \tag{5.7}$$

By applying an inverse Laplace transform $\mathscr{L}^{-1}[\cdot]$ to both sides of Eqs. (5.6) and (5.7), we can obtain $\mathbf{x}_n(t)$ for any n. A semi-analytical solution of Eq. (5.1) is defined as the sum of the first N terms of $\mathbf{x}_n(t)$ as shown in Eq. (5.8):

$$\mathbf{x}^{SAS}(t) = \sum_{n=0}^{N-1} \mathbf{x}_n(t) \tag{5.8}$$

5.1.2 Solving Stochastic Differential Equations

With S stochastic variables $y_1(t), \dots, y_S(t)$, which could be stochastic loads following S different distributions. Each $y_i(t)$ can be transformed by function $g_i(\cdot)$ in Eq. (5.9) from some ε_i in a normal distribution.

$$\mathbf{y}(t) = \left[g_1(\varepsilon_1)\ g_2(\varepsilon_2)\ \cdots\ g_S(\varepsilon_S)\right]^T \tag{5.9}$$

The Ornstein–Uhlenbeck process is utilized to generate each ε_i from Eq. (5.10).

$$\dot{\boldsymbol{\varepsilon}}(t) = -\mathbf{a} \circ \boldsymbol{\varepsilon}(t) + \mathbf{b} \circ \mathbf{W}(t) \tag{5.10}$$

where $\mathbf{W}(t)$ is the white noise vector whose dimension equals the number of load buses, **a** and **b** parameters are drifting and diffusion parameters of the stochastic differential equations, operator "∘" is the Hadamard Product, i.e. element-wise multiplication.

Consider a nonlinear system modeled by stochastic differential equations (5.11) having M deterministic state variables $x_1, \ldots, x_M$, such as the state variables of generators, exciters and speed governors, and S stochastic variables $y_1, \ldots, y_S$.

$$\dot{\mathbf{x}}(t) = \mathbf{f}(\mathbf{x}(t), \mathbf{y}(t)) \tag{5.11}$$

The Adomian polynomials can be calculated using Eqs. (5.12) and (5.13).

$$f_k(\mathbf{x}, \mathbf{y}) = \sum_{n=0}^{\infty} A_{k,n}(\mathbf{x}_0, \mathbf{x}_1, \ldots, \mathbf{x}_n, \mathbf{y}) \tag{5.12}$$

$$A_{k,n} = \frac{1}{n!}\left[\frac{\partial^n}{\partial \lambda^n} f_k\left(\sum_{k=0}^{n} \mathbf{x}_k \lambda^k, \mathbf{y}\right)\right]\Bigg|_{\lambda=0} \tag{5.13}$$

Then the semi-analytical solution (5.14) can be derived from the same recursive formulas (5.6) and (5.7) as in the deterministic case. In the resulting semi-analytical solution, stochastic variables in **y** appear explicitly as symbolic variables.

$$\mathbf{x}^{SAS}(t, \mathbf{y}) = \sum_{n=0}^{N} \mathbf{x}_n(t, \mathbf{y}) \tag{5.14}$$

5.2 Adomian Decomposition of Deterministic Power System Models

5.2.1 Applying Adomian Decomposition Method to Power Systems

For a power system having K synchronous generators, consider the fourth-order two-axis model (5.15) to model each generator with saliency ignored [7]. All generators are coupled through nonlinear algebraic equations in (5.16) about the network. In Eqs. (5.15) and (5.16), ω_R is the rated angular frequency; δ_k, ω_k, H_k, and D_k are respectively the rotor angle, rotor speed, inertia, and damping coefficient of the machine k; $\mathbf{Y}_k$ is the kth row of the reduced admittance matrix **Y**; **E** is the column vector of all generator's electromotive forces and E_k is the kth element; P_{mk} and P_{ek} are the mechanical and electric powers; E_{fdk} is the internal field voltage; e'_{qk}, e'_{dk}, i_{qk}, i_{dk}, T'_{q0k}, T'_{d0k}, x_{qk}, x_{dk}, x'_{qk}, and x'_{dk} are transient voltages, stator currents, open-circuit time constants, synchronous reactances, and transient reactances in q- and d-axes, respectively; V_k is the terminal bus voltage magnitude.

$$\begin{cases} \dot{\delta}_k = \omega_k - \omega_R \\ \dot{\omega}_k = \dfrac{\omega_R}{2H_k}\left(P_{mk} - P_{ek} - D_k\dfrac{\omega_k - \omega_R}{\omega_R}\right) \\ \dot{e}'_{qk} = \dfrac{1}{T'_{d0k}}\left[E_{fdk} - e'_{qk} - \left(x_{dk} - x'_{dk}\right) i_{dk}\right] \\ \dot{e}'_{dk} = \dfrac{1}{T'_{q0k}}\left[-e'_{dk} + \left(x_{qk} - x'_{qk}\right) i_{qk}\right] \end{cases} \tag{5.15}$$

$$\begin{aligned} &E_k = e'_{dk}\sin\delta_k + e'_{qk}\cos\delta_k + j\left(e'_{qk}\sin\delta_k - e'_{dk}\cos\delta_k\right) \\ &I_{tk} = i_{Rk} + ji_{Ik} \overset{def}{=} \mathbf{Y}_k^{*}\mathbf{E} \\ &P_{ek} = e_{qk}i_{qk} + e_{dk}i_{dk} \\ &i_{qk} = i_{Ik}\sin\delta_k + i_{Rk}\cos\delta_k, \quad i_{dk} = i_{Rk}\sin\delta_k - i_{Ik}\cos\delta_k \\ &e_{qk} = e'_{qk} - x'_{dk}i_{dk}, \qquad e_{dk} = e'_{dk} + x'_{qk}i_{qk} \\ &V_k = \sqrt{e_{dk}^2 + e_{qk}^2} \end{aligned} \tag{5.16}$$

In addition, consider the following first-order exciter and governor models [8]:

$$\dot{E}_{fdk} = \frac{1}{T_{Ak}}[-E_{fdk} + K_{Ak}(V_{refk} - V_k)] \tag{5.17}$$

$$\dot{P}_{mk} = \frac{1}{T_{gk}}\left(-P_{mk} + P_{refk} - \frac{\omega_k - \omega_R}{R_k}\right) \tag{5.18}$$

where T_{Ak} and K_{Ak} are respectively the time constant and gain in voltage regulation with the exciter, V_{refk} is the reference voltage value, T_{gk} is total time constant of the governor and turbine, P_{refk} is the setting point of the mechanical power output, R_k is the speed regulation factor.

In the following context, the fourth-order model is utilized as an example to illustrate the derivation of a semi-analytical solution for simplicity of description. A similar procedure is applied to the sixth-order differential equation in (5.15), (5.17), and (5.18) and other differential equations. Substitute algebraic equations (5.16) into differential equations (5.15) to eliminate i_{qk}, i_{dk}, and P_{ek}. Then, the differential-algebraic equations (5.15) and (5.16) are transformed into the form of Eq. (5.1), where state vector $\mathbf{x} = \left[\delta_1\ \omega_1\ e'_{q1}\ e'_{d1}\ \cdots\ \delta_K\ \omega_K\ e'_{qK}\ e'_{dK}\right]^T$ has $M = 4K$ state variables as the elements. Then, a semi-analytical solution of this set of differential equations can be derived by formulas (5.6) and (5.7) [9], as illustrated below about the generator speed ω of a single machine infinite bus system modeled by Eq. (5.15). Assume that the infinite bus has voltage $V_\infty = 1$ p.u. Let $\mathbf{x} = \left[\delta, \omega, e'_q, e'_d\right]^T$ and $\mathbf{f} = [f_1, f_2, f_3, f_4]^T$, which are the nonlinear functions in four differential equations. From Eq. (5.3),

$$\delta(t) = \sum_{n=0}^{\infty} \delta_n(t) \quad \omega(t) = \sum_{n=0}^{\infty} \omega_n(t) \quad e'_d(t) = \sum_{n=0}^{\infty} e'_{d,n}(t) \quad e'_q(t) = \sum_{n=0}^{\infty} e'_{q,n}(t) \tag{5.19}$$

Then, Eq. (5.2) about ω becomes:

$$\mathscr{L}[\omega] = \frac{\omega(0)}{s} + \frac{\mathscr{L}\left[f_2\left(\delta, \omega, e'_q, e'_d\right)\right]}{s} \tag{5.20}$$

From Eqs. (5.4) and (5.5), the first two Adomian polynomials for f_2 are given in

$$\begin{aligned}
A_{2,0} &= \frac{\omega_R}{2H}\left[-e'_{q,0}\gamma_1 - e'_{d,0}\gamma_2 + \gamma_1\gamma_2\left(x'_d - x'_q\right) + P_m - D\frac{\omega_0 - \omega_R}{\omega_R}\right] \\
A_{2,1} &= \frac{\omega_R}{2H}\left[Y_o x'_d\left(e'_{q,1}\gamma_2 + e'_{d,1}\gamma_1\right) - \delta_1 Y_\infty^2\left(x'_d - x'_q\right)\cos 2\delta_0\right. \\
&\quad - 2Y_\infty\left(e'_{q,0}e'_{q,1} + e'_{d,0}e'_{d,1}\right) - Y_\infty\left(e'_{d,1}\cos\delta_0 - e'_{q,1}\sin\delta_0\right) \\
&\quad \left. -x'_d Y_o\gamma_4 + Y_o Y_\infty \delta_1 x'_q \gamma_4 + Y_\infty\delta_1\gamma_3 - Y_o x'_q\left(e'_{d,1}\gamma_1 + e'_{q,1}\gamma_2\right) - \frac{D\omega_1}{\omega_R}\right]
\end{aligned}$$

where

$$\begin{aligned}
&\gamma_1 = Y_o e'_{q,0} - Y_\infty \sin\delta_0, \quad \gamma_2 = Y_o e'_{d,0} + Y_\infty\cos\delta_0, \\
&\gamma_3 = e'_{q,0}\cos\delta_0 + e'_{d,0}\sin\delta_0, \quad \gamma_4 = e'_{d,0}\cos\delta_0 + e'_{q,0}\sin\delta_0
\end{aligned} \tag{5.21}$$

where Y_o and $Y_\infty = |Y_\infty|\angle\beta$ are respectively the admittances from the generator's electromotive force to the ground and to the infinite bus. Note that E_{fd}, which is constant in this fourth-order differential equation model, only explicitly appears in the Adomian polynomials about e'_q.

An important note about the semi-analytical solution derived from ADM is that although the Adomian decomposition of nonlinear functions is unique, there are more than one ways to incorporate the Adomian polynomials into the semi-analytic solutions. For example, by defining the first term of the semi-analytical solution differently, two different forms of semi-analytical solutions can be derived. This can be demonstrated through a single-machine infinite bus system considering only the rotor angle and speed as state variables. For the second order single machine infinite bus system,

$$2H\ddot{\delta} + D\dot{\delta} = P_m - P_{max}\sin\delta \tag{5.22}$$

where H is the inertia, D is damping, P_m is the mechanical power setpoint, and P_{max} is the maximum electrical power output. If instead of defining the first term of the semi-analytical solution as in Eq. (5.6), making it also include the information of the initial value of the derivative of rotor angle (i.e. rotor speed),

$$\mathscr{L}[\delta_0] = \delta(0)/s + \dot{\delta}(0)/s^2 \tag{5.23}$$

The resulting semi-analytical solution will contain sinusoidal functions of time [10]. This suggests that in addition to numerical simulations, semi-analytical solutions may also have the potential to provide a means to embed some nonlinear dynamic information (e.g. oscillation frequency, etc.) for direct analysis of the system behavior after disturbances. In this case, the semi-analytical solution of rotor angle with three terms is,

$$\delta(t) = \sum_{n=0}^{2} \delta_n(t) \tag{5.24}$$

$$\delta_0 = bt + a$$

$$\delta_1 = \frac{-Db + P_m}{4H}t^2 - \frac{P_{max}\cos a}{2Hb}t + \frac{P_{max}[-\sin a + \sin(bt + a)]}{2Hb^2}$$

$$\begin{aligned}\delta_2 = {} & \frac{D^2b - DP_m}{24H^2}t^3 + \frac{Db(\cos a - \cos c) + P_m \cos c}{8H^2b^2}P_{max}t^2 \\ & + \frac{P_{max}\left[8Db(\sin a + \sin c) - 4P_m(\sin a + 2\sin c) - P_{max}(2\cos d + 2\cos bt + \cos 2a + 4)\right]}{16H^2b^3}t \\ & - \frac{P_{max}\left[32Db(\cos a - \cos c) - 24P_m(\cos a - 2\cos c) - P_{max}(4\sin d + \sin 2c + 12\sin bt - 5\sin 2a)\right]}{32H^2b^4}\end{aligned}$$

where, $a = \delta(0)$, $b = \dot{\delta}(0)$, $c = bt + a$, and $d = bt + 2a$. The sinusoidal functions of time t can be observed in terms $n = 1$ and above.

Since the accuracy of a semi-analytical solution defined by Eq. (5.8) only lasts for a limited time window T [11–14], a multistage strategy [15–18]) is adopted to extend the accuracy of the same semi-analytical solution to an expected simulation period by these two steps:

Step-1: Partition the simulation period into sequential windows of T each able to keep an acceptable accuracy of the semi-analytical solution.

Step-2: Evaluate the semi-analytical solution at desired time points in the first T using the given initial state and the values of other symbolic variables; starting from the second window T, evaluate the semi-analytical solution by taking the final state of the previous T as the initial state.

As long as the final state of each window is accurate enough, the accuracy of the next window will be ensured. To apply this approach to simulate a contingency, we may first perform the numerical approach until the contingency is cleared to obtain the initial state for the initial value problem about the post-contingency simulation period, and then the Multistage ADM can be performed.

5.2.2 Convergence and Time Window of Accuracy

Consider a single-machine infinite bus system having a second-order classical model generator connected to the infinite bus by an impedance. Thus, Y_o is zero and the electromotive force E of the generator has a constant magnitude so as to eliminate two differential equations on e_d' and e_q' in Eq. (5.15). System parameters and initial conditions are listed in Table 5.1. Mechanical power P_m determines the operating condition. V_∞ is the voltage magnitude of the infinite bus, whose phase angle is considered zero. $\delta(0)$ and $\omega(0)$ are the initial rotor angle and speed of the generator, which are initial state variables.

Figure 5.1 plots the trajectories of six different semi-analytical solutions with $N = 3$–8, respectively, and compare them with the numerical integration result from the R-K 4.

Table 5.1 Parameters of the SMIB system.

H	D	$Y_\infty = \lvert Y_\infty \rvert \angle\beta$	Y_o	P_m	$\lvert E \rvert$	V_∞	ω_R	$\delta(0)$	$\omega(0)$
3 s	0 s	0.9∠90° p.u.	0 p.u.	0.8 p.u.	1.1 p.u.	1	377 rad/s	0.06 rad	2.05 rad/s

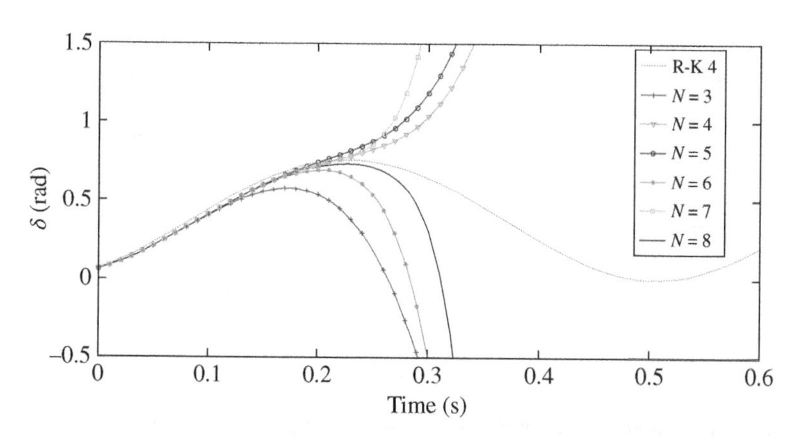

Figure 5.1 Comparison of semi-analytical solutions with numerical results.

For $N = 5$, five terms of the semi-analytical solution are given in Eq. (5.25) as an example and its trajectory and the trajectories of individual terms are shown in Figure 5.2.

$$\delta(t) = \sum_{n=0}^{4} \delta_n = 11.39t^8 + 4.50t^7 + 1353.32t^6 + 361.02t^5 - 240.09t^4 - 47.72t^3 + 20.81t^2 + 2.05t + 0.06 \quad (5.25)$$

where

$$\delta_0 = 0.06$$
$$\delta_1 = 20.81t^2 + 2.05t$$
$$\delta_2 = -241.61t^4 - 47.72t^3$$
$$\delta_3 = 1184.67t^6 + 350.95t^5 + 1.52t^4$$
$$\delta_4 = 11.39t^8 + 4.50t^7 + 168.66t^6 + 10.07t^5$$

In Figure 5.2, T_{max} denotes a limit of the time window of accuracy. Also define the absolute value of the last term, i.e. $|x_{N-1}|$, as a divergence indicator I_D, which is close to zero within T_{max} and sharply increases the magnitude, otherwise. T_{max} can be estimated by selecting an appropriate threshold $I_{D,max}$ for I_D. For instance in Figure 5.2, $I_{D,max}$ is set at 0.01 rad to determine T_{max}. There are two observations from Figure 5.2:

- The semi-analytical solution from the ADM matches well the R-K 4 result within 0.2 seconds, i.e. a time window of accuracy.

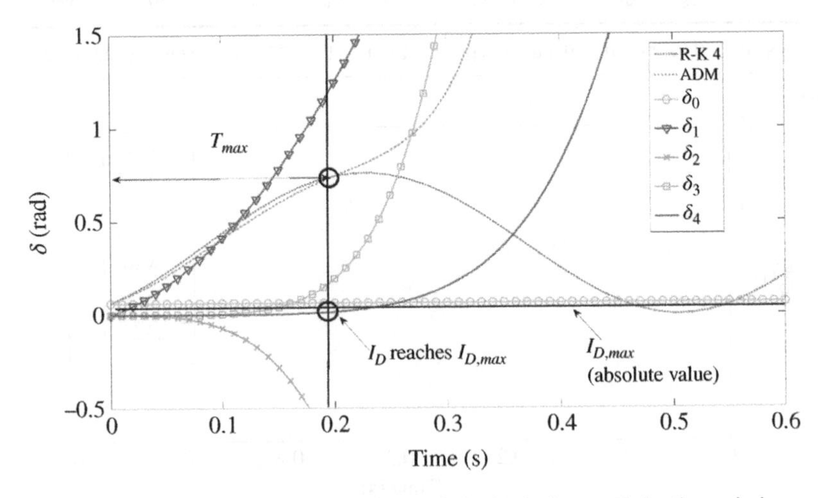

Figure 5.2 Different terms of the semi-analytical solutions and the time window of accuracy.

- The higher order of a term, the less contribution it has and the faster it diverges to infinity. The last term δ_4 diverges quickly outside 0.2 seconds.

To unveil the relation between T_{max} and time constants of a multi-machine system, the IEEE 3-generator 9-bus system in [19] is studied. Gradually decrease H_3, the inertia of generator 3, from 4.5 to 1.0 seconds while keeping the other two unchanged at original 23.64 and 6.4 seconds, such that eight system models are yielded as shown in Table 5.2. Because the system has two oscillation modes and their oscillation periods T_1 and T_2 may be important time constants influencing T_{max}, T_1, and T_2 are estimated from each linearized model of the system and are listed in Table 5.2. A three-phase fault at bus 7 cleared by tripping lines 5–7 is simulated on each model by both the R-K 4 and the ADM with $N = 3$ (using the post-fault state from the R-K 4 as its initial state). Using 0.01 rad as $I_{D,max}$, the estimated T_{max} for each model is given in the table. Figure 5.3 illustrates that T_1, T_2, and T_{max} monotonically increase with H_3. The bigger time constant T_1 does not change significantly with H_3. Figure 5.4 shows values of T_{max} for

Table 5.2 T_{max} vs. time constants of the system.

No.	H_3 (s)	T_1 (s)	T_2 (s)	T_{max} (s)
1	4.5	0.9510	0.5516	0.2546
2	4.0	0.9438	0.5280	0.2342
3	3.5	0.9369	0.5014	0.2131
4	3.0	0.9304	0.4718	0.1905
5	2.5	0.9241	0.4365	0.1662
6	2.0	0.9183	0.3961	0.1410
7	1.5	0.9128	0.3479	0.1137
8	1.0	0.9076	0.2881	0.0845

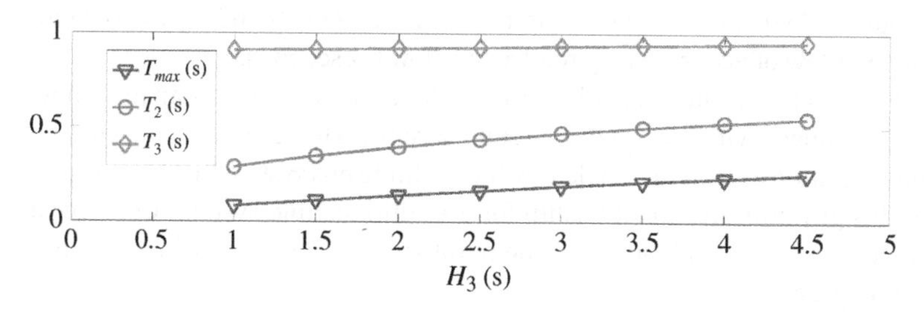

Figure 5.3 Relationships between T_{max}, T_2, T_1, and H_3.

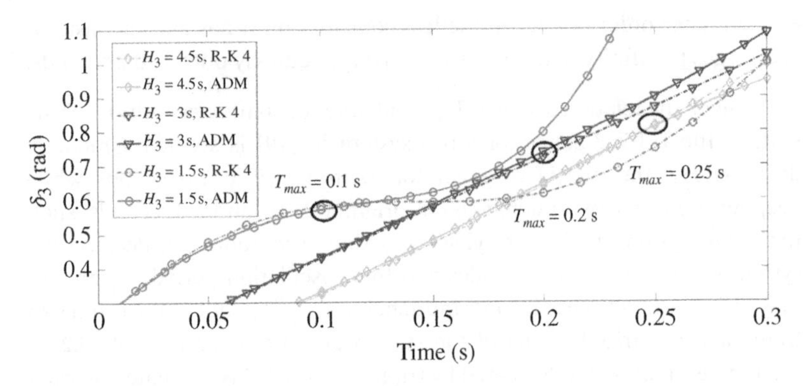

Figure 5.4 T_{max}'s with respect to selected H_3's.

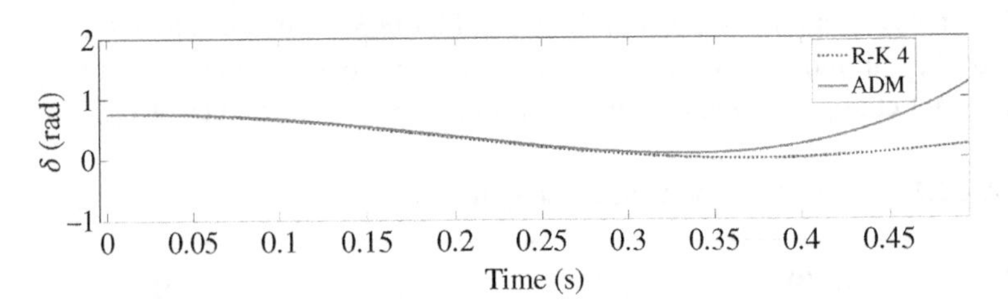

Figure 5.5 Using an initial state with $\omega(0) = 0$ rad/s and $\delta(0) = 0.76$ rad.

$H_3 = 1.5$, 3, and 4.5 seconds, beyond which the ADM result starts diverging from the R-K 4 result. A hypothesis for a multi-machine power system is that T_{max} is mainly influenced by the smallest time constant.

If the initial state varies, the time of accuracy may change as well. For the single machine infinite bus system, different values of $\delta(0)$ and $\omega(0)$ will lead to different T_{max}'s. As illustrated by Figures 5.5 and 5.6, the semi-analytical solution evaluated starting from an initial state with $\omega(0) = 0$ rad/s and $\delta(0) = 0.76$ rad keeps its accuracy for a time window around 0.25 seconds while for a larger $\omega(0) = 1.38$ rad/s and $\delta(0) = 0.04$ rad, the window of accuracy may reduce to below 0.2 seconds.

For a general multi-machine system, it can be difficult to analyze how T_{max} changes with $I_{D,max}$ about a state variable. However, we may analyze their relationship on the single-machine infinite bus system first to help gain an insight into their relationship for a multi-machine system. Consider a three-term semi-analytical solution of rotor angle δ, whose last term δ_2 has this expression

$$\delta_2 = c_1 t^4 + c_2 t^3 \tag{5.26}$$

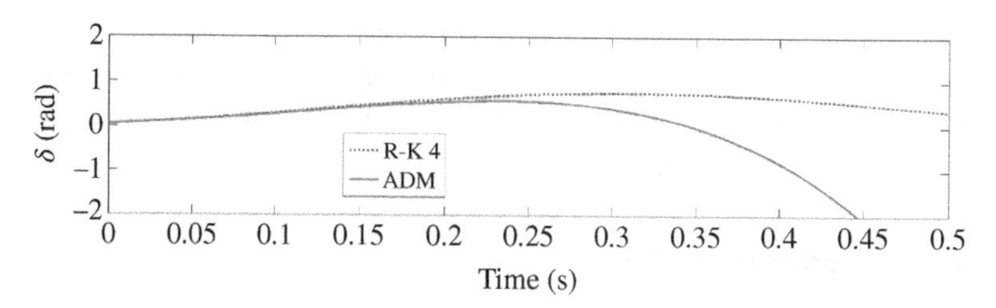

Figure 5.6 Using an initial state with $w(0) = 1.38$ rad/s and $\delta(0) = 0.04$ rad.

where

$$c_1 = \frac{\omega_0^2 V_\infty |Y_\infty E|^2 \sin(\beta - \delta(0))}{96H^2} \left[|E| \cos\beta + V_\infty \cos(\beta - \delta(0)) - \frac{P_m}{|Y_\infty E|} \right]$$

$$c_2 = \frac{-\omega_0 V_\infty \mid Y_\infty E \mid (\omega(0) - \omega_R) \sin(\beta - \delta(0))}{12H}$$

Define divergence indicator I_D as δ_2 and let $t = T_{max}$ and $\delta_2 = I_{D,max}$ in Eq. (5.26) to obtain

$$I_{D,max} = c_1 T_{max}^4 + c_2 T_{max}^3 \tag{5.27}$$

T_{max} has four roots as given in

$$T_{max} = -\frac{c_2}{4c_1} \pm \frac{p_4}{2} \mp \frac{\sqrt{p_5 \pm p_6}}{2} \text{ or } -\frac{c_2}{4c_1} \pm \frac{p_4}{2} \pm \frac{\sqrt{p_5 \pm p_6}}{2} \tag{5.28}$$

where

$$p_1 = -27c_2^2 I_{D,max}, \quad p_2 = p_1 + \sqrt{4 \cdot (12c_1 I_{D,max})^3 + p_1^2}$$

$$p_3 = -\frac{4I_{D,max}}{\sqrt[3]{2/p_2}} + \frac{\sqrt[3]{p_2/2}}{3c_1}, \quad p_4 = \sqrt{\frac{c_2^2}{4c_1^2} + p_3}$$

$$p_5 = \frac{c_2^2}{2c_1^2} - p_3, \quad p_6 = \frac{-c_2^3}{4c_1^3 p_4}$$

Since $T_{max} > 0$, the smallest positive root should be selected as an estimate of T_{max}. For a multi-machine system, Eqs. (5.27) and (5.28) can also be applied to approximately analyze the relationship of $I_{D,max}$ and T_{max} for state variables of each machine by means of a single-machine infinite bus equivalent about that machine against the rest of the system.

The studies above show that, for a semi-analytical solution, its T_{max} depends on time constants of the system, the initial state starting the evaluation and the contingency as well. Therefore, we may either choose a fixed time window less than the most conservative T_{max} observed offline based on many simulations on probable contingency scenarios or allow the window T to change adaptively as long as divergence indicator I_D remains below a preset threshold $I_{D,max}$ for each state variable.

5.2.3 Adaptive Time Window

The convergence of the semi-analytical solutions for a general nonlinear system is still an open question [20], and no sufficient condition for convergence has been proved yet. A necessary condition [21] is proven to be the satisfaction of a ratio test: $\|\mathbf{x}_{n+1}\|_2 < \alpha\|\mathbf{x}_n\|_2$ holds for $n = 0, 1, \ldots, N-1$, where $0 < \alpha < 1$ is a constant depending on the system. However, α is difficult to derive analytically for a high-dimensional system.

A practical approach for evaluation of an ADM-based N-term semi-analytical solution is using an adaptive time window. The approach compares divergence indicator I_D with a preset threshold $I_{D,max}$ to adaptively judge the end of the current window for evaluation and proceed to the next window until the entire simulation period is made up. $I_{D,max}$ is estimated by the following procedure for a list of scenarios that each have a contingency simulated under a specific operating condition:

Step-1: For each scenario, use the post-contingency state from the R-K 4 as the initial state to run the Multistage-ADM using a small enough fixed time window T.

Step-2: Find the maximum per unit absolute value that the last semi-analytical solution term, i.e. $|x_{k,N-1}|$, of any state variable may reach over the entire simulation period. Use that value as a guess of $I_{D,max}$.

Step-3: Add a small random variation to the post-contingency state and repeat *Step-2* for a number of times. Take the smallest guess of $I_{D,max}$.

Step-4: After finishing *Steps 1–3* for all contingencies, choose the smallest $I_{D,max}$ as the final threshold.

Remarks: (i) *Step-2* on guessing an $I_{D,max}$ may exclude $\delta_{k,N-1}$, i.e. the last semi-analytical solution term for each rotor angle δ_k since its divergence can be detected through the divergence of the last semi-analytical solution term of ω_k; (ii) *Step-2* finds the maximum value of all last terms rather than the minimum value in order to provide a necessary condition for convergence rather than an overconservative, sufficient condition causing loss of the advantage of using an adaptive time window; (iii) the random variation in *Step-3* is added to make the $I_{D,max}$ more independent of the post-contingency state.

The above procedure can be performed offline for potential contingencies and operating conditions. Based on our tests, $I_{D,max}$ does not vary significantly with contingencies, so in practice, the list of scenarios does not have to be large to find an effective $I_{D,max}$.

5.2.4 Simulation Scheme

A two-stage scheme is presented for power system simulation using the Multistage-ADM, which comprises an offline stage to derive the semi-analytical solutions and an online stage to evaluate the semi-analytical solutions as shown in Figure 5.7.

5.2.4.1 Offline Stage

Assuming a constant impedance load at each bus, a semi-analytical solution is derived by the ADM for each generator with symbolic variables from, e.g. one of these two groups:

- **Group-1**: Time, the initial state, and the operating condition (e.g. generator outputs and load impedances)
- **Group-2**: *Group-1* plus selected symbolized elements (symbolized parameters of system that subject to changes) in the system admittance matrix.

Group-1 assumes a specific post-contingency system topology (i.e. a constant system admittance matrix) but relaxes the system operating condition so as to enable one semi-analytical solution to simulate multiple loading conditions. *Group-2* additionally relaxes selected elements in the admittance matrix and hence enables one semi-analytical solution suitable for simulating multiple contingencies. Other symbolic variables can also be added as undetermined parameters but the more symbolic variables the more complex expression of the semi-analytical solution. All semi-analytical solutions derived in the offline stage will be saved in storage for later online use.

If an adaptive time window for semi-analytical solution evaluation is used, the offline stage also needs to estimate $I_{D,max}$. If a fixed window is adopted, T can be chosen less than the minimum T_{max} estimated by a procedure similar to that for the determination of $I_{D,max}$ using a list of scenarios.

5.2.4.2 Online Stage

For a specific contingency scenario, this stage evaluates the corresponding semi-analytical solutions of every generator consecutively over time windows T, fixed or adaptive, until making up the expected simulation period. The first time window needs to know the post-contingency initial system state, which can be obtained from numerical integration for the fault-on

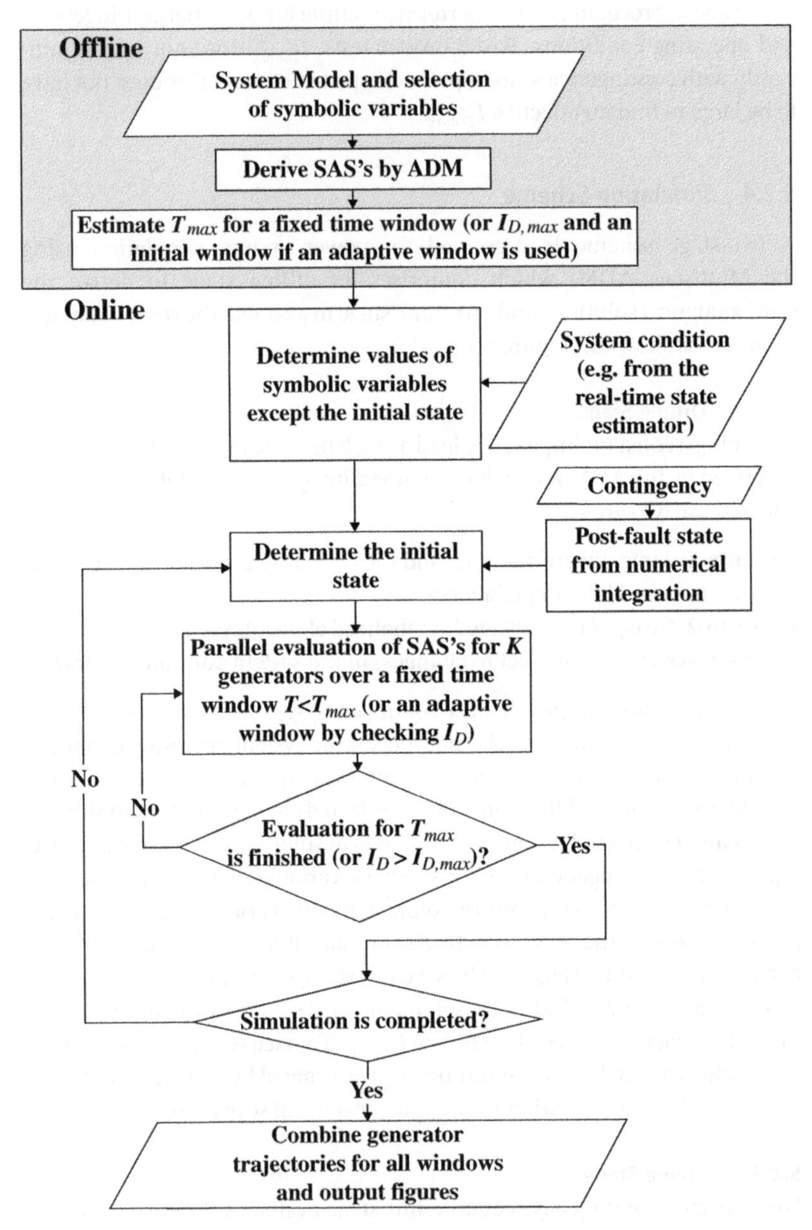

Figure 5.7 Flowchart of the semi-analytical scheme for power system simulation.

period until the fault is cleared. Starting from the second window, the initial state takes the final state of the previous window.

If an adaptive time window is applied, an initial window may be chosen less than the estimated T_{max} for a fixed window. Then, during each window, the divergence indicator I_D for each state variable is calculated and compared with the threshold $I_{D,max}$ acquired in the offline stage in order to decide when to proceed to the next window, i.e. the end of the current window. Thus, even if the initial window is not small enough, comparison of I_D and $I_{D,max}$ will enable self-adaptive adjustment of the window.

Within each window, because semi-analytical solutions are independent expressions, their evaluations can be performed simultaneously on parallel computers. In expression, each semi-analytical solution is the sum of terms in this form

$$C \cdot \underbrace{x_i \dots x_j}_{h} t^n \underbrace{f_k(x_k) \dots f_l(x_l)}_{m} \quad \text{where } f(\cdot) \text{ is } \sin(\cdot) \text{ or } \cos(\cdot) \tag{5.29}$$

where C is a constant that depends on system parameters, t is time, i, j, k, and l are integer indices of state variables. For different numbers of semi-analytical solution terms and different systems, the ranges of h, m, and n are different. Expression Eq. (5.29) is defined as one computing unit in this paper. All computing units can be evaluated simultaneously on parallel processors to accelerate the online stage.

The semi-analytical solution approach may be applied for fast power system simulation in the real-time operating environment: in the offline stage, a semi-analytical solution is derived that symbolizes a group of uncertain parameters like *Group-2*; then, in the online stage, whenever the real-time state estimation is finished (typically, every one to three minutes) to give the current power-flow solution and network topology, the semi-analytical solution will be evaluated to provide simulation results on a given contingency. However, if a change in the network topology or any parameter about the operating condition is detected in real time by, e.g. the SCADA system [22] and makes the most recent state estimation result invalid, the semi-analytical solution evaluation should wait until the state estimator gives a new estimation result. Thus, online power system simulation using the semi-analytical approach can be performed synchronously with real-time state estimation.

5.2.5 Examples

IEEE 10-generator, 39-bus system, as shown in Figure 5.8, is used to demonstrate the semi-analytical solution approach for power system simulation.

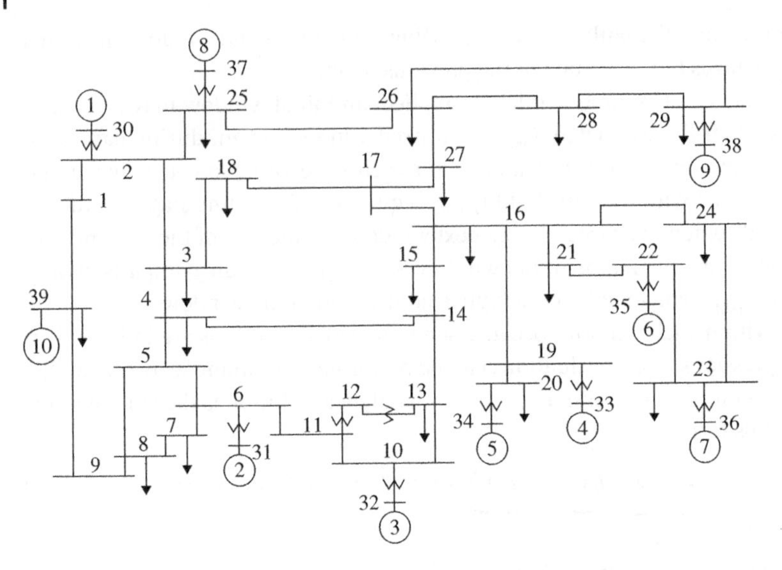

Figure 5.8 IEEE 10-generator 39-bus system.

Generator 39 has the largest inertia and its rotor angle is defined as the reference. The two-stage scheme is demonstrated using both a fixed time window and an adaptive time window. To achieve the fastest simulation, only one data point on each trajectory is evaluated within each time window, which is enough for transient stability assessment.

5.2.5.1 Fixed Time Window

A permanent three-phase fault lasting for 0.08 seconds is applied to lines 3–4 at bus 3. We preset $I_{D,max} = 0.005$ p.u. (per unit) for all state variables except for the rotor angle. If all generators are represented by the fourth-order model in (5.15), our tests show that when a semi-analytical solution with two terms is evaluated over a time window of 0.002 seconds, the largest second semi-analytical solution term of the state variables is 0.0047 p.u. $< I_{D,max}$, which means $T_{max} \approx 0.002$ seconds for a two-term semi-analytical solution. Figure 5.9 gives the results from the Multistage-ADM (dash lines) using a 0.001 seconds window and the results from the R-K 4 (solid lines) with a 0.001 seconds integration step, which are identical. The relatively small time step (0.001 seconds) required by the fixed time window Multistage-ADM approach is due to its explicit nature similar to R-K 4. While implicit methods like Trapezoid Method can use a larger integration step, they involve solving their integration equations iteratively. If the time window and integration step are both increased to 0.01 seconds (>T_{max}), the simulation results from the R-K 4 and Multistage-ADM have slight, noticeable differences as shown in Figure 5.10.

Although including more terms is expected to increase T_{max} as indicated by Figure 5.1, using a semi-analytical solution with three terms does not extend T_{max} significantly in this case. For example, use a 0.01 seconds time window to run a three-term semi-analytical solution for the same contingency, there are still obvious mismatches between the R-K 4 and Multistage-ADM results. Moreover, a three-term semi-analytical solution has a more complex expression, so it takes longer to evaluate than a two-term semi-analytical solution.

When a semi-analytical solution is evaluated over a fixed time window T for power system simulation, the last terms, i.e. divergence indicator I_D's, of

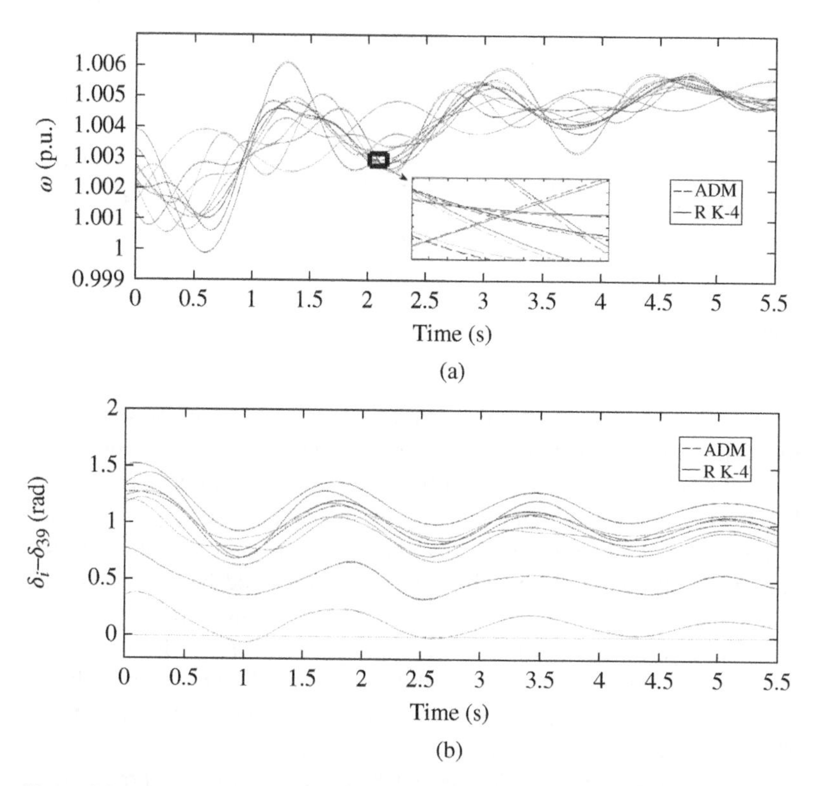

Figure 5.9 Comparison of the simulation results given by the R-K 4 and the two-term semi-analytical solution using a fixed time window of 0.001 seconds. (a) Rotor speeds, (b) rotor angles, (c) q-axis transient voltages, (d) d-axis transient voltages.

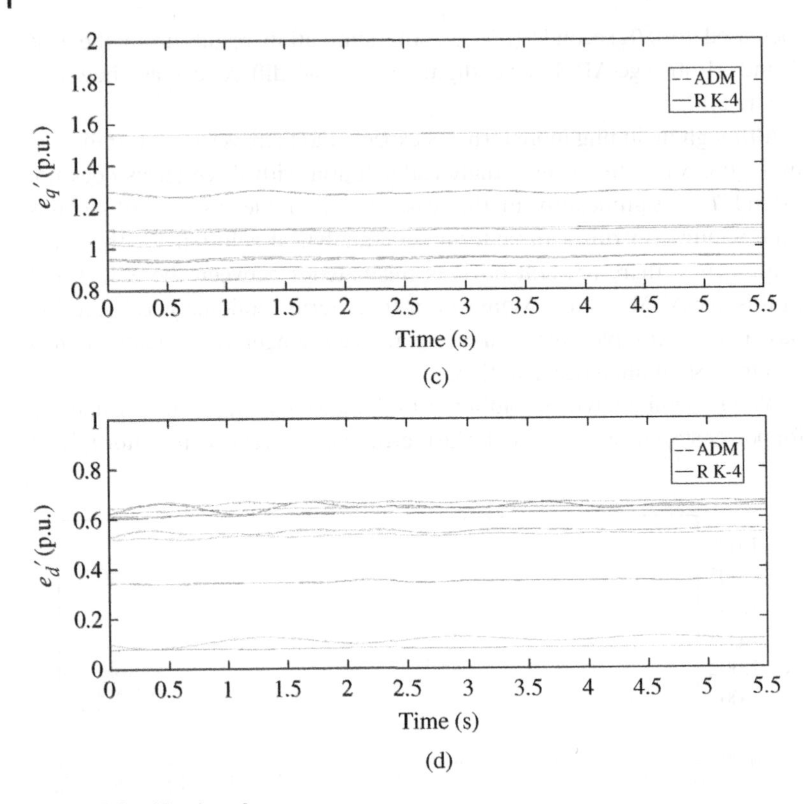

Figure 5.9 *(Continued)*

all state variables can distinguish numerical instability from power system instability: if the simulated system trajectory becomes unstable while all I_D's are still small, e.g. much less than the predefined $I_{D,max}$, it is very likely to be power system instability; if some I_D also increases drastically to approach or exceed $I_{D,max}$ when the system trajectory appears to be unstable, numerical instability may happen. Thus, a smaller T should be used to reevaluate the semi-analytical solution for verification of numerical instability. For example, if T is increased to 0.02 seconds $\approx 10 \times T_{max}$, the simulation results diverge with numerical instability introduced on purpose as shown in Figure 5.11, where the results from the R-K 4 method are still stable. That numerical instability can be detected by I_D's $> I_{D,max}$ for many windows. From the results of Figures 5.9–5.11, as T increases from 0.001 to 0.01 seconds and then to 0.02 seconds, the largest I_D of all states variables increases from 0.0023 to 0.0279 p.u. (i.e. 12.1 times) and then to 0.1051 p.u. (i.e. 45.7 times), which indicates the occurrence of numerical instability. I_D

can be utilized to avoid numerical instability by changing the time window adaptively.

The Multistage-ADM is also tested on the system having each generator represented by the sixth-order differential equation model in (5.15), (5.17), and (5.18) containing the exciter and governor. The parameters of exciters and governors are set up as $T_{Ak} = 0.02$ seconds, $K_{Ak} = 5$, $T_{gk} = 0.5$ seconds, $R_k = 0.01$ for all machines. A two-term semi-analytical solution is derived for each of the six state variables, and the time window is selected to be 0.001 seconds within the estimated T_{max}. Under the same contingency on line 3–4. The R-K 4 simulation indicates the frequency oscillation is better damped than that without a governor. Figure 5.12 compares the results from the Multistage-ADM (dash lines) and R-K 4 (solid lines) for each state variable, which match well.

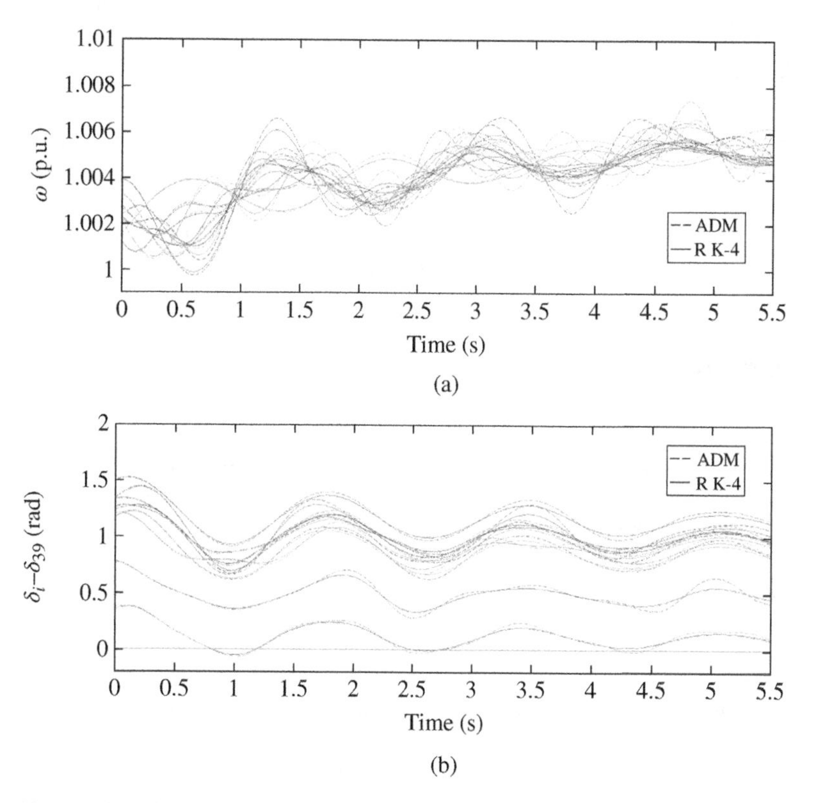

Figure 5.10 Comparison of the simulation results given by the R-K 4 and the two-term semi-analytical solution using a fixed time window of 0.01 seconds. (a) Rotor speeds, (b) rotor angles, (c) q-axis transient voltages, (d) d-axis transient voltages.

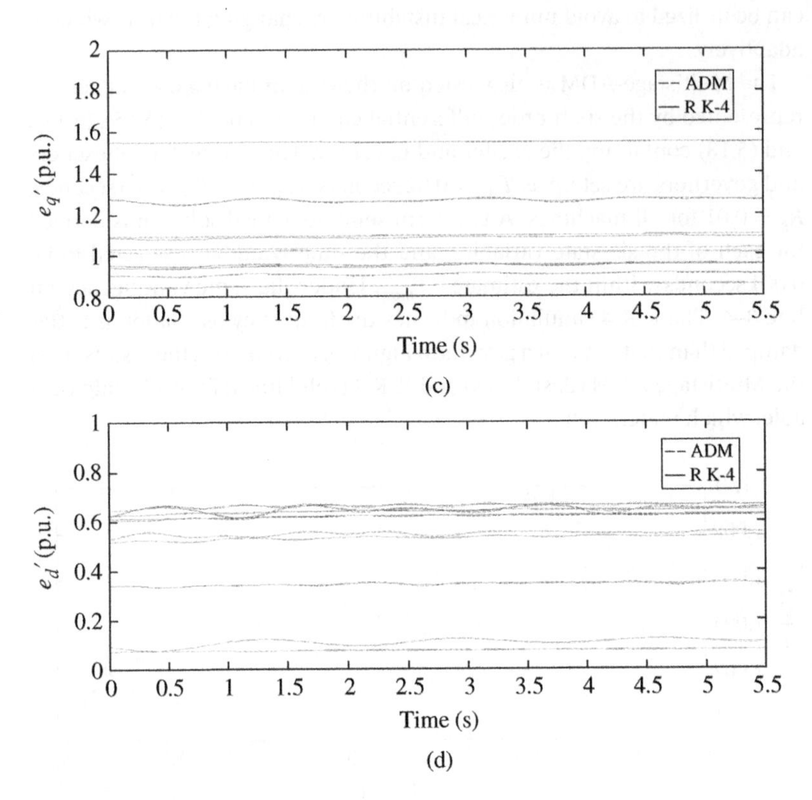

Figure 5.10 *(Continued)*

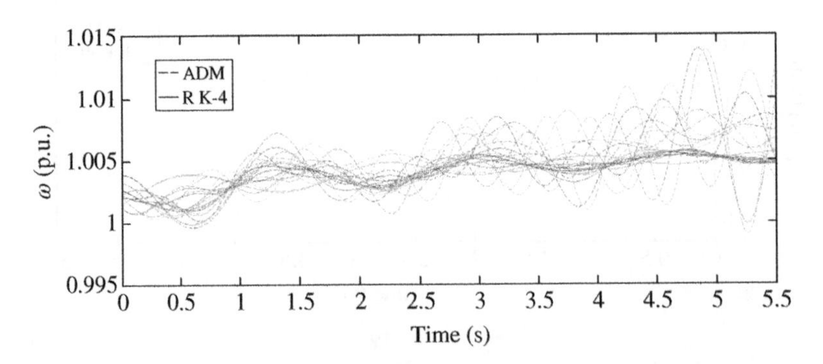

Figure 5.11 Comparison of the simulation results of rotor speeds given by the R-K 4 and the two-term semi-analytical solution using a fixed time window of 0.02 seconds.

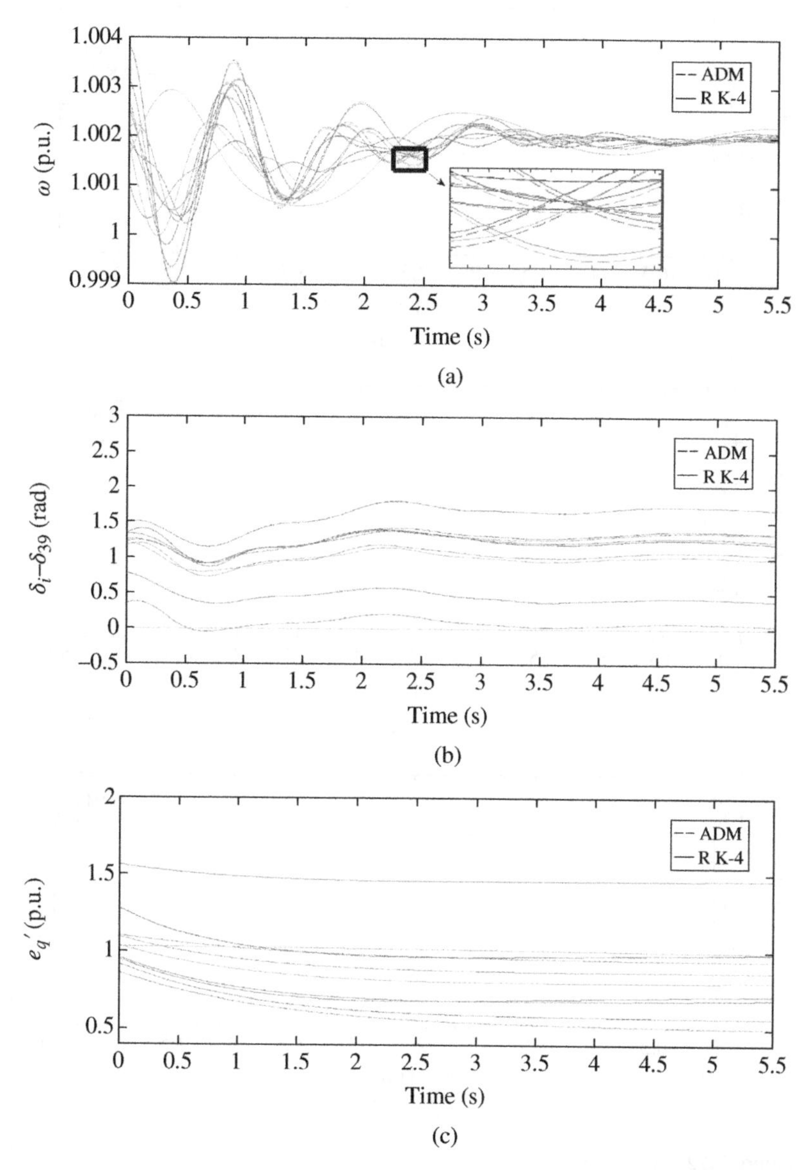

Figure 5.12 Comparison of the simulations using the sixth-order generator model by the R-K 4 and the two-term semi-analytical solution using a fixed time window of 0.001 seconds. (a) Rotor speeds, (b) rotor angles, (c) q-axis transient voltages, (d) d-axis transient voltages, (e) field voltages, (f) governor outputs.

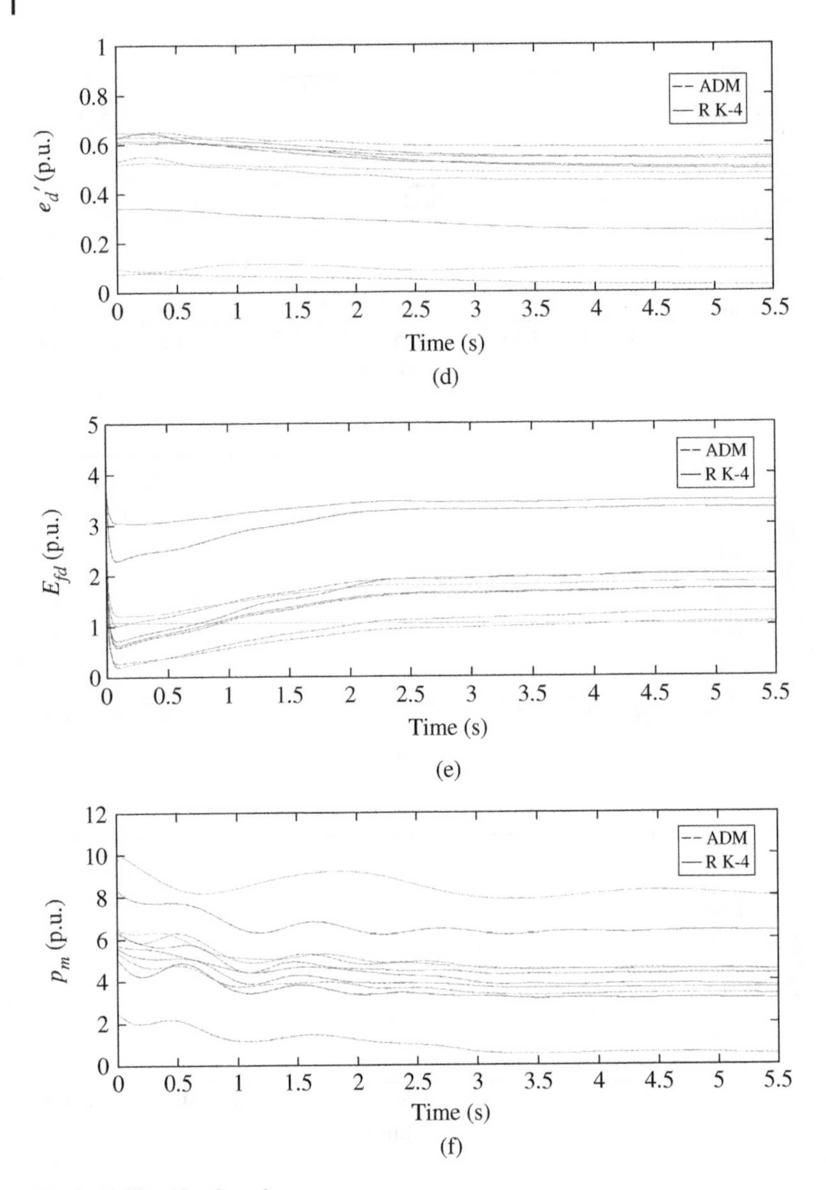

Figure 5.12 *(Continued)*

5.2.5.2 Adaptive Time Window

The first step is to use a list of contingencies to determine an $I_{D,max}$ that can guarantee the accuracy of a semi-analytical solution and avoid numerical instability in simulation by the Multistage-ADM. For the illustration

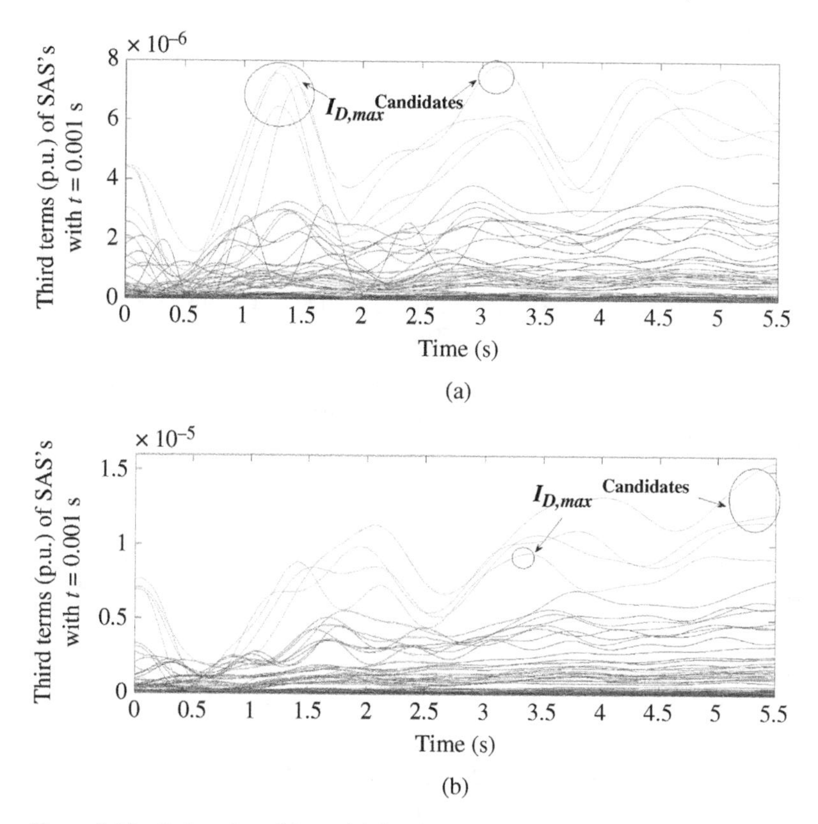

Figure 5.13 Estimation of $I_{D,max}$. (a) Contingency 1, (b) Contingency 2.

purpose, the above contingency on lines 3–4 and a second contingency adding a three-phase fault lasting 0.08 seconds on lines 15–16 at bus 15 are considered. Consider the third term of each state variable (except the rotor angle) in per unit as an I_D. Figure 5.13 plots the I_D's for all those state variables of 10 generators, where 3 random variations are added and the resulting trajectories are also plotted in the same figure. The effective $I_{D,max}$ for two contingencies are found both associated with $|e'_{d5,2}|$, which are 6.5×10^{-6} and 9.4×10^{-6} (p.u.), respectively. Figure 5.14 gives the result from a three-term semi-analytical solution evaluated over an adaptive time window, which is identical to the R-K 4 result.

Figure 5.15 plots how the length of the time window changes with time during a 5.5-seconds simulation for three cases: (1) the two-term semi-analytical solution with an initial $T = 0.001$ seconds, (2) the same semi-analytical solution with an initial $T = 0.01$ seconds, and (3) the three-term semi-analytical solution with an initial $T = 0.001$ seconds. The

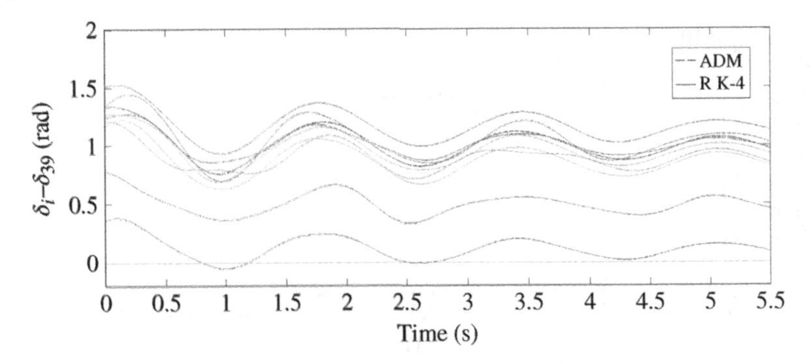

Figure 5.14 Comparison of rotor angles given by the R-K 4 and the three-term semi-analytical solution using an adaptive time window initiated from 0.001 seconds.

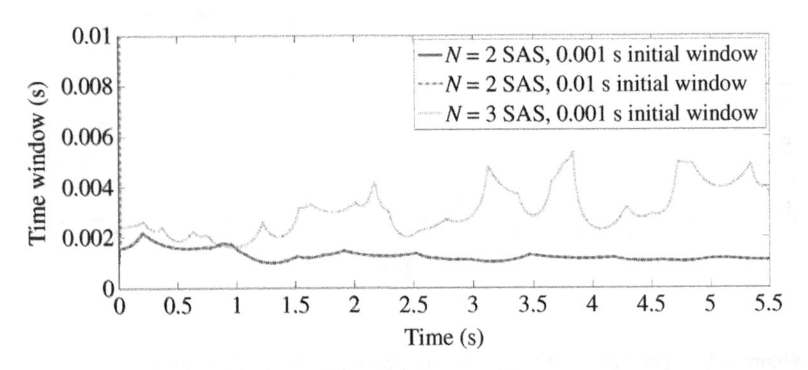

Figure 5.15 The comparison of adaptive changing of time window length.

comparison of the cases (1) and (2) in Figure 5.15 verifies that, if an adaptive time window is used, the accuracy of simulation is independent of the choice of the initial time window since the T of the case (2) adaptively decreases below 0.002 seconds soon after simulation starts. For the cases (1) and (2), the largest T reaches 0.0022 seconds. A main advantage of using an adaptive time window is that the total number of windows for evaluation is effectively reduced. The Multistage-ADM using a fixed 0.001 seconds window evaluates 5500 windows to finish 5.5-seconds simulation while the case (1) using an adaptive time window only takes 4500 windows (i.e. 4500/5500 = 81.8%) to finish the same simulation period. For the case (3), the reduction of time windows is even more significant. As shown in Figure 5.15, the largest T reaches 0.005 seconds, which is more than twice the largest T for the two-term semi-analytical solution. Also, the total number of windows drops to 2000 (i.e. 2000/5500 = 36.4%). Thus, a conclusion is that using an adaptive time window enables the Multistage-ADM to better exploit the advantage with a higher-order semi-analytical solution in terms of the reduction of the window number.

5.2.5.3 Time Performance

To demonstrate the time performance of the semi-analytical approach, the following three cases are tested:

- **Case-A**: only symbolizing time t and initial state variables, i.e. for one specific simulation.
- **Case-B:** beside *Case-A*, also symbolizing the reduced admittance matrix **Y** about 10 generator EMFs, i.e. for simulating different faults under one specific loading condition. Magnitudes and angles of elements of the reduced admittance matrix are symbolized separately to generate two symmetric symbolic 10×10 matrices.
- **Case-C**: beside *Case-B*, also symbolizing generators' mechanical powers to make the semi-analytical solution be also good for simulating various loading conditions.

Here, the load at each bus is represented by a constant impedance load model and is embedded in the reduced admittance matrix **Y**. In the online stage, for a given power-flow condition with all loads known, load impedances will first be calculated, and then with the knowledge of the post-fault network topology, all elements of **Y** can be calculated in order to evaluate the semi-analytical solution.

The offline stage is implemented in MAPLE and the online stage is performed in MATLAB. For fourth-order and sixth-order generator models, the numbers of computing units comprising the three-term semi-analytical solutions of each state variable are given in Tables 5.3 and 5.4, respectively, for three cases.

For *Case-A*, it only takes less than 3 μs to evaluate one computing unit. If all computing units are evaluated simultaneously on parallel processors,

Table 5.3 The number of computing units for the fourth-order model system.

State variable	Case-A	Case-B	Case-C
ω_k	4269	11 430	11 430
δ_k	150	150	150
e'_{qk}	225	301	301
e'_{dk}	223	299	299

Table 5.4 The number of computing units for the sixth-order model system.

State variable	Case-A	Case-B	Case-C
ω_k	4272	11 434	11 434
δ_k	150	150	150
e'_{qk}	227	303	303
e'_{dk}	223	299	299
E_{fd}	2644	5234	5234
P_m	153	155	155

it takes about 3 μs to evaluate one semi-analytical solution for each time window plus the time costs for communication in parallel computing. Because summating the values of all computing units for a state variable is essentially the addition of constants, it is extremely fast. The additions for different state variables can also be performed in parallel. Thus, the final time for summating all computing units equals the time for the most complex semi-analytical solution, often on a rotor speed, which only takes 7 μs. Therefore, the ideal total time cost for evaluations of state variables of one generator is $3 + 7 = 10$ μs per time window T. If evaluations for various generators are also done simultaneously on an unlimited number of parallel processors, that time is also the time cost τ for semi-analytical solution evaluation over each time window T. The R-K 4 method takes 0.37 seconds to finish a 5.5-seconds simulation with all generators represented by the fourth-order model on one computer processor. (It takes 0.48 seconds if all generators are represented by the sixth-order model.) Given the fact that a three-term semi-analytical solution only needs 2000 adaptive time windows for a 5.5-seconds simulation, it can be concluded that the online stage ideally only takes $0.00001 \times 2000 = 0.02$ seconds to finish simulation on parallel processors, which is about 18 times faster than the time cost of the R-K 4. Ratio $T/\tau = 5.5/0.02 = 275$, i.e. the number of times faster than wall-clock time. For *Case-B* and *Case-C*, $T/\tau = 137.5$ as given by Table 5.5, which indicates how many times the simulation can be faster than the wall-clock time.

By comparing Tables 5.3 and 5.4, it can be easily noticed that even after the exciter and governor models are added, the state variables that have the most computing units are still rotor speeds. Meanwhile, the number of computing units of each rotor speed's semi-analytical solution only increases very slightly (by 3 for *Case-A* and 4 for *Case-B* and *Case-C*.) when

Table 5.5 Time performance on the fourth-order model system.

	Case-A	Case-B	Case-C
Offline time cost (s)	198.05	682.18	711.17
Online time cost (s)	0.02	0.04	0.04
Ratio T/τ	275.0	137.5	137.5

Table 5.6 Time Performance on the sixth-order model system.

	Case-A	Case-B	Case-C
Offline time cost (s)	6215.51	13 472.91	16 339.71
Online time cost (s)	0.02	0.04	0.04
Ratio T/τ	275.0	137.5	137.5

the generator model changes from the fourth-order to the 6sixth-order. Basically, adding those details or controllers to each generator does not influence the online performance of the approach.

The time performance of the offline stage is not as critical as the online stage, so it is evaluated in a sequential computing manner. Tables 5.5 and 5.6 summarize the time performances of both offline and online stages for two systems respectively using the fourth and sixth order generator models under the assumption of an ideal parallel computing capability.

Considering that the number of parallel processors cannot be infinite in practice, we also studied how the time performance changes with the number of available processors. As theoretical estimates, ideal parallelism among all available processors is assumed. Thus, all processors are assumed to take equal computational burdens. The results are listed in Table 5.7 for *Case-A* using three-term semi-analytical solutions. From the table, when the number of processors drops to 100, the simulation time increases to 0.3 seconds, which is close to 0.37 seconds of the R-K 4. If the number of parallel processors is further decreased, the simulation using the Multistage-ADM becomes slower than the R-K 4.

When a long list of contingencies needs to be simulated, parallel processors may simulate multiple contingencies simultaneously, so power system simulation using the semi-analytical solution approach will be parallelized also at the contingency level besides the computing unit level.

Table 5.7 Influence of parallel capability on time performance.

Number of parallel processors	Time cost of each time window (s)	Time cost for a 5.5-s simulation (s)
∞	1.0×10^{-5}	2.0×10^{-2}
1000	1.4×10^{-5}	2.8×10^{-2}
100	1.5×10^{-4}	3.0×10^{-1}
10	1.5×10^{-3}	3.0

5.2.5.4 Simulation of a Contingency With Multiple Disturbances

The semi-analytical approach can be used to simulate a contingency containing multiple disturbances, e.g. "n–1–1" and even "n–k" contingencies, which involve one or more disturbances during the simulation period. The same semi-analytical solution can be used for the entire simulation period if all parameters that may change during the simulation period are defined as symbolic variables like a semi-analytical solution from *Case-B* or *Case-C*.

In the following, we demonstrate how to use the semi-analytical solutions of *Case-B* to perform an "n–1–1" simulation involving a topological change of the system during the simulation period. The sixth-order generator models are adopted. The initial contingency is still the same as that in Figures 5.9–5.12 except that at $t = 3$ seconds, the line 22–35 is opened, making the system have a different topology in the remaining 2.5 seconds. The semi-analytical solutions derived for *Case-B* treat all elements of reduced $\mathbf{Y}$ matrix as symbolic variables. Therefore, at $t = 3$ seconds, the time when topology changes, new values of the elements in the reduced $\mathbf{Y}$ matrix should be plugged into the semi-analytical solutions. The simulation results are shown in Figure 5.16. Generator 35 loses its stability. The online time cost is 0.04 seconds with ideal parallelism on sufficient processors.

5.3 Adomian Decomposition of Stochastic Power System Models

For power systems simulated as stochastic differential equations, the semi-analytical nature of a semi-analytical solution yielded by the ADM can be utilized to embed stochastic processes, e.g. a stochastic load model, into the semi-analytical solution. Evaluation of a semi-analytical solution with the stochastic model whose parameters are represented symbolically will not increase many computational burdens compared to evaluating a semi-analytical solution for deterministic simulation.

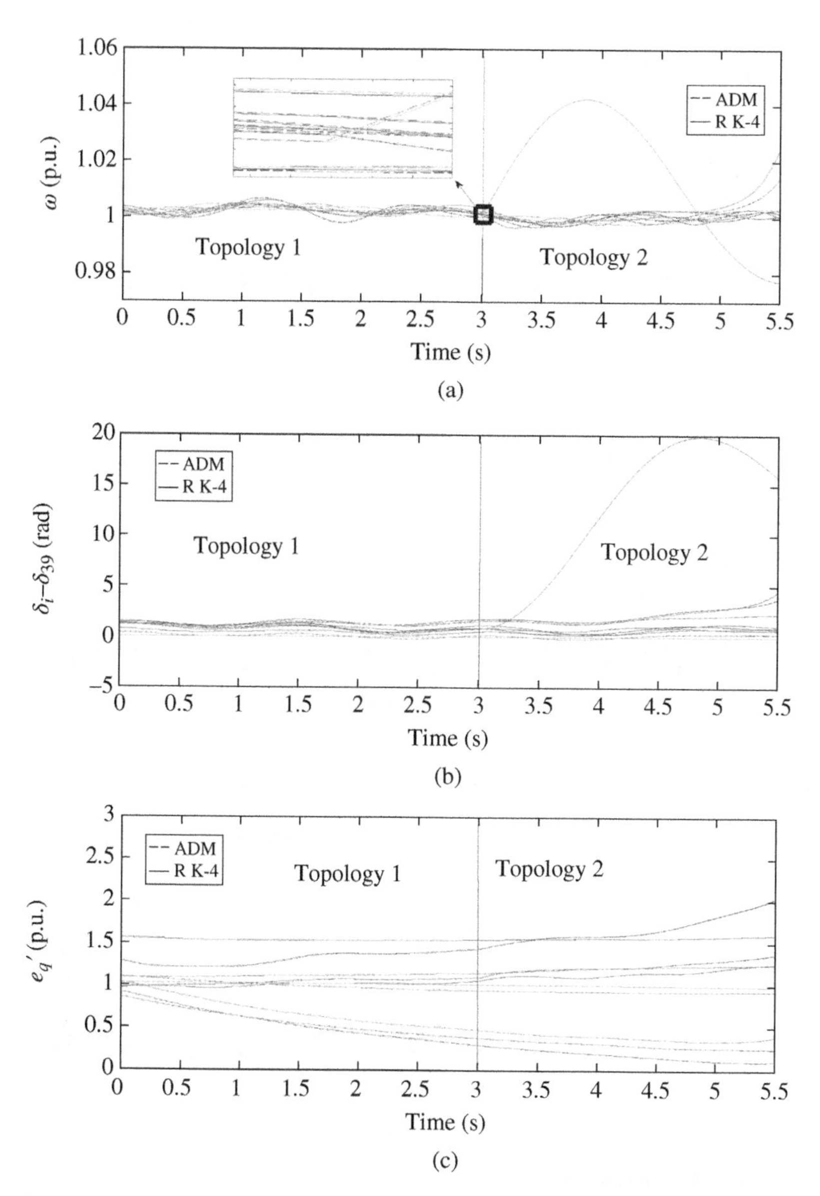

Figure 5.16 Comparison of the simulation results with a topology change at $t = 3$ seconds given by the R-K 4 and a three-term semi-analytical solution using an adaptive time window. (a) Rotor speeds, (b) rotor angles, (c) q-axis transient voltages, (d) d-axis transient voltages, (e) field voltages, (f) governor outputs.

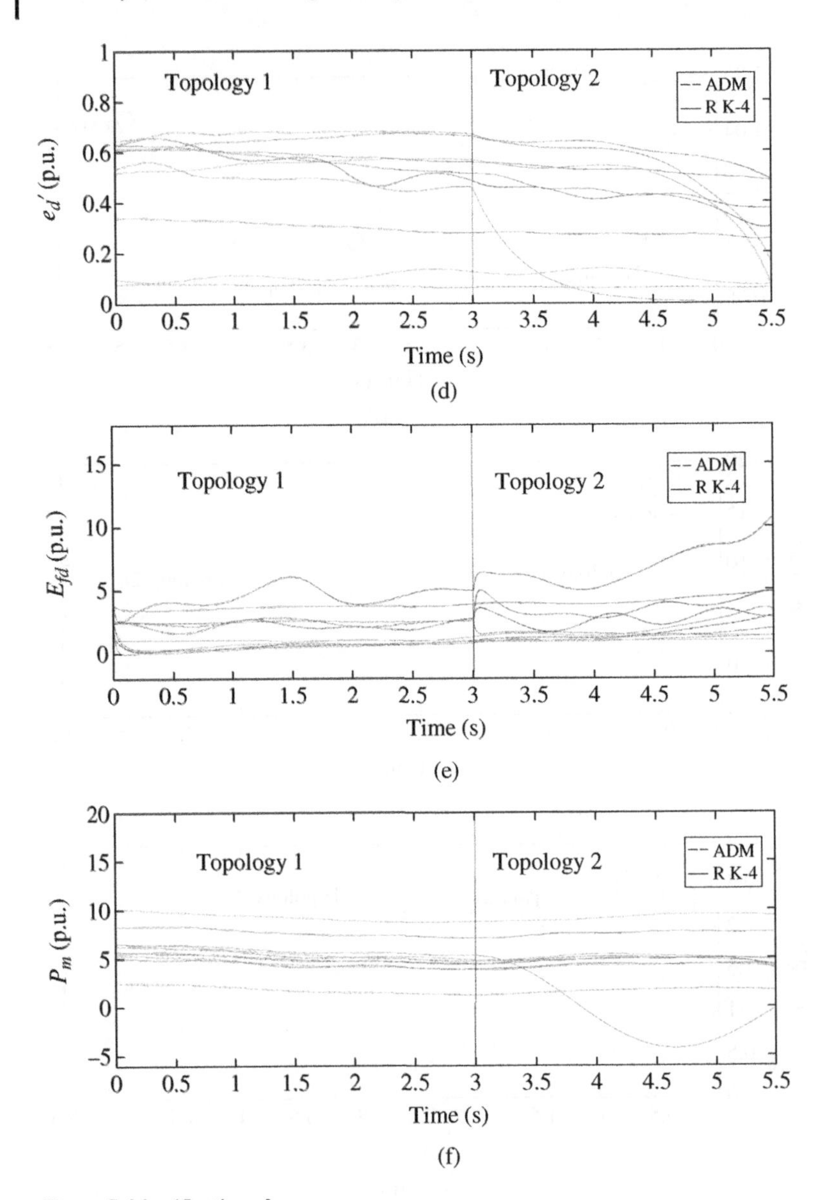

Figure 5.16 *(Continued)*

5.3.1 Single-Machine Infinite Bus System With a Stochastic Load

A single-machine infinite bus system with a stochastic load shown in Figure 5.17 can be modeled as (5.30) The stochastic load is connected to

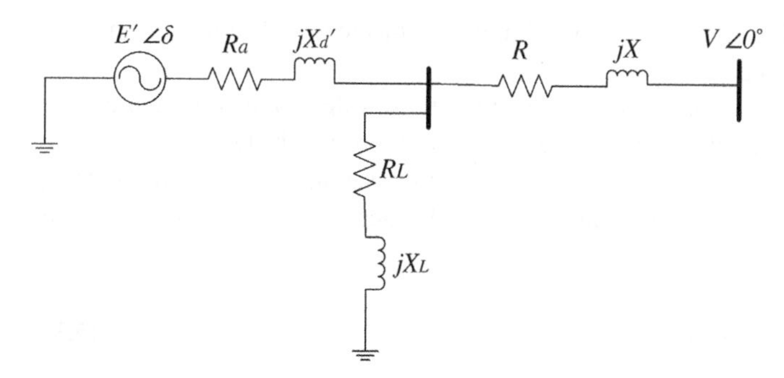

Figure 5.17 Single-machine infinite bus system with constant impedance load at generator bus.

the generator bus and has its resistance R_L and reactance X_L modeled by stochastic variables.

$$\begin{cases} \dot{\delta} = \omega - \omega_R \\ \dot{\omega} = \dfrac{\omega_R}{2H}\left(-D\dfrac{\omega-\omega_R}{\omega} + P_m - \left(k_3 + \dfrac{E'V}{k_1k_2}(k_4\cos(\delta) + k_5\sin(\delta))\right)\right) \\ \dot{R}_L = -a_1R_L + b_1W(t) \\ \dot{X}_L = -a_2X_L + b_2W(t) \end{cases} \tag{5.30}$$

where

$$\begin{aligned}
G_L + jB_L &= \frac{1}{R_L + jX_L} \\
G_S + jB_S &= \frac{1}{R_a + jX_d'} \\
G_R + jB_R &= \frac{1}{R + jX} \\
k_1 &= \frac{(G_L + G_R + G_S)^2}{(B_L + B_R + B_S)} + (B_L + B_R + B_S) \\
k_2 &= (G_L + G_R + G_S) + \frac{(B_L + B_R + B_S)^2}{(G_L + G_R + G_S)} \\
k_3 &= E'^2\left(\frac{G_S(B_L+B_R)+B_S(G_L+G_R)}{k_1} - \frac{B_S(B_L+B_R)-G_S(G_L+G_R)}{k_2}\right) \\
k_4 &= -k_2(B_SG_R + G_SB_R) + k_1(B_SB_R - G_SG_R) \\
k_5 &= -k_2(B_SB_R - G_SG_R) - k_1(B_SG_R + G_SB_R)
\end{aligned}$$

G_S, B_S, G_R, B_R, G_L, and B_L are the conductances and susceptances at the generator sending side, the infinite bus receiving side, and the load side, respectively. Since R_L and X_L change stochastically, G_L and B_L cannot be treated as constants.

The variances of R_L and X_L depend on the values of drifting parameters a_1 and a_2 and diffusion parameters b_1 and b_2, respectively.

To find the semi-analytical solution of this system, the first step is to apply ADM to Eq. (5.30). The resulting second order semi-analytical solution for rotor speed ω is,

$$\omega(t) = \sum_{n=0}^{2} \omega_n(t) \tag{5.31}$$

where,

$$\omega_0(t) = \omega(0)$$

$$\omega_1(t) = -\frac{t\omega_R}{2H}\left[\frac{D(\omega(0)-\omega_R)}{\omega_R} - P_m + k_3 + \frac{k_4E'V}{k_1k_2}\cos(\delta(0)) + \frac{k_5E'V}{k_1k_2}\sin(\delta(0))\right] \tag{5.32}$$

$$\begin{aligned}\omega_2(t) = \frac{t^2\omega_R}{8H^2}\Bigg[&\frac{D^2(\omega(0)-\omega_R)}{\omega_R} \\ &+ D\left(-P_m + k_3 + \frac{k_4E'V}{k_1k_2}\cos(\delta(0)) + \frac{k_5E'V}{k_1k_2}\sin(\delta(0))\right) \\ &+ 2H\omega_R\frac{k_5E'V}{k_1k_2}\cos(\delta(0)) - 2H\omega_R\frac{k_4E'V}{k_1k_2}\sin(\delta(0)) \\ &-2H(\omega(0)-\omega_R)\frac{E'V}{k_1k_2}\cos(\delta(0))\Bigg]\end{aligned} \tag{5.33}$$

Once the semi-analytical solution of the system differential equations is derived, the semi-analytical solution of the stochastic differential equations can be derived and incorporated into it. For example, the second order semi-analytical solution of R_L can be derived using ADM as,

$$R_L(t) = \sum_{n=0}^{2} R_{L,n}(t) \tag{5.34}$$

where,

$$R_{L,0}(t) = R_L(0) + b_1B(t) \tag{5.35}$$

$$R_{L,1}(t) = -a_1R_L(0)t - a_1b_1\int_0^t B(s_1)ds_1 \tag{5.36}$$

$$R_{L,2}(t) = a_1^2R_L(0)\frac{t^2}{2!} + a_1^2b_1\int_0^t\int_0^{s_1} B(s_2)ds_2ds_1 \tag{5.37}$$

$B(t)$ is the Brownian motion starting at origin and $dB(t) = W(t)dt$. The second order semi-analytical solution of X_L can be derived in similar manner as in Eqs. (5.34)–(5.37).

To derive the semi-analytical solution of the entire system considering both the deterministic and stochastic differential equations, replace the symbolic variables in the deterministic differential equations' semi-analytical solution representing the stochastic variables with the stochastic differential equations' semi-analytical solution, i.e. the second order semi-analytical solution of the system (5.30) can be derived by replacing the symbolic variables R_L and X_L in Eq. (5.31) with their semi-analytical solutions.

For some forms of stochastic differential equations, the analytical solution may exist. In such cases, the stochastic differential equations' analytical solution instead of the semi-analytical solution also can be incorporated into the deterministic differential equations' semi-analytical solution to derive the semi-analytical solution of the entire system.

For example, the general expression of the semi-analytical solution terms of the load resistance in Eq. (5.30) can be written as,

$$R_{L,n}(t) = (-1)^n a_1^n R_L(0)\frac{t^n}{n!} + (-1)^n a_1^n b_1 \int_0^t \int_0^{s_1} \cdots \int_0^{s_{n-1}} B(s_n)ds_n \ldots ds_2 ds_1 \tag{5.38}$$

Therefore the infinite order semi-analytical solution of it becomes,

$$R_L(t) = R_L(0)\sum_{i=0}^{\infty}\frac{(-a_1 t)^i}{i!} + b_1 B(t) + b_1 \sum_{i=1}^{\infty}(-a_1)^i \int_0^t \int_0^{s_1} \cdots \int_0^{s_{i-1}} B(s_i)ds_i \ldots ds_2 ds_1 \tag{5.39}$$

Apply Maclaurin expansion of an exponential function and Lemma 2.3 in [23] to Eq. (5.39), the solution becomes,

$$R_L(t) = R_L(0)e^{-a_1 t} + b_1 B(t) - a_1 b_1 \int_0^t e^{a_1 s - a_1 t} B(s)ds \tag{5.40}$$

Then apply the integration by parts formula,

$$\int_0^t e^{a_1 s} dB(s) = e^{a_1 t}B(t) - \int_0^t a_1 e^{a_1 s} B(s)ds \tag{5.41}$$

The close-form solution can be found as,

$$R_L(t) = e^{-a_1 t}\left[R_L(0) + b_1 \int_0^t e^{a_1 s} dB(s)\right] \tag{5.42}$$

In this case, the symbolic variable R_L in Eq. (5.28) can be replaced by Eq. (5.39) instead of Eq. (5.31).

5.3.2 Examples

The IEEE 10-machine 39-bus system in the deterministic case shown in Figure 5.8 is modified to include stochastic behavior of loads. All loads are assumed to change stochastically while all generators are represented by deterministic models. In each example, the stochastic simulation result by the Euler–Maruyama approach is used as the benchmark, and the second-order semi-analytical solutions (i.e. $N = 2$) are used and evaluated every 0.001 seconds. The value of each stochastic variable is changed every 0.1 seconds. To make the studies repeatable and the comparisons fair, a same pseudorandom number generating scheme is used for both approaches. For each case, 100 sample trajectories are generated. The fault applied in cases A and B is a self-clearing 4-cycle 3-phase fault at bus 1. All simulations are performed in MATLAB R2017b on a laptop computer with an Intel Core i5-6300U 2.40 GHz CPU and 4 GB RAM.

5.3.2.1 Stochastic Loads with Low Variances

In the first case, the variances of the loads are 0.1% of their mean values. The results of 100 runs from the semi-analytical approach are shown in Figure 5.18. Among all the generators, generator 1 has the shortest the electrical distance to bus 1, hence its rotor speed deviation from 377 rad/s is presented in the following results.

As shown in Figure 5.18, under the assumption of low load variances, the mean trajectory of 100 trajectories is a sufficient representation of the system dynamic. The $\mu + \sigma$ and $\mu + 2\sigma$ envelopes (which enclose 68% and 95% of the sample trajectories, respectively) are very close to the mean trajectory. To benchmark the accuracy, the mean, variance, and skewness of the trajectories from the semi-analytical approach are compared with those

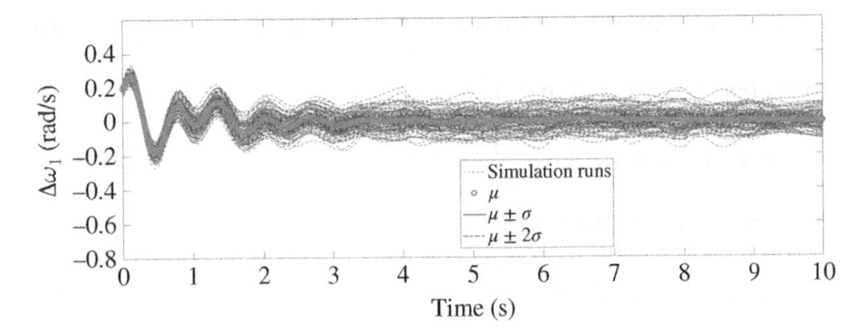

Figure 5.18 Semi-analytical stochastic simulation results of generator 1 rotor speed deviation (low load variances case).

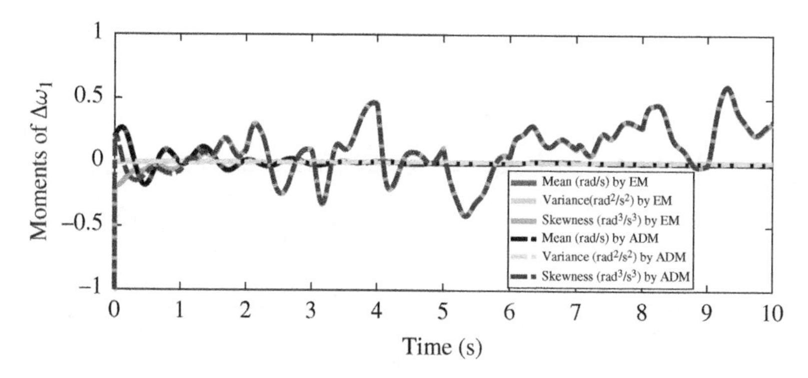

Figure 5.19 Mean, variance, and skewness comparisons of the generator 1 rotor speed deviation from the semi-analytical approach and the Euler–Maruyama approach (low load variance case).

from Euler–Maruyama approach. The comparisons show good agreement in Figure 5.19.

5.3.2.2 Stochastic Loads with High Variances

In the second case, the variances of the loads are increased to 1% of their mean values. As the uncertainty in the system increases, the mean trajectory can no longer represent the system dynamically.

As shown in Figure 5.20, the $\mu + \sigma$ and $\mu + 2\sigma$ envelopes are noticeably apart from the mean trajectory, they even show an undamped behavior. It implies that the probability of the system becoming marginally unstable is no longer negligible. That justifies the necessity of using stochastic load

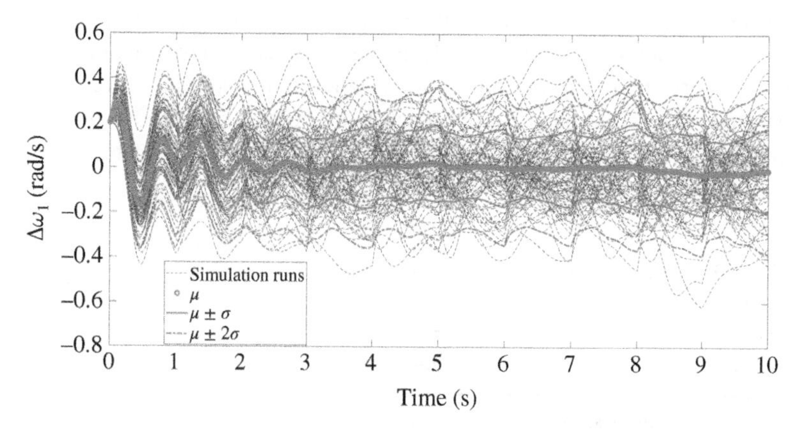

Figure 5.20 Semi-analytical stochastic simulation results of generator 1 rotor speed deviation (high load variances case).

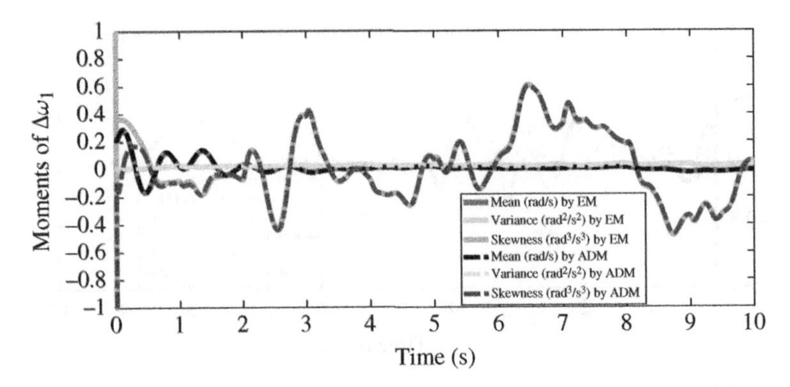

Figure 5.21 Mean, variance, and skewness comparisons of the generator 1 rotor speed deviation from the semi-analytical approach and the Euler–Maruyama approach (high load variance case).

models to study the stability of power systems with a high penetration of stochastic loads.

The accuracy of the semi-analytical approach is not affected by the increase in the uncertainty. As shown in Figure 5.21, the mean, variance, and skewness from the semi-analytical approach still match those from the Euler–Maruyama approach very well.

5.3.2.3 Comparison of Time Performances

The time performances for the semi-analytical approach and the Euler–Maruyama approach are compared in Table 5.8, from which the semi-analytical approach only takes 75% of the time cost of the Euler–Maruyama approach. The advantage of the semi-analytical approach in time performance is more prominent when many simulation runs are required. The semi-analytical approach is inherently suitable for parallel implementation, which could help further improve the time performance if high-performance parallel computers are available.

Table 5.8 Time performance comparison of stochastic load cases

Methods	Time costs (s)
Euler–Maruyama 100 runs	424.8
ADM 100 runs	317.0

5.3.2.4 Control Informed by Stochastic Simulation

Energy management systems (EMSs) and distribution management systems (DMSs) are two essential components that inform the operation of power transmission and distribution systems. They are traditionally considered independent, but more and more efforts are being made to integrate them for more reliable and efficient control and planning. The faster stochastic simulation enabled by ADM provides an opportunity for EMS and DMS to have transactional communication and achieve cooperative control.

Under deterministic assumption, to mitigate instability, the most common remedial action is load shedding. But the economic impact of load shedding cannot be overlooked. With stochastic simulation, the uncertainty of load becomes a new attribute based on which new control strategies can be designed.

The following case demonstrates a scenario where the DMS has sufficient capability to adjust load scheduling, renewable energy generation, and energy storage operation to reduce the uncertainty of the aggregative behavior of the loads under a distribution feeder [24]. Such uncertainty reduction strategy is associated with certain costs therefore undesirable under normal operation when the power transmission system can maintain its stability.

However, when potential instability is indicated by the stochastic simulation performed by EMS, instead of applying load shedding, load uncertainty reduction strategy can be applied to alleviate the stress on the system, which comes with a much lower cost than load shedding. As shown in Figure 5.22, a 30-cycle self-clearing 3-phase fault is applied at bus 1. With all loads having 2% of mean value variance, the system survives first a few swings, but the oscillation is poorly damped. In Figure 5.23, the DMSs at all the load buses receive the control signal from the EMS to apply load uncertainty reduction strategy at $t = 6$ seconds. The system is

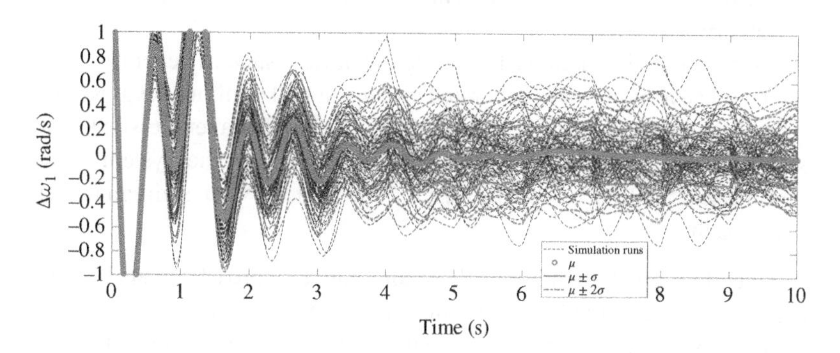

Figure 5.22 Results of the semi-analytical stochastic simulation without control.

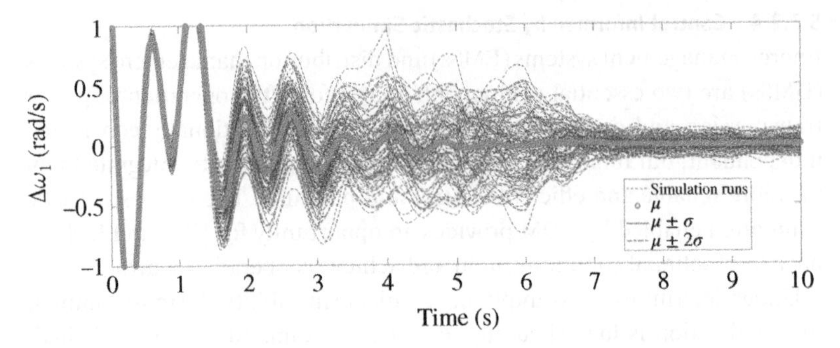

Figure 5.23 Results of the semi-analytical stochastic simulation with the EMS-DMS cooperative control taking effect at $t = 6$ seconds.

stabilized after that because of the cooperative control of EMS and DMSs. The foundation of such uncertainty-based control action is the advanced stochastic simulation approaches such as the semi-analytical approach.

5.4 Large-Scale Power System Simulations Using Adomian Decomposition Method

Several design choices can be made to scale the ADM when applying to realistic power systems with more than 1000 buses. The main challenge is the large memory the semi-analytical solution utilizes for a large power system. To mitigate this issue, we suggest three adjustments:

- The semi-analytical solutions can be limited to the differential equations only and leave the algebraic equations to traditional numerical solvers [25]. This alternative approach avoids embedding a large amount of network information in the semi-analytical solutions and helps save memory.
- The differential equations describing the devices such as generator, exciter, and governor, can be separated into linear and nonlinear expressions. Semi-analytical solutions can be limited to the nonlinear expressions only.
- Hybrid approaches that simulate some portion of the system using numerical methods and the rest of the system with semi-analytical approach.

Depending on the situation, these alternative approaches can be used together or individually to improve the simulation speed and reduce memory usage. However, the potential alternatives are not limited to these three suggestions.

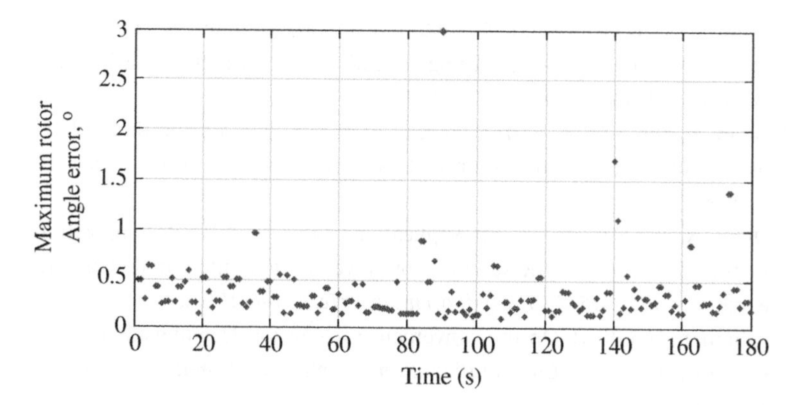

Figure 5.24 ADM maximum rotor angle error w.r.t Trapezoidal Method for IEEE 327-generator 2383-bus Polish system.

The IEEE Polish System with 327 generators and 2383 buses are simulated using multistage-ADM [26]. The maximum rotor angle error with respect to Trapezoidal method is shown in Figure 5.24.

References

1 Adomian, G. (2013). *Solving Frontier Problems of Physics: The Decomposition Method*, vol. 60. Springer Science & Business Media.

2 Adomian, G. (1976). Nonlinear stochastic differential equations. *Journal of Mathematical Analysis and Applications* 55 (2): 441–452.

3 Jafari, H., Khalique, C.M., and Nazari, M. (2011). Application of the Laplace decomposition method for solving linear and nonlinear fractional diffusion–wave equations. *Applied Mathematics Letters* 24 (11): 1799–1805.

4 Khuri, S.A. (2001). A Laplace decomposition algorithm applied to a class of nonlinear differential equations. *Journal of Applied Mathematics* 1 (4): 141–155.

5 Biazar, J. and Shafiof, S.M. (2007). A simple algorithm for calculating Adomian polynomials. *International Journal of Contemporary Mathematical Sciences* 2 (20): 975–982.

6 Wazwaz, A.M. (1999). The modified decomposition method and Padé approximants for solving the Thomas–Fermi equation. *Applied Mathematics and Computation* 105 (1): 11–19.

7 Qi, J., Sun, K., and Kang, W. (2014). Optimal PMU placement for power system dynamic state estimation by using empirical observability Gramian. *IEEE Transactions on Power Systems* 30 (4): 2041–2054.

8 Chow, J.H. and Cheung, K.W. (1992). A toolbox for power system dynamics and control engineering education and research. *IEEE Transactions on Power Systems* 7 (4): 1559–1564.

9 Duan, N. and Sun, K. (2016). Power system simulation using the multistage Adomian decomposition method. *IEEE Transactions on Power Systems* 32 (1): 430–441.

10 Duan, N. and Sun, K. (2015). Application of the Adomian decomposition method for semi-analytic solutions of power system differential algebraic equations. In: *2015 IEEE Eindhoven PowerTech*, 1–6. IEEE.

11 Adomian, G. (1984). On the convergence region for decomposition solutions. *Journal of Computational and Applied Mathematics* 11 (3): 379–380.

12 Duan, J.S. and Rach, R. (2011). New higher-order numerical one-step methods based on the Adomian and the modified decomposition methods. *Applied Mathematics and Computation* 218 (6): 2810–2828.

13 Holmquist, S.M. (2007). *An Examination of the Effectiveness of the Adomian Decomposition Method in Fluid Dynamic Applications.* University of Central Florida.

14 Bougoffa, L., Rach, R.C., and El-Manouni, S. (2013). A convergence analysis of the Adomian decomposition method for an abstract Cauchy problem of a system of first-order nonlinear differential equations. *International Journal of Computer Mathematics* 90 (2): 360–375.

15 Razali, N.I., Chowdhury, M.S.H., and Asrar, W. (2013). The multistage adomian decomposition method for solving chaotic lü system. *Middle-East Journal of Scientific Research* 13: 43–49.

16 Kolebaje, O.T., Akinyemi, P., and Adenodi, R.A. On the application of the multistage Laplace Adomian decomposition method with pade approximation to the Rabinovich-Fabrikant system. 4 (3): 232–243.

17 Duan, J.S., Rach, R., Baleanu, D., and Wazwaz, A.M. (2012). A review of the Adomian decomposition method and its applications to fractional differential equations. *Communications in Fractional Calculus* 3 (2): 73–99.

18 Duan, J.S., Rach, R., and Wazwaz, A.M. (2014). A reliable algorithm for positive solutions of nonlinear boundary value problems by the multistage Adomian decomposition method. *Open Engineering* 5 (1): 59–74.

19 Anderson, P.M. and Fouad, A.A. (2008). *Power System Control and Stability*. Wiley.

20 Ramos, J.I. (2009). Piecewise-adaptive decomposition methods. *Chaos, Solitons & Fractals* 40 (4): 1623–1636.

21 Mahmood, A.S., Casasús, L., and Al-Hayani, W. (2005). The decomposition method for stiff systems of ordinary differential equations. *Applied Mathematics and Computation* 167 (2): 964–975.

22 Kezunovic, M. (2006). Monitoring of power system topology in real-time. In: *Proceedings of the 39th Annual Hawaii International Conference on System Sciences (HICSS'06)*, vol. 10, 244b–244b. IEEE.

23 Nouri, K. (2016). Study on stochastic differential equations via modified Adomian decomposition method. *UPB Scientific Bulletin, Series A* 78 (1): 81–90.

24 Fenglei, G., Tao, X., Hao, X. et al. (2016). An uncertainty reduction strategy to schedule and operate microgrids with renewable energy sources. In: *2016 IEEE PES Asia-Pacific Power and Energy Engineering Conference (APPEEC)*, 1191–1199. IEEE.

25 Gurrala, G., Dimitrovski, A., Sreekanth, P. et al. (2015). Application of adomian decomposition for multi-machine power system simulation. In: *2015 IEEE Power & Energy Society General Meeting*, 1–5. IEEE.

26 Gurrala, G., Dinesha, D.L., Dimitrovski, A. et al. (2017). Large multi-machine power system simulations using multi-stage adomian decomposition. *IEEE Transactions on Power Systems* 32 (5): 3594–3606.

6

Application of Homotopy Methods in Power Systems Simulations

Gurunath Gurrala and Francis C. Joseph

Department of Electrical Engineering, Indian Institute of Science, Bangalore, Karnataka, India

6.1 Introduction

Power system time-domain simulations are used to assess the stability of the system under varying operating conditions and disturbances. The time-domain simulation for stability consists of the solution of differential equations comprised by the machines and its control systems as well as the transmission network modeled as algebraic equations. This entire system of equations comes under the class of differential-algebraic equations (DAE). Two methods of solution of such DAE systems used in power systems are [1, 2]:

1. Discretize the differential equations using appropriate numerical schemes to covert them to algebraic form and solve them along with the network algebraic equations. This is called the simultaneous solution approach.
2. Solve the differential equations and algebraic equations in an alternating fashion. This is called the partitioned solution approach.

Due to the nonlinear nature of the final algebraic equations arising from the composite load models and saturation modeling, the simultaneous approach uses a flavor of Newton–Raphson method [2]. Such methods may perform frequent inversions in each time step. However, the partitioned approach may develop interface errors, which can be reduced using multiple iterations within each time step. The partitioned approach can use both explicit and implicit methods of integration, though the latter requires appropriate sequencing while solving [1]. The semi-analytic methods provide an approximate solution to a set of nonlinear equations in the form

Power System Simulation Using Semi-Analytical Methods, First Edition. Edited by Kai Sun.

Published 2024 by John Wiley & Sons, Inc.

of a power series or closed-form solutions. These methods are widely used to solve both ordinary differential equation (ODE) and DAE problems [3], but these require derivation of new mathematical expressions, by applying the method for the intended system. These methods produce solutions that resemble explicit numerical integration. In [4], a Multistage Adomian Decomposition Method (MADM) is introduced for power system dynamic simulations and it was shown that a few terms of the Adomian polynomial was sufficient to compute the solution. Homotopy Analysis Method (HAM) is a widely used semi-analytical method in Science and Engineering that breaks down the nonlinear equations into a series of solutions. What sets HAM apart from other semi analytical methods, is that it creates a family of solutions for the given problem by varying an auxiliary linear operator and an auxiliary parameter [5]. Homotopy-based methods' application to continuation power flow was reported in [6–9]. The application of HAM to find bifurcation points for transient stability analysis was reported in [10] for a Single Machine Infinite Bus (SMIB) system using the classical model. In [11], a Multistage Homotopy Analysis Method (MHAM) was introduced for the power system stability simulations. It was shown that a family of solutions could be obtained by varying the nonzero auxiliary parameter, which allows different convergence characteristics for the system. It was also shown that MHAM could be considered as a general formulation for MADM and Homotopy Perturbation Method (HPM). The execution performance of both the semi-analytic methods [4, 11] was shown to be better than its conventional counterparts, Modified Euler (ME) and Midpoint Trapezoidal. In [12], MHAM was used to provide initial condition solver and serial correction (Corase Propagator) in a time-parallel solver called the Parareal in time.

6.2 The Homotopy Method

This section illustrates the concept of HAM as per [13]. It was first propounded in 1992 [14] and a special case called Homotopy Perturbation Method was presented in 1998 [15]. Both will be outlined in this section. Consider a general nonlinear differential equation:

$$\mathcal{A}[u(t)] = 0 \tag{6.1}$$

where, $\mathcal{A}$ is a nonlinear operator, t is time and $u(t)$ is an unknown variable. Let $u_0(t)$ denote an initial approximation of $u(t)$. Consider an auxiliary linear operator $\mathcal{L}$ with the following property:

$$\mathcal{L}[f(t)] = 0 \text{ when } f(t) = 0 \tag{6.2}$$

Let $q \in [0,1]$ be an embedding parameter, c be a nonzero auxiliary parameter and $H(t)$ be a nonzero auxiliary function of time t. Now, one can construct a homotopic relationship as follows:

$$\mathcal{H}(\phi; q, c, H) = (1 - q)\mathcal{L}[\phi(t; q, c, H) - u_0(t)] - qH(t)\mathcal{A}[\phi(t; q, c, H)] \tag{6.3}$$

The solution of the homotopic equation can be obtained by enforcing it to zero, which results in the following equation:

$$(1 - q)\mathcal{L}[\phi(t; q, c, H) - u_0(t)] = qH(t)\mathcal{A}[\phi(t; q, c, H)] \tag{6.4}$$

where $\phi(t; q, c, H)$ is a function of t, q, c and H. This equation is called the zero-order deformation equation. When $q = 0$, the above equation gives:

$$\mathcal{H}(\phi(t; q, c, H), q)|_{q=0} = \mathcal{L}[\phi(t; 0, c, H) - u_0(t)] = 0 \tag{6.5}$$

Since $\mathcal{L}$ is a linear operator

$$\begin{aligned}\mathcal{L}[\phi(t; 0, c, H) - u_0(t)] = 0\\ \phi(t; 0, c, H) = u_0(t)\end{aligned} \tag{6.6}$$

Similarly, for $q = 1$, one will get:

$$\mathcal{H}(\phi(t; q, c, H), q)|_{q=1} = H(t)\mathcal{A}[\phi(t; 1, c, H)] = 0 \tag{6.7}$$

Since $c \neq 0$ and $H(t) \neq 0$

$$\begin{aligned}\mathcal{A}[\phi(t; 1, c, H)] = 0\\ \phi(t; 1, c, H) = u(t)\end{aligned} \tag{6.8}$$

At $q = 1$, the solution still depends on the c and $H(t)$. Hence, the solution $\phi(t; q, c, H)$ of the generalized homotopy function given in (6.3) approximates the solution of (6.1) from initial condition $u_0(t)$ to the exact solution $u(t)$ as the embedding parameter q varies from 0 to 1. Such a kind of continuous variation is called deformation in homotopy. A family of solution curves can be obtained depending on the values of c and $H(t)$. These parameters give a simple way to adjust and control the convergence region and rates of approximation series [5]. The solution $u(t)$ can be expressed by a series of polynomial functions, fractional functions, or exponential functions. For polynomial functions, $H(t)$ is uniquely defined as unity, i.e. $H(t) = 1$ (Ref. Chapter 2, Section 3.5 in [5]). Using Taylor's theorem, $\phi(t; q, c)$ can be expanded as follows:

$$\phi(t; q, c) = u_0(t) + \sum_{m=1}^{+\infty} u_m q^m \tag{6.9}$$

where, u_m is defined as:

$$u_m = \frac{1}{m!}\frac{\partial^m \phi(t;q)}{\partial q^m}\bigg|_{q=0} \tag{6.10}$$

If the power series given in (6.9) converges at $q = 1$ and the deformation derivative u_m exists for $n = 1,2, \ldots,$ then according to (6.8),

$$\phi(t;1,c) = u(t) = u_0(t) + \sum_{n=1}^{+\infty} u_m(t) \tag{6.11}$$

This equation gives a relationship between the exact solution of $u(t)$ and the initial approximation of $u_0(t)$ by means of terms $u_m(t)$, which can be determined by the high-order deformation equations described as follows:

Differentiating the zero-order deformation equation, (6.4), m times w.r.t. the embedding parameter, q, and then dividing it by $m!$ and setting $q = 0$, one will obtain the following mth-order deformation equation:

$$\mathcal{L}[u_m(t) - \chi_m u_{m-1}(t)] = c\mathcal{R}_m(t) \tag{6.12}$$

where,

$$\chi_m = \begin{cases} 0, & m \leq 1 \\ 1, & m > 1 \end{cases}$$

and

$$\mathcal{R}_m(t) = \frac{1}{(m-1)!}\frac{\partial^{m-1} A[\phi(t;q,c)]}{\partial q^{m-1}}\bigg|_{q=0} \tag{6.13}$$

In [13], the HPM is shown to be same as HAM for $c = -1.0$. Adomian decomposition method (ADM) is shown to be equivalent to HPM in [16]. Hence, using HAM one can derive both HPM and ADM models for power system.

6.2.1 Multi-stage MHAM

HAM has limited convergence when moved farther from the initial conditions [17–19]. Let the total simulation time be divided into K equal time intervals. Consider one such interval $[t_{i-1}, t_i]$ where t_{i-1} denotes the time at the end of the previous interval while t_i denotes the time at the end of the current interval.

The HAM solution in a time subinterval is obtained based on the following transformation $u(t) \to u^{(i)}(T)$ which is the approximated solution in the ith subinterval. Then the new time shifted variable will satisfy the following condition: $0 \leq \mathrm{T} \leq (t_i - t_{i-1} = h)$. If the linear operator is selected as $\mathcal{L}(u(t)) = \frac{du(t)}{dt}$ over the interval $[t_{i-1}, t_i]$, then it will be transformed to $\mathcal{L}(u^{(i)}(T)) = \frac{du^{(i)}(T)}{dT}$ over the interval $[0, h]$. Hence, all terms which are function of "t" can be considered to be constant during that time interval since the differentiation is being carried out w.r.t. "T." In this approach, the solution $u^{(i)}(T)$ of the previous interval will become the initial condition for the next iteration. Hence HAM will be applied again for the new interval, and the process is repeated till all K time segments are solved. This revised method is called Multistage Homotopy Analysis Method (MHAM).

6.2.2 Stability of Homotopy Analysis

ODE are described in the form

$$\frac{dx}{dt} = f(x, t) \tag{6.14}$$

where, $x(0)$ is the initial condition.

Generally the stability of numerical method is tested out using the following first order ODE [20].

$$\frac{dx}{dt} = \lambda x \tag{6.15}$$

where, $x(0) = 1$ is the initial condition. The test equation, (6.15), has the solution $x(t) = e^{\lambda t}$. The discretized version of $e^{\lambda t}$ is $e^{h\lambda n}$ where h is the step size. The value at each discrete point can be shown as summation as shown in (6.16).

$$\begin{aligned} e^{zn} &= 1 + \frac{zn}{1!} + \frac{(zn)^2}{2!} + \frac{(zn)^3}{3!} + \cdots \\ e^{zn} &= \sum_{i=0}^{\infty} \frac{(zn)^i}{i!} \\ &\textit{where } z = h\lambda \end{aligned} \tag{6.16}$$

Let us consider 3 explicit numerical solvers

1. **Forward Euler** is a first order solver defined by (6.17) for a system of differential equations given by (6.14).

$$x_{n+1} = x_n + hf\left(x_n, t\right) \tag{6.17}$$

The general solution of Forward Euler when applied to (6.15) is given in (6.18).

$$x_n = (1 + h\lambda)^n \ \forall \ n \geq 0 \tag{6.18}$$

The solution is stable if $|1 + h\lambda| < 1$.

2. **Modified Euler** is a second order method defined by (6.19) for a system of differential equations given by (6.14).

$$\begin{aligned} \text{Predictor}: \; x_{n+1}^{P} &= x_n + hf\left(x_n, t\right) \\ \text{Corrector}: \; x_{n+1}^{C} &= x_n + \frac{h}{2}\left(f\left(x_n, t\right) + f\left(x_{n+1}^{P}, t+h\right)\right) \end{aligned} \tag{6.19}$$

The first step of ME, given in (6.19), is to compute a predicted value of x_{n+1}^{p}. This step is the same as Forward Euler. The general solution of (6.15) by ME is

$$x_n = \left(1 + h\lambda + \frac{h^2\lambda^2}{2}\right)^n \; \forall \, n \geq 0 \tag{6.20}$$

The solution given by (6.20) is stable if $|1 + h\lambda + \frac{h^2\lambda^2}{2}| < 1$.

3. **Runge–Kutta of Order 3 (RK3)** is a third order method defined by (6.21) for a system of differential equations given by (6.14).

$$\begin{aligned} k_1 &= hf\left(x_n, t\right) \\ k_2 &= hf\left(x_n + k_1/2, t + h/2\right) \\ k_3 &= hf\left(x_n - k_1 + 2k_2, t + h/2\right) \\ x_{n+1} &= x_n + \frac{1}{6}(k_1 + 4k_2 + k_3) \end{aligned} \tag{6.21}$$

The general solution of (6.15) by RK3 is

$$x_n = \left(1 + h\lambda + \frac{h^2\lambda^2}{2} + \frac{h^3\lambda^3}{6}\right)^n \; \forall \, n \geq 0 \tag{6.22}$$

The solution given by (6.22) is stable if $|1 + h\lambda + \frac{h^2\lambda^2}{2} + \frac{h^3\lambda^3}{6}| < 1$.

Comparing the first step result of (6.18) and (6.20) with the actual solution (6.16), it can be seen that the lowest order term of error is $\frac{h^2\lambda^2}{2}$ for Forward Euler and $\frac{h^3\lambda^3}{6}$ for ME. It is observed that after the second step of ME, the solution has higher accuracy than the predicted state.

Consider a system given in (6.14), it is rewritten as

$$\begin{aligned} \dot{x} - f(x, t) &= 0 \\ \mathcal{A}[x(t)] = \dot{x} - f(x, t) &= 0 \end{aligned} \tag{6.23}$$

where $A[\cdot]$ denotes the operator being used. Semi-analytical methods such as HAM provides the solution as a sum of infinite series, which can be approximated by considering a few starting terms. The solution of integration from a period $[t, t+h]$ is given by

$$x(t+h) = \phi(t; q)|_{q=1} = u_0 + \sum_{i=1}^{\infty} u_i q^i \Big|_{q=1} = u_0 + \sum_{i=1}^{\infty} u_i \tag{6.24}$$

where, $u_0 = x(t)$.

The remaining terms of $u_i(t)$ can be computed using (6.13) and (6.12). Lets consider upto 4 MHAM terms for (6.15), $x(t+h) = u_0 + u_1 + u_2 + u_3$

1. Rearranging the equations to the form in (6.23), $\dot{x} - \lambda x = 0$
2. Defining 4 term power series, from (6.24) $\phi(t;q) = u_0 + u_1 q + u_2 q^2 + u_3 q^3$
3. **Finding the term $\mathbf{u_0}$** From (6.5), $u_0 = x(t_0)$ as this forms the initial condition for the solution.
4. **Finding the term $\mathbf{u_1}$** For the next set of terms we substitute $\phi(t;q) = u_0 + u_1 q + u_2 q^2 + u_3 q^3$ into (6.12) and (6.13). Therefore for u_1

$$\mathcal{L}[u_1 - 0 * u_0] = c\mathcal{R}_1 = \frac{c}{(1-1)!}$$
$$\frac{\partial^{(1-1)}(\dot{u}_0 + \dot{u}_1 q + \dot{u}_2 q^2 + \dot{u}_3 q^3 - \lambda(u_0 + u_1 q + u_2 q^2 + u_3 q^3))}{\partial^{(1-1)} q}$$

Here we choose the linear operator $\mathcal{L}[\cdot]$ as $\frac{d(\cdot)}{dT}$ $\forall\ 0 \leq T \leq h$. Since no partial derivative is needed, here q is set to zero.

$$\frac{du_1}{dT} = c\mathcal{R}_1 = c(\dot{u}_0 - \lambda u_0)$$

Setting $\dot{u}_0 = 0$ and integrating with respect to T from 0 to h.

$$u_1 = \int_0^h (-c\lambda u_0) dT = -hc\lambda u_0$$

5. **Finding the term $\mathbf{u_2}$** As the previous step, we substitute $\phi(t;q) = u_0 + u_1 q + u_2 q^2 + u_3 q^3$ into (6.12) and (6.13)

$$\mathcal{L}[u_2 - 1 * u_1] = c\mathcal{R}_2 = \frac{c}{1!}$$
$$\frac{\partial(\dot{u}_0 + \dot{u}_1 q + \dot{u}_2 q^2 + \dot{u}_3 q^3 - \lambda(u_0 + u_1 q + u_2 q^2 + u_3 q^3))}{\partial q}$$

$$\mathcal{L}[u_2 - u_1] = c\mathcal{R}_2$$
$$= c(\dot{u}_1 + 2\dot{u}_2 q + 3\dot{u}_3 q^2 - \lambda(u_1 + 2u_2 q + 3u_3 q^2))|_{q=0}$$
$$\mathcal{L}[u_2 - u_1] = c(\dot{u}_1 - \lambda(u_1))$$

Integrating with respect to T

$$u_2 - u_1 = \int_0^h (c\dot{u}_1 - c\lambda(u_1)) dT$$

$$u_2 = (1+c)u_1 - c\lambda \int_0^h u_1\, dT$$

From step 4 u_1 is function of T, hence

$$\int_0^h u_1\, dT = \int_0^h \int_0^T (-c\lambda u_0) dT\, dT = -\frac{h^2}{2} c\lambda u_0 = \frac{hu_1}{2}$$

Substituting

$$u_2 = (1 + c)u_1 - c\lambda \frac{hu_1}{2}$$

6. **Finding the term $\mathbf{u_3}$**

$$\mathcal{L}[u_3 - 1 * u_2] = cR_3 = \frac{c}{2!} \frac{\partial^2(\dot{u}_0 + \dot{u}_1 q + \dot{u}_2 q^2 + \dot{u}_3 q^3 - \lambda(u_0 + u_1 q + u_2 q^2 + u_3 q^3))}{\partial^2 q}$$

$$\mathcal{L}[u_3 - u_2] = cR_3 = \frac{c}{2}(2\dot{u}_2 + 6\dot{u}_3 q - \lambda(2u_2 + 6u_3 q))|_{q=0}$$

$$\mathcal{L}[u_3 - u_2] = (\dot{u}_2 - \lambda(u_2))$$

Integrating with respect to T

$$u_3 - u_2 = \int_0^h \left(c\dot{u}_2 - c\lambda(u_2)\right) dT$$

$$u_3 = (1 + c)\, u_2 - c\lambda \int_0^h u_2\, dT$$

From Steps 4 and 5 u_2 and u_1 are functions of T, hence

$$\int_0^h u_2\, dT = \int_0^h (1 + c)\dot{u}_1\, dT - c\lambda \int_0^h \int_0^T (-c\lambda u_0) dT\, dT$$

$$\int_0^h u_2\, dT = (1 + c)\frac{hu_1}{2} + \frac{h^3}{6} c^2 \lambda^2 u_0$$

$$\int_0^h u_2\, dT = (1 + c)\frac{hu_1}{2} - \frac{h^2 c\lambda u_1}{3}$$

$$\int_0^h u_2\, dT = \frac{h}{2}\left((1 + c)\frac{u_1}{2} - \frac{hc\lambda u_1}{3}\right)$$

$$\int_0^h u_2\, dT = \frac{h}{2}\left(u_2 + \frac{c\lambda hu_1}{2} - \frac{hc\lambda u_1}{3}\right)$$

$$\int_0^h u_2\, dT = \frac{h}{2}\left(u_2 + \frac{c\lambda hu_1}{6}\right)$$

Substituting

$$u_3 = (1 + c)u_2 - \frac{ch}{2}\left(u_2 + \frac{h\lambda u_1}{6}\right)$$

From the computed terms, the numerical solution for (6.15) would be

- **Using two terms**:

$$x(t + h) = u_0 + u_1$$

$$= x(t) - ch\lambda u_0$$
$$= (1 - ch\lambda)x(t) \tag{6.25}$$

- **Using three terms**:

$$x(t+h) = u_0 + u_1 + u_2$$
$$= (1 - ch\lambda)x(t) + (1+c)u_1 - \frac{ch\lambda u_1}{2}$$
$$= \left(1 - ch\lambda - (1+c)ch\lambda + \frac{ch\lambda(ch\lambda)}{2}\right)x(t)$$
$$= \left(1 - (2c + c^2)h\lambda + \frac{c^2h^2\lambda^2}{2}\right)x(t) \tag{6.26}$$

- **Using four terms**:

$$x(t+h) = u_0 + u_1 + u_2 + u_3$$
$$= \left(1 - (2c+c^2)h\lambda + \frac{c^2h^2\lambda^2}{2}\right)x(t) + (1+c)u_2 - \frac{ch}{2}\left(u_2 + \frac{h\lambda u_1}{6}\right)$$
$$= \left(1 - (3c + 3c^2 + c^3)h\lambda + h^2\lambda^2\left(\frac{3c^2}{2} + c^3\right) - \frac{c^3h^3\lambda^3}{6}\right)x(t) \tag{6.27}$$

Setting the auxiliary parameter $c = -1.0$ and generalizing the solution in the form of (6.20)

- **Using two terms**:

$$x_n = (1 + h\lambda)^n \tag{6.28}$$

- **Using three terms**:

$$x_n = \left(1 + h\lambda + \frac{h^2\lambda^2}{2}\right)^n \tag{6.29}$$

- **Using four terms**:

$$x_n = \left(1 + h\lambda + \frac{h^2\lambda^2}{2} + \frac{h^3\lambda^3}{6}\right)^n \tag{6.30}$$

It can be seen that HAM is able to generate FE (comparing (6.18) and (6.28)), ME (comparing (6.20) and (6.29)) and RK3 (comparing (6.22) and (6.30)) sequences. Also observe that N terms of HAM gives the N terms of the actual solution given in (6.16) for a time step $n = 1$.

Lets consider the stability of the method, when λ is real and h is positive nonzero real value. The value of $\phi(z)$ is computed for $-\infty < z < +\infty$. As long as $|\phi(z)| < 1$ the method should not diverge,

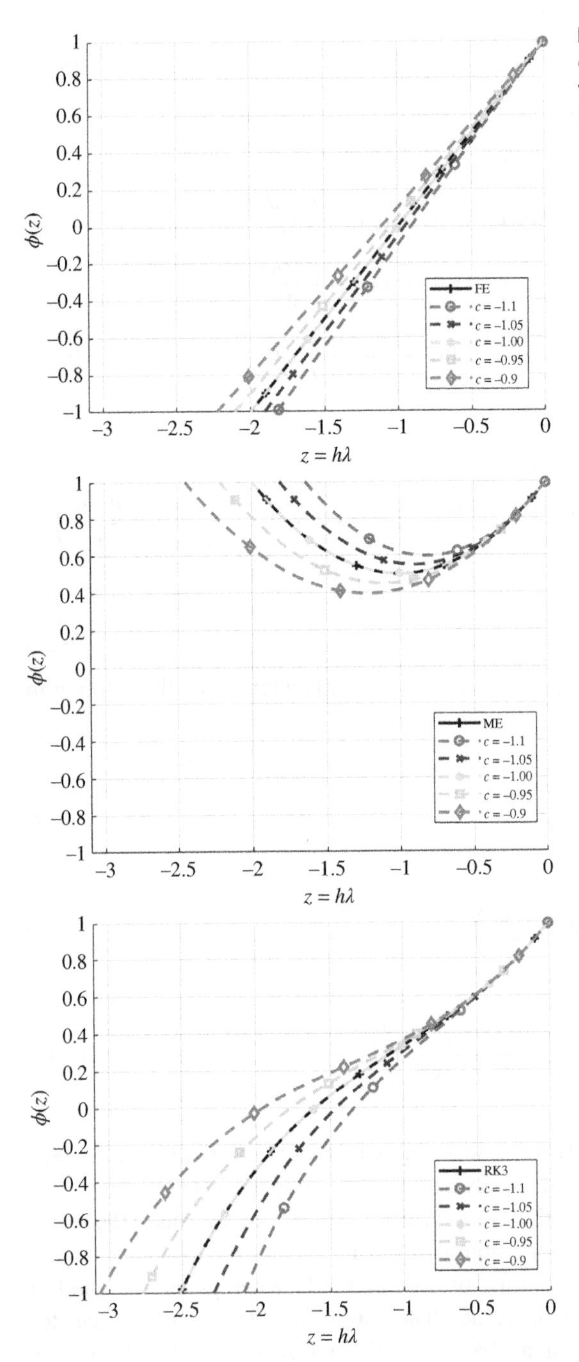

Figure 6.1 Stability of Homotopy Method with Forward Euler, Modified Euler, and RK3 methods.

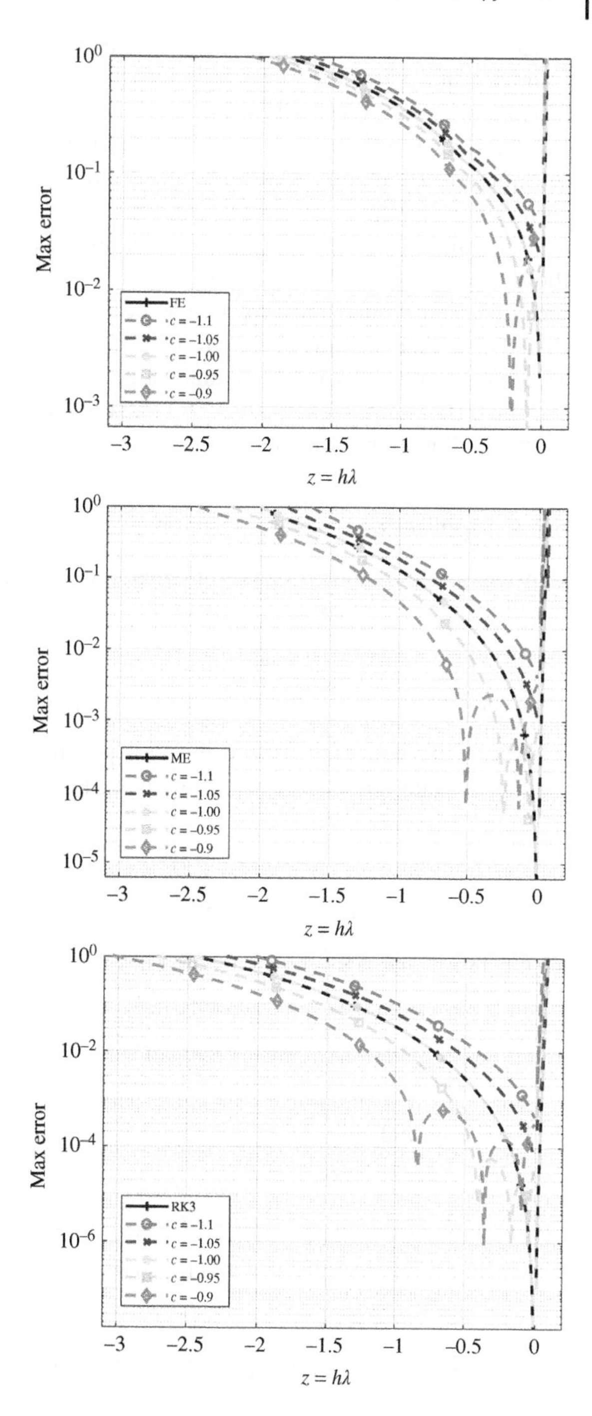

Figure 6.2 Error of Homotopy Method with respect to Forward Euler, Modified Euler, and RK3 Methods.

at $|\phi(z)| = 1$, $x(n) = 1\ \forall n$ and $|\phi(z)| = 0$, $x(n) = 0\ \forall n$. The value $\phi(z)$ is computed for $c = \{-0.9, -0.95, -1.0, -1.05, -1.1\}$ for 2, 3, and 4 terms of HAM and compared with a numerical method of the same accuracy class as shown in Figure 6.1. It can be seen that $c = -1.0$ the HAM tracks the numerical method. HAM has the similar characteristics that of the corresponding numerical method. The first and third order method has $\phi(z)$ crossing zero, indicating that results obtained from this series with these values of z will have alternating "+−" signs whereas the second order method would not have that issue. It can be seen in Figure 6.1 that, the auxiliary parameter c can influence the range of z that can be used. The maximum error with the actual solution is shown in Figure 6.2. For the test ODE (6.15), higher values of c shows better performance than the corresponding numerical method.

6.2.3 Application to a Linear System

Consider a step response of a second order system as given in (6.31)

$$\frac{Y(s)}{X(s)} = \frac{w_0^2}{s^2 + 2\zeta w_0 s + w_0^2} \tag{6.31}$$

Let us consider a three term HAM expansion of the solution.

1. Rewrite the equation in HAM form (as per (6.23)), using U_1 and U_2 as the two state variables associated with (6.31) and $x(t_0) = x_0$ as the input. Output $y(t) = U_1$.

$$\dot{U}_1 - U_2 = 0$$
$$\dot{U}_2 - w_0^2 x + (w_0^2 U_1 + 2\zeta w_0 U_2) = 0$$

2. The power series solution as per (6.24) is

$$\phi_{U_1}(t;q) = U_{1,0} + U_{1,1}q + U_{1,2}q^2$$
$$\phi_{U_2}(t;q) = U_{2,0} + U_{2,1}q + U_{2,2}q^2$$

   where $U_{1,0} = U_1(t_0)$ and $U_{2,0} = U_2(t_0)$. Hence first term is done
3. For the second term,

$$\mathcal{L}[U_{1,1}] = c(\dot{U}_{1,0} + \dot{U}_{1,1}q + \dot{U}_{1,2}q^2 - U_{2,0} - U_{2,1}q - U_{2,2}q^2)$$
$$\mathcal{L}[U_{2,1}] = c(\dot{U}_{2,0} + \dot{U}_{2,1}q + \dot{U}_{2,2}q^2 - w_0^2 x_0 + (w_0^2 U_{1,0} + 2\zeta w_0 U_{2,0}))$$

   Setting $\dot{U}_{1,0}$, $\dot{U}_{2,0}$, and q to 0 and integrating from 0 to step time h we get

$$U_{1,1} = c(-U_{2,0})h$$
$$U_{2,1} = c(-w_0^2 x_0 + (w_0^2 U_{1,0} + 2\zeta w_0 U_{2,0}))h$$

4. for the third term.

$$\mathcal{L}[U_{1,2} - U_{1,1}] = c(\dot{U}_{1,1} + 2\dot{U}_{1,1}q - U_{2,1} - 2U_{2,1}q)$$
$$\mathcal{L}[U_{2,2} - U_{2,1}] = c\big(\dot{U}_{2,1} + 2\dot{U}_{2,2}q + w_0^2(U_{1,1} + 2U_{1,2}q) + 2\zeta w_0(U_{2,1} + 2U_{2,2}q)\big)$$

Considering input to be a constant under integration and setting $q = 0$, and integrating from 0 to step time h we get

$$U_{1,2} = (1 + c)U_{1,1} - cU_{2,1}\frac{h}{2}$$
$$U_{2,2} = (1 + c)U_{2,1} + c\left(w_0^2 U_{1,1} + 2\zeta w_0 U_{2,1}\right)\frac{h}{2}$$

Starting from zero initial conditions the step response can be computed for various values of $c = \{-1.1 \text{ to } -0.9\}$. The output results are shown in Figure 6.3. We can see that with the auxiliary parameter at $c = -1.0$, the output of step function and homotopy based approach are same. When $c \neq -1.0$ there is deviations from the expected solution. These deviations are seen at initial sections of the waveforms where the oscillations exist. For all c, the step response converges to the same steady state value.

6.2.4 Application to a Nonlinear System

Consider a SMIB system comprised of a double circuit transmission line with each line's reactance $X = j0.2$ p.u. The generator operating at 1.05 p.u. terminal voltage and 1.0 p.u. active power with an internal reactance of $X_g = j0.01$ p.u. and inertia of $H = 3.5$MJ/MVA. Let us simulate a 3-phase fault in the middle of a line and clear the fault by removing the faulted

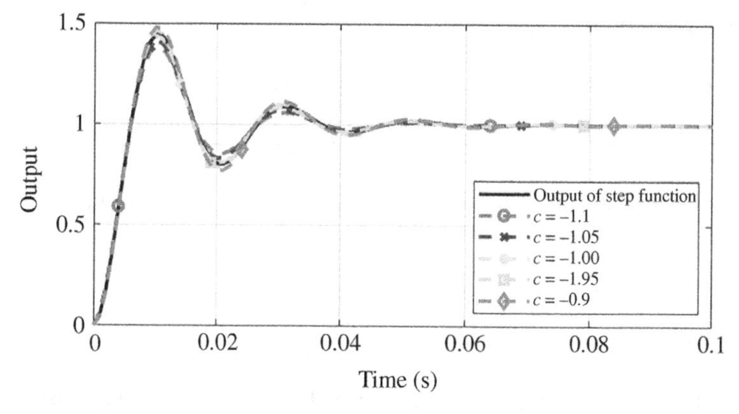

Figure 6.3 Second order system step response $w_0 = 314$, $\zeta = 0.25$ and $h = 1$ ms.

line after 3 cycles. Let us find the swing curve using HAM. Assume infinite bus voltage to be 1.0 p.u. and 50 Hz system. Let's consider a three term HAM expansion of the solution. Here we use the classical model of the generator [1].

Step 1: Rewrite the equations in HAM form (as per (6.23)), using δ and S_m as the two state variables δ, S_m.

$$\dot{\delta} - w_B S_m = 0$$

$$\dot{S}_m - \frac{P_m - P_e}{2H} = 0$$

$$P_e = P_{MAX} \sin(\delta)$$

Step 2: The power series solution as per (6.24) is

$$\phi_\delta(t;q) = \delta_0 + \delta_1 q + \delta_2 q^2$$

$$\phi_{S_m}(t;q) = S_{m,0} + S_{m,1} q + S_{m,2} q^2$$

where $\delta_0 = \delta(t_0)$ and $S_{m,0} = S_m(t_0)$. Hence first term is done

Step 3: For the second terms,

$$\mathcal{L}[\delta_1] = c\left(\dot{\delta}_0 + \dot{\delta}_1 q + \dot{\delta}_2 q^2 - w_B(S_{m,0} + S_{m,1} q + S_{m,2} q^2)\right)$$

$$\mathcal{L}[S_{m,1}] = c\frac{\dot{S}_{m,0} + \dot{S}_{m,1} q + \dot{S}_{m,2} q^2 - P_m + P_{MAX} * \sin(\delta_0 + \delta_1 q + \delta_2 q^2)}{2H}$$

Setting $\dot{\delta}_0$, $\dot{S}_{m,0}$, q to 0 and integrating from 0 to h,

$$\delta_1 = c(-w_B S_{m,0})h$$

$$S_{m,1} = c\frac{(-P_{m,0} + P_{MAX} \sin(\delta_0))}{2H}h$$

Step 4: For the third terms.

$$\mathcal{L}[\delta_2 - \delta_1] = c\left((\dot{\delta}_1) + 2(\dot{\delta}_2)q - w_B(S_{m,1} + 2S_{m,2} q)\right)$$

$$\mathcal{L}[S_{m,2} - S_{m,1}] = c\frac{\left(\dot{S}_{m,1} + 2\dot{S}_{m,2} q + P_{MAX} \cos(\delta_0 + \delta_1 q + \delta_2 q^2)(\delta_1 + 2\delta_2 q)\right)}{2H}$$

Considering input to be a constant under integration and setting $q = 0$, and integrating from $0 \to h$, we get

$$\delta_2 = (1 + c)\delta_1 - c w_B S_{m,1} h/2$$

$$S_{m,2} = (1 + c)S_{m,1} + c\frac{P_{MAX}\delta_1 \cos(\delta_0)}{2H}h/2$$

In order to solve the swing curve, the initial value of $\delta(0)$ and the $P_{MAX}(n)$ at pre-fault, fault and post-fault is needed.

1. Power transferred from the generator is 1.0 p.u. via an inductive circuit ($X_{eq} = j0.1$ p.u.) to grid. Therefore, the generator bus angle θ can be computed as

$$\theta = \sin^{-1}\left(\frac{1.0 * 0.1}{1.05 * 1.00}\right) = 5.4650^{\circ}$$

2. The generator current is

$$I_T = \frac{1.05\angle\theta - 1.0}{j0.1} = 1.0975\angle - 24.3359^{\circ}$$

3. The generator internal voltage and angle can be computed using $E_g\angle\delta = 1.05\angle\theta + j0.01 * I_T$

$$E_g = |1.05\angle\theta + j0.01 * I_T| = 1.0555$$
$$\delta = arg(1.05\angle\theta + j0.01 * I_T) = 5.9820^{\circ}$$

4. There are three network configurations hence three values of $P_{MAX}(n)$
   (a) **Pre-fault**: The generator impedance is in series with the double circuit line as shown in Figure 6.4.

$$P_{MAX}(0 \leq t < t_f) = \frac{1.0555 * 1.0}{(0.2/2 + 0.01)} = 9.5954$$

   (b) **Fault**: The double circuit line now resembles a π network due to the fault with shunt impedance of $j0.1$ on both sides as shown in Figure 6.4. The generator impedance is in series with this π section. Hence using $Y - \Delta$ conversion the equivalent reactance connecting both the buses can be found.

$$P_{MAX}(t_f \leq t < t_p) = \frac{1.0555 * 1.0}{(0.2 + 3 * 0.01)} = 4.5891$$

   (c) **Post-fault**: The generator impedance is in series with only a single circuit line as shown in Figure 6.4.

$$P_{MAX}(t \geq t_p) = \frac{1.0555 * 1.0}{(0.2 + 0.01)} = 5.0262$$

The swing curve can be computed for various value $c = \{-1.1 \text{ to} - 0.9\}$. The output results are shown in Figure 6.5. The output for $c = -1.0$ follows the Modified Euler trajectory exactly. But for the other values, $c \neq -1.0$ there exists deviations form the expected values.

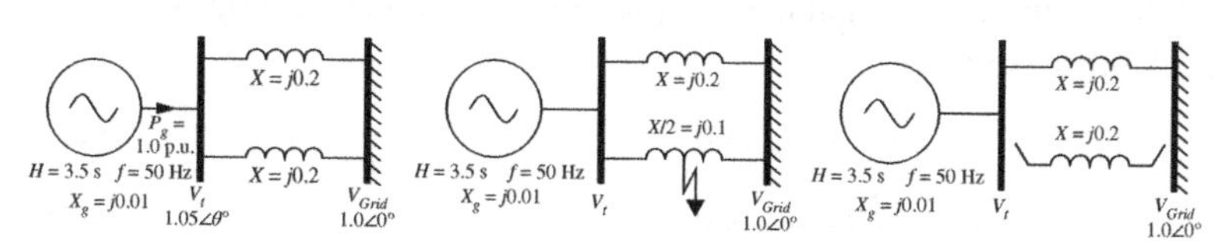

Figure 6.4 The prefault, fault, and post fault circuits.

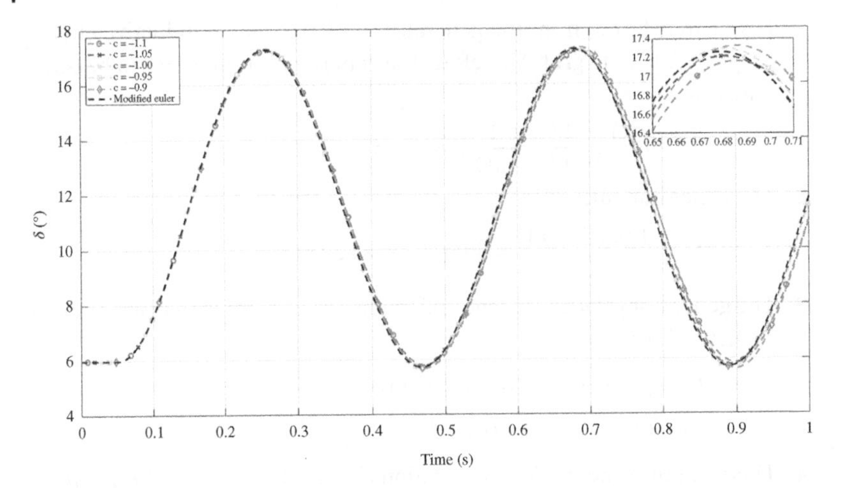

Figure 6.5 Swing Curve for a SMIB, 3-phase midline fault cleared by tripping the faulted line after 3 cycles.

6.3 Application of Homotopy Methods to Power Systems

Power system dynamics can be modeled as DAEs of the following form:

$$\dot{X} = f(X, Y) \tag{6.32}$$

$$g(X, Y) = 0 \tag{6.33}$$

where, the differential equations represent the dynamic models of the devices and the algebraic equations represent the network. Partitioned approach has been used along with MHAM for the time domain simulations in this chapter [2]. In Section 6.3.1 the development of MHAM models of IEEE 1.1 and IEEE 2.2 synchronous machine models are discussed.

6.3.1 Generator Model for Transient Stability

A generator possesses two sets of equations, mechanical and electrical. Usually a single rotational mass, representing the kinetic energy, forms the mechanical equations.

$$\dot{\delta} - w_B S_m = 0 \tag{6.34}$$

$$\dot{S}_m - \frac{T_m - K_D S_m - T_e}{2H} = 0 \tag{6.35}$$

where, the mechanical equations have two states δ – the rotor angle and S_m – the slip speed. Table 6.1 shows the 3 terms for HAM for (6.34).

Table 6.1 MHAM terms for the mechanical equations of the generator.

New state $X(n+1)$	First term X_0	Second term X_1	Third term X_2
$\delta(n+1)$	$\delta(n)$	$-cw_B S_{m,1} h$	$(1+c)\delta_1 - cw_B S_{m,1} h/2$
$S_m(n+1)$	$S_m(n)$	$-c\dfrac{(T_{m,0} - T_{e,0} - K_D S_{m,0})}{2H}h$	$(1+c)S_{m,1} - c\dfrac{(T_{m,1} - T_{e,1} - K_D S_{m,1})}{2H}h/2$

The procedure is similar to the swing curve problem discussed in Section 6.2.4.

The generator can be modeled with a field winding and damper winding (IEEE 1.1 model) with two states E'_q and E'_d. The generator current and voltage is an algebraic quantity, obtained by solving (6.33), and can be treated as input quantity. Therefore there is no HAM terms of i_d, i_q, v_d, and v_q.

$$\frac{dE'_q}{dt} - \frac{1}{T'_{do}}\left(-E'_q + \left(x_d - x'_d\right) i_d + E_{fd}\right) = 0 \tag{6.36}$$

$$\frac{dE'_d}{dt} - \frac{1}{T'_{qo}}\left(-E'_d - \left(x_q - x'_q\right) i_q\right) = 0 \tag{6.37}$$

Equations (6.36) and (6.37) resemble a first order transfer function similar to second order transfer function solution of Section 6.2.3, therefore same procedure is followed. The MHAM terms are given in Table 6.2.

Table 6.2 MHAM terms for the field and damper windings.

New state $X(n+1)$	First term X_0	Second term X_1	Third rerm X_2
$E'_q(n+1)$	$E'_q(n)$	$-c\dfrac{\left(-E'_{q,0} + \left(x_d - x'_d\right) i_d + E_{fd,0}\right)}{T'_{do}}h$	$(1+c)E'_{q,1} - c\dfrac{\left(-E'_{q,1} + E_{fd,1}\right)}{T'_{do}}h/2$
$E'_d(n+1)$	$E'_d(n)$	$-c\dfrac{\left(-E'_{d,0} - \left(x_q - x'_q\right) i_q\right)}{T'_{qo}}h$	$(1+c)E'_{d,1} - c\dfrac{\left(-E'_{d,1}\right)}{T'_{qo}}h/2$

Table 6.3 MHAM terms for the second damper windings.

New state $X(n+1)$	First term X_0	Second term X_1	Third term X_2
$E''_q(n+1)$	$E''_q(n)$	$-c\dfrac{\left((x'_d - x''_d)\, i_d + E'_{q,0} - E''_{q,0}\right)}{T''_{do}}h$	$(1+c)E''_{q,1} - c\dfrac{\left(E'_{q,1} - E''_{q,1}\right)}{T''_{do}}h/2$
$E''_d(n+1)$	$E''_d(n)$	$-c\dfrac{\left(-(x'_q - x''_q)\, i_q + E'_{d,0} - E''_{d,0}\right)}{T''_{qo}}h$	$(1+c)E''_{d,1} - c\dfrac{\left(E'_{d,1} - E''_{d,1}\right)}{T''_{qo}}h/2$

Generator can also be modeled with two additional damper windings (IEEE 2.2 model) having two states E''_q and E''_d.

$$\frac{dE''_q}{dt} - \frac{1}{T''_{do}}\left((x'_d - x''_d)\, i_d + E'_q - E''_q\right) = 0 \tag{6.38}$$

$$\frac{dE''_d}{dt} - \frac{1}{T''_{qo}}\left(-(x'_q - x''_q)\, i_q + E'_d - E''_d\right) = 0 \tag{6.39}$$

the MHAM terms are given in Table 6.3. The electrical torque T_e depends on the number of windings present

For IEEE 1.1 Model: $T_e = E'_d i_d + E'_q i_q + (X'_d - X'_q) i_d i_q$. Hence $T_{e,0} = E'_{d,0} i_d + E'_{q,0} i_q + (X'_d - X'_q) i_d i_q$ and $T_{e,1} = E'_{d,1} i_d + E'_{q,1} i_q$

For IEEE 2.2 Model: $T_e = E''_d i_d + E''_q i_q + (X''_d - X''_q) i_d i_q$ Hence $T_{e,0} = E''_{d,0} i_d + E''_{q,0} i_q + (X''_d - X''_q) i_d i_q$ and $T_{e,1} = E''_{d,1} i_d + E''_{q,1} i_q$

6.3.1.1 Single Machine Infinite Bus with IEEE Model 1.1

Consider a SMIB system comprise a single circuit transmission line with X = $j0.2$ p.u. Generator Parameters: $Eb = 1$ p.u., $Ra = 0.000$ p.u., $X_d = 1.93$ p.u., $X_q = 1.77$ p.u., $X'_d = 0.23$ p.u., $X'_q = 0.5$ p.u., $T'_{do} = 5.2$ seconds, $T'_{qo} = 0.81$ seconds, $H = 3.74$ MVA/MJ, $D = 0$. The Generator operating at 0.3 p.u. active power and 0.02 p.u. reactive power. The Generator terminal voltage is $1.0022\angle 3.29488°$. Let us consider the following two cases for simulation:

1. The second circuit is turned on at 1 second, as shown in Figure 6.6.
2. A 3 cycle self-clearing fault occurs at the mid point of the line as shown in Figure 6.7.

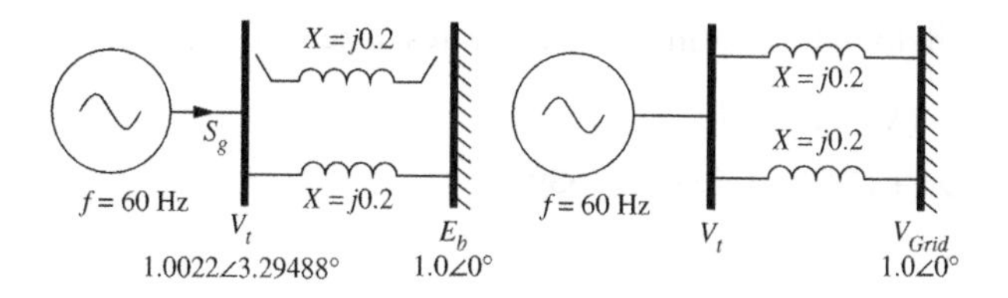

Figure 6.6 The pre- and post-switch on circuits.

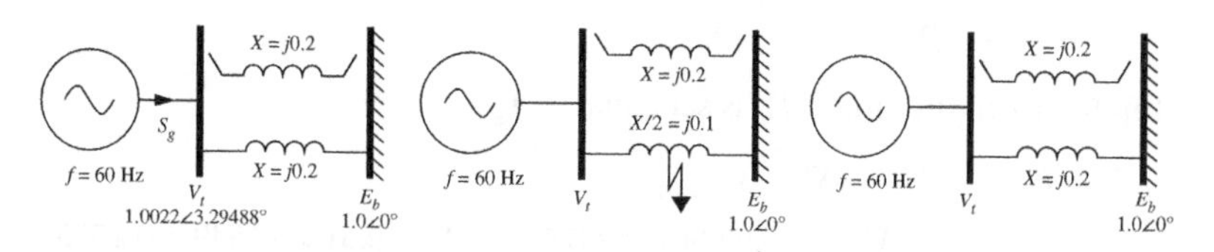

Figure 6.7 The pre-fault, fault, and post-fault SMIB networks.

Both the cases start with the same initial condition, hence the four state variables (IEEE 1.1 Model) δ, S_m, E'_q, and E'_d need to be found. Also the inputs to the generator T_m and E_{fd} need to be computed.

Step 1: The terminal current I_T is given by

$$
\begin{aligned}
I_T &= \left(\frac{P + jQ}{V_t\angle\theta}\right)^* \\
&= \left(\frac{0.3 + j0.02}{1.0022\angle 3.29488^\circ}\right)^* \\
&= 0.3\angle - 0.5192^\circ
\end{aligned}
$$

Step 2: The internal voltage E_{q0} is given by

$$
\begin{aligned}
E_{q0} &= Vt\angle\theta + I_T * (R_a + jX_q) \\
&= 1.0022\angle 3.29488^\circ + 0.3\angle - 0.5192^\circ * (0 + j1.77) \\
&= 1.1650\angle 30.3480^\circ
\end{aligned}
$$

Hence $\delta = 30.3480^\circ$ and $S_m = 0$, indicating the system is running at rated frequency.

$$
i_q + ji_d = e^{-j\delta} * I_t = 0.2575 - j0.1539
$$

Step 3: To compute the other states, E_{fd0} the excitation voltage is to be computed as,

$$
\begin{aligned}
E_{fd0} &= |E_{q0}| - (X_d - X_q) * i_d \\
&= 1.1650 + (1.93 - 1.77) * 0.1539 \\
&= 1.1896
\end{aligned}
$$

Step 4: E'_q and E'_d are found by setting their derivates to zero

$$\begin{aligned} E'_q &= E_{fd0} - (X_d - X'_d) * i_d \\ &= 1.1896 + (1.93 - 0.23) * 0.1539 \\ &= 0.9279 \\ E'_d &= -(X_q - X'_q) * i_q \\ &= -(1.77 - 0.5) * 0.2575 \\ &= -0.3270 \end{aligned}$$

Step 5: The Input torque T_m is set equal to T_e

$$\begin{aligned} T_e &= E'_d i_d + E'_q i_q + (X'_d - X'_q) i_d i_q \\ &= 0.3270 * 0.1539 + 0.9279 * 0.2575 - (0.23 - 0.5) * 0.1539 * 0.2575 \\ &= 0.300 \end{aligned}$$

T_m is equal to the active power since the generator power loss is zero.

The network equations are given by,

$$Y_{BUS} V_{BUS} = I_{BUS}$$

Considering the generator is Bus 1 and Grid is Bus 2, we can consider only the first row of Y_{BUS} relating grid and the generator voltages with Grid (Bus 2) as the reference [1].

$$y_{11} V_T + y_{12} E_b = I_T$$

Rearranging with respect to V_T

$$\begin{aligned} V_T &= \frac{1}{y_{11}} \left(I_T - y_{12} E_b \right) \\ &= h_1 I_T - h_2 E_b \end{aligned}$$

$$where,\ h_1 = \frac{1}{y_{11}} \text{ and } h_2 = \frac{y_{12}}{y_{11}}$$

The generator stator equation in its reference frame is given as

$$\begin{aligned} E'_q + X'_d i_d - R_a i_q &= v_q \\ E'_d - X'_q i_q - R_a i_d &= v_d \end{aligned}$$

where $v_q + j v_d = V_T e^{-j\delta}$.

From the network equations, the expression for the voltages can be derived

$$\begin{aligned} v_q + j v_d &= h_1 I_T e^{-j\delta} - h_2 E_b e^{-j\delta} \\ v_q + j v_d &= h_1 (i_q + j i_d) - E_b h_2 e^{-j\delta} \end{aligned}$$

$$v_q = Real(h_1)i_q - Imag(h_1)i_d - E_b\left(Real(h_2)\cos(\delta) + Imag(h_2)\sin(\delta)\right)$$
$$v_d = Real(h_1)i_d + Imag(h_1)i_q - E_b\left(-Real(h_2)\sin(\delta) + Imag(h_2)\cos(\delta)\right)$$

Solving these equations i_q and i_d can be found for each time step. For both the cases the initial system configuration is same. Hence

$$y_{11} = \frac{1}{X} = \frac{1}{j0.2} = -j5$$
$$y_{12} = -\frac{1}{X} = -\frac{1}{j0.2} = j5$$

Hence $h_1 = \frac{1}{y_{11}} = j0.2$ and $h_2 = \frac{y_{12}}{y_{11}} = \frac{j5}{-j5} = -1$. For each case, the values of h_1 and h_2 are computed as follows:

- **For Case 1**: At 1 second, the line become a double circuit, hence the impedance halves as shown in Figure 6.6.

$$y_{11} = \frac{2}{X} = -j10$$
$$y_{12} = -\frac{2}{X} = j10$$

  Hence $h_1 = \frac{1}{y_{11}} = j0.1$ and $h_2 = \frac{y_{12}}{y_{11}} = \frac{j10}{-j10} = -1$ after the switch on.
- **For Case 2**: At 1 second, the line becomes a shunt due to fault as shown in Figure 6.7. There is no connectivity between both the buses since a zero voltage point is created at the fault location. Hence,

$$y_{11} = \frac{2}{X} = -j10$$
$$y_{12} = 0$$

  Hence $h_1 = \frac{1}{y_{11}} = j0.1$ and $h_2 = \frac{y_{12}}{y_{11}} = 0$ during fault.
  When the fault clears, it returns to the initial value of $h_1 = j0.1$ and $h_2 = -1$.

The time domain solution is computed by evaluating each term of the generator given in Tables 6.1 and 6.2. The results for Cases 1 and 2 are shown in Figures 6.8 and 6.9 respectively. A comparison using Forward Euler method (with half the step time) is also provided. It can be seen in both cases, that the homotopy based solution does follow the reference exactly for $c = -1.0$. But the deviations observed from reference are not significant for the values of $c \neq -1.0$.

6.4 Multimachine Simulations

As mentioned, Transient stability is used to assess the stability of a system post a disturbance. Hence MHAM is used to solve the differential part of the

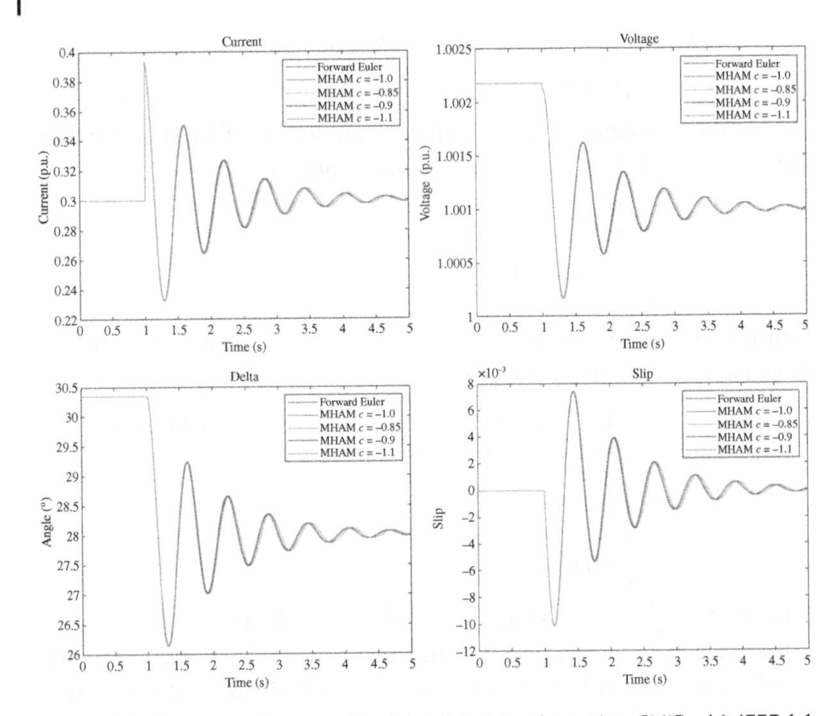

Figure 6.8 Generator Response to a Line Switching Operation, SMIB with IEEE 1.1 model.

DAE system given in (6.32). MHAM is formulated as an explicit method and hence the partitioned method is used, where the algebraic and differential are solved alternatively. For the application of MHAM to such time domain solutions, the steps can be summarized as follows:

1. Read the power system data including control systems, load flow outputs, and simulation parameters.
2. Initialize the machines and its control systems states. For example, Padiyar [1] has outlined initialization procedure for generators and its control blocks.
3. Formulate the network admittance matrix.
4. In every time step, from $t = 0$ to $t = T_{end}$, the flow as shown in Figure 6.10 is performed.
   1. Firstly, check if any disturbance is there. The disturbances can be faults, line outages, etc., which result in network change, or reference input changes such as exciter reference change or power reference change or a combination of both. Network change might require an

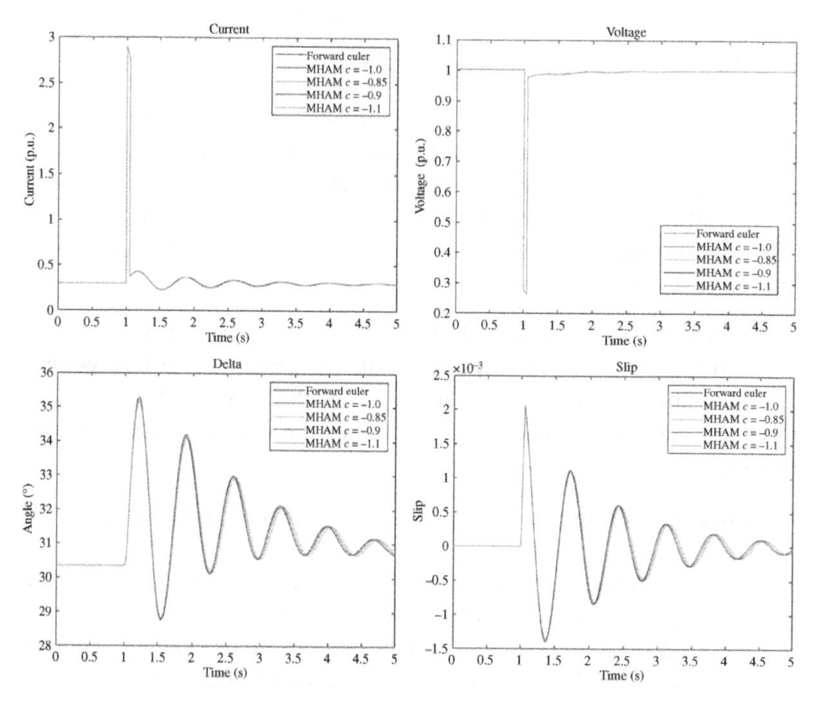

Figure 6.9 Generator Response to a self-cleared Line Fault, SMIB with IEEE 1.1 model.

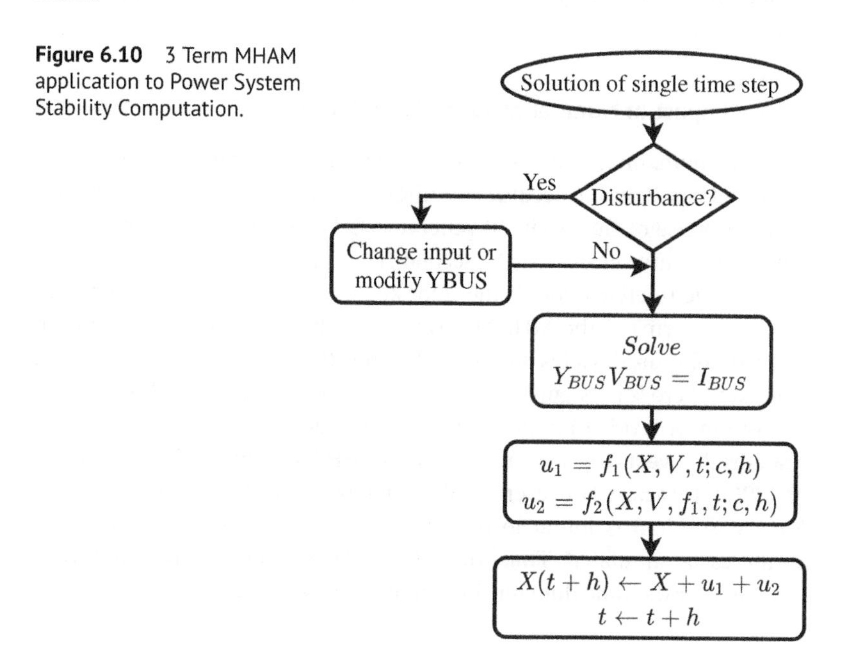

Figure 6.10 3 Term MHAM application to Power System Stability Computation.

update of the Y_{BUS} and its factorization. Reference change requires update to the input vectors appropriately.

2. Secondly, an algebraic solution of network equations is performed. If the injected bus currents can be computed solely based on the states X and the equipment parameters then this solution is a direct inverse. If not, an iterative solver would be used.
3. Lastly, compute the MHAM terms for each generator and its control system. Then the next state can be easily computed as a vector sum.

In [11], MHAM is used to solve the ODEs of 7 widely used test systems ranging from 10 generators 39 bus systems to 4092 generators 13 659 bus systems. The simulations are carried out on MATLAB® R2015a. The following disturbances have been simulated for evaluating the stability and accuracy of MHAM:

1. **Case 1**: 3ϕ close-in, self-cleared faults on all the buses with 4 and 6 cycles fault duration
2. **Case 2**: 3ϕ self-cleared faults on all the branches, fault at 30% distance from one end, with 4 and 6 cycles fault duration
3. **Case 3**: 3ϕ close-in faults on one of the buses of all the branches followed by tripping of the branch with 4 and 6 cycles fault duration
4. **Case 4**: 3ϕ faults on all the branches, fault at 30% distance from one end, followed by tripping of the branch with 4 and 6 cycles fault duration

6.4.1 Impact of Number of Terms Considered

MHAM approximates the solution as an infinite series, which can be truncated based on the accuracy requirements. Figure 6.11 shows the rotor angle response of Gen-10 in 10 generator 39 bus system, for a 3-ϕ, 4 cycle self-cleared fault on Bus-20 at 1 second. The center of inertia (COI [21]) is taken as the reference for all the rotor angles. The results are shown with 2, 3, and 4 terms of the MHAM series. It can be observed that the 2-term MHAM solution deviates significantly from the 3-term MHAM solution. However, there is no significant difference between the 3-term MHAM and the 4-term MHAM solutions. The maximum deviation in the rotor angle response between the 2-term and 3-term MHAM solution is found to be 0.0179° in this case. However, the deviation between the 3-term and 4-term MHAM solution is found to be 0.0024°. Similar errors are observed for all the cases simulated. Thus, the 3-term MHAM solution is found to be sufficient for the time domain simulations of power systems.

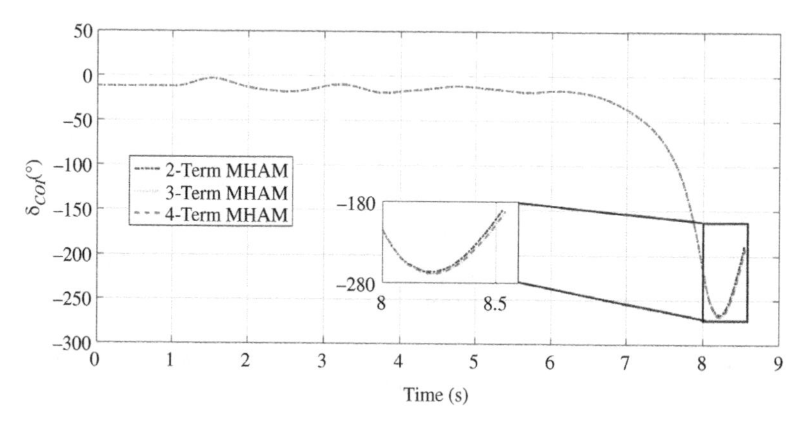

Figure 6.11 Impact of Number of Terms, δ_{COI} of Gen-10, 10 Generator 39 Bus System, 3ϕ Self-Cleared Fault at Bus-20, 4 Cycles. Source: [11] (©[2019] IEEE).

6.4.2 Effect of *c*

The nonzero auxiliary parameter, c, sets MHAM apart from other semi-analytical methods as it allows the construction of a family of solutions. Thus, the convergence region and the rate of solution can be varied depending on c. By plotting the c-curves, that is the curve of state variable w.r.t. c, one can easily see the proper value of c that can be employed. Figure 6.12(a,b) shows the rotor angle, slip speed, field voltage, and mechanical torque responses of Gen-9 for a self cleared $3-\phi$ fault of 4 cycles duration on Bus 5 for the 10 generator 39 bus system. A response with Midpoint Trapezoidal method using 1 ms time step is shown as the reference. The responses of MHAM are plotted with 1 ms time step by varying c from -1.1 to -0.8. It can be observed that as the value of c deviates from -1.0, the curves also start deviating farther from the reference. The solution curve with $c = -1.0$ is seen to be closest to the reference, whereas the curve with $c = -0.8$ is the most deviated one. In MHAM, the errors occur due to truncation of the terms as well as with the change in the value of c. These errors get accumulated over the time steps and the solution starts deviating significantly after the cumulative errors become significant. From Figure 6.12(a,b), one can observe that the cumulative errors become significant after 4 seconds and the deviations from the reference are observable. Figures 6.13 and 6.14 show the maximum rotor angle errors w.r.t. Midpoint Trapezoidal method for all the disturbances described in Cases 1–4 for 10 generator 39 bus system. The errors are obtained for all the cases by varying c between -1.1 and -0.9. Since this system is inherently an unstable system, the errors are calculated at the instant of instability defined by

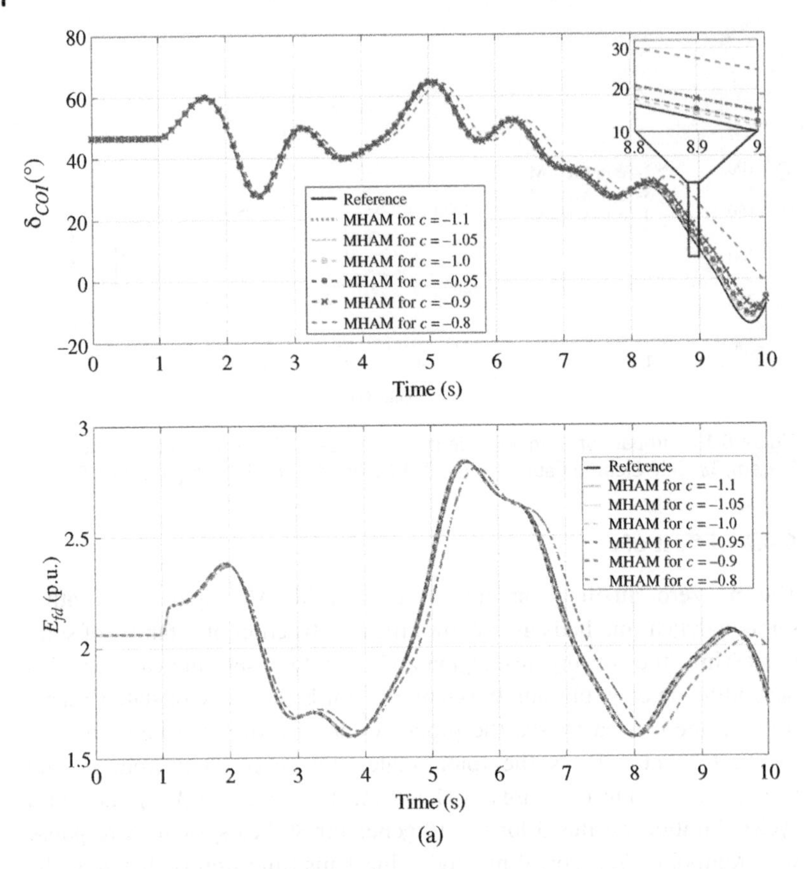

Figure 6.12 (a) δ and E_{fd} response of Generator-9 for Self Clearing 3ϕ Fault, 4 Cycles on Bus 5 for c varying between −1.1 to −0.8, 10 generator 39 bus system. Source: [11] (©[2019] IEEE).

the time instant $minimum\{t[|\delta_{i,COI}| \geq 100°]\}$ where $i = 1,2,3,\ldots,N$ where N = number of generators. It can be observed that the errors obtained with $c = -1.1$ and $c = -0.9$ are very close to each other while the errors obtained with $c = -1.05$ and $c = -0.95$ are almost the same. It can be observed in Figure 6.13 that the majority of errors of MHAM with $c = -1.0$ are within 1° and 0.05 p.u. respectively. Errors between 2°–5° and 0.05–0.2 p.u. respectively are noticed for cases that show instability within 10 seconds of simulation time during the period of loss of synchronism. For all the other state variables similar error patterns are observed. From the above simulations on 10 generator 39 bus system, it is found that the errors are minimum with $c = -1.0$. The errors patterns are symmetric about $c = -1.0$. It is also necessary to test the consistency of the errors obtained with variation of c for different systems. Thus, a 3-ϕ, 4 cycles, self-cleared fault is

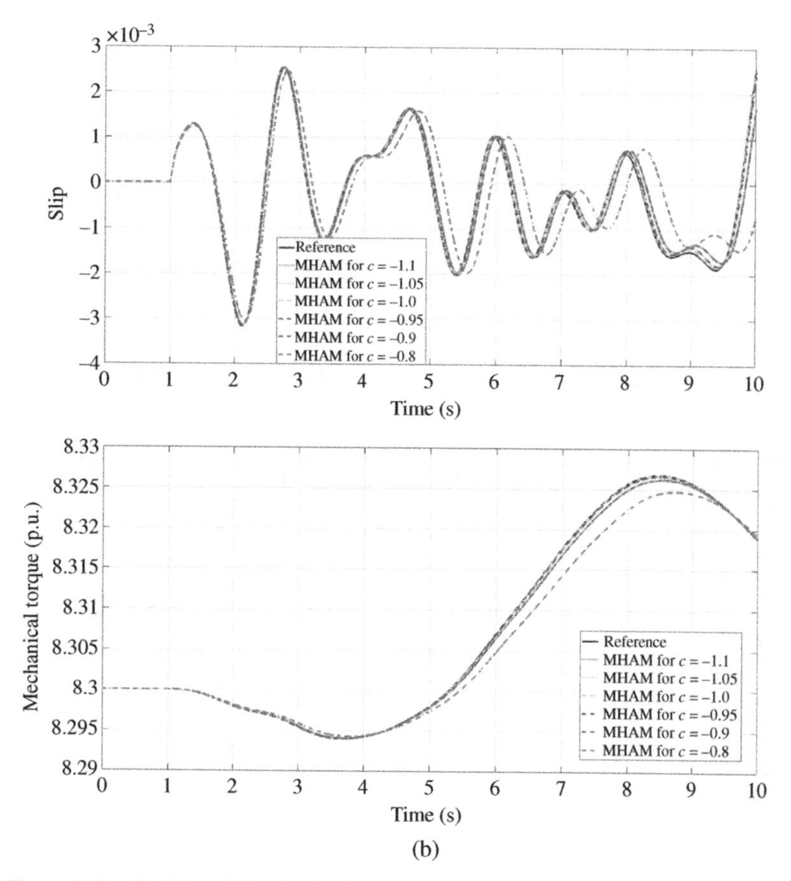

Figure 6.12 (b) Slip and Mechanical Torque response of Generator-9 for Self Clearing 3ϕ Fault, 4 Cycles on Bus 5 for c varying between −1.1 to −0.8, 10 generator 39 bus system. Source: [11] (©[2019] IEEE).

created on Bus-5 for the 10 generator 39 bus, the 19 generator 118 bus, and 948 generator 6468 bus; on Bus-8 for the 70 generator 300 bus; on Bus-10 for the 327 generator 2383 bus; on Bus-6 for the 1445 generator 9421 bus and on Bus-12 for the 4092 generator 13659 bus systems. Figure 6.15 shows the maximum rotor angle (w.r.t. COI) errors obtained in each of these cases for 10 seconds simulation w.r.t. Midpoint Trapezoidal method. The errors are plotted for $c = [-1.1, -1.05, -1.0, -0.95, -0.9]$. Time step, $\Delta t = 1$ ms, is used in these simulations. From the figure, it can be observed that the errors are less than 6° for c between −1.1 to −0.9. The maximum rotor angle errors are obtained for the simulation length of 10 seconds for all the systems. The errors for all systems are minimum when $c = -1.0$. The errors are more or less symmetric about $c = -1.0$. The execution time for 10 seconds simulation with 1 ms time step for change in c values for all the above test

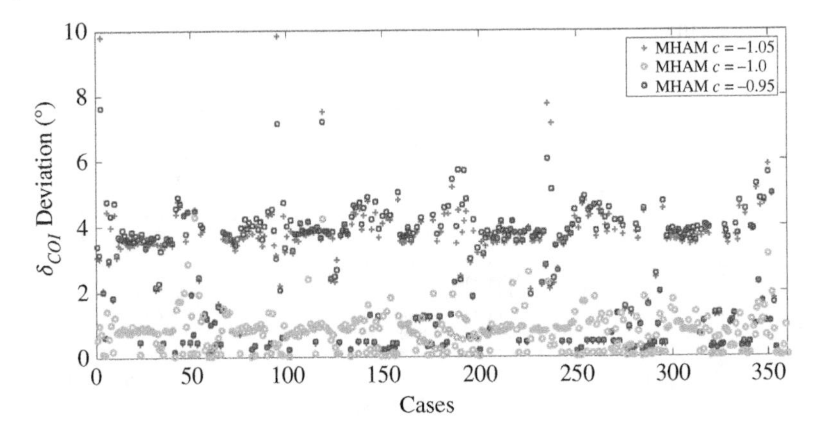

Figure 6.13 Maximum Rotor Angle Error using MHAM w.r.t. Midpoint Trapezoidal Method for all cases with $c = \{-1.05, -1.0, -0.95\}$, 10 generator 39 bus system. Source: [11] (©[2019] IEEE).

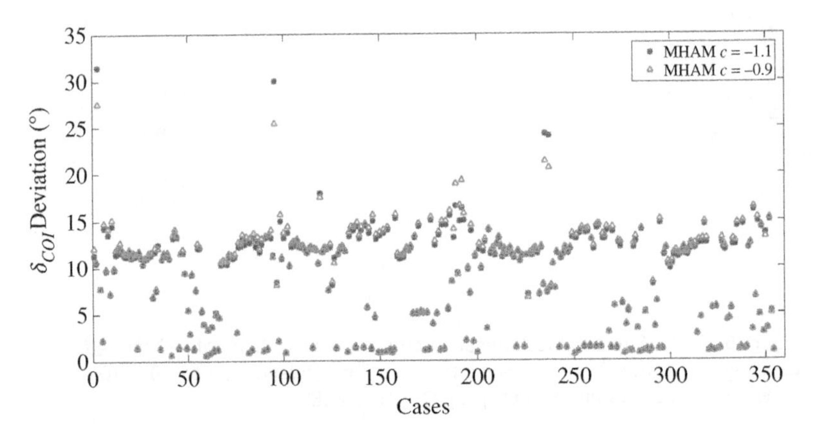

Figure 6.14 Maximum Rotor Angle Error using MHAM w.r.t. Midpoint Trapezoidal Method for all cases with $c = \{-1.1, -0.9\}$, 10 generator 39 bus system. Source: [11] (©[2019] IEEE).

cases are provided in Table 6.4. It can be observed that the execution time of MHAM is faster than ME and Midpoint Trapezoidal. As the system size increases, the speed up is significant with MHAM. Also, it can be observed that the execution time of MHAM is unaffected by the value of c. Figure 6.16 shows the rotor angle w.r.t. COI of Gen-7 for a self cleared $3-\phi$ fault of 4 cycles duration at Bus-5 in 10 generator 39 bus system. The responses are plotted for the maximum time step beyond which the solution blows up due to numerical instability. Solution of Midpoint Trapezoidal method

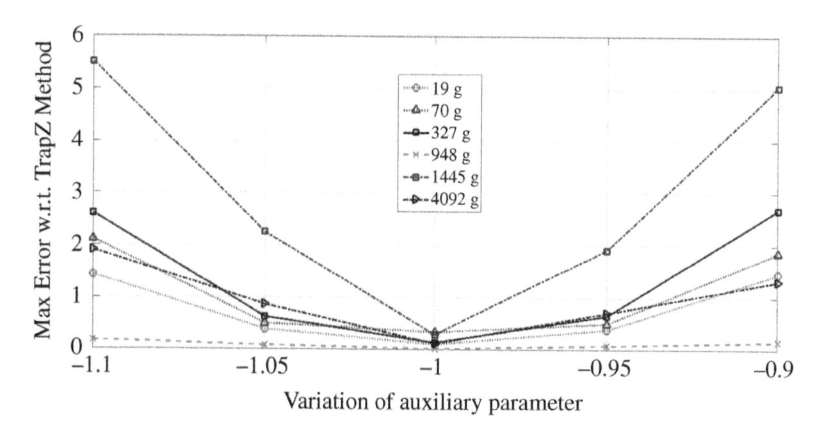

Figure 6.15 Maximum Rotor Angle Error w.r.t. Midpoint Trapezoidal Method for various networks. Source: [11] (©[2019] IEEE).

Table 6.4 Execution time (s) for all test systems [11].

	Execution time (s)						
Method	**10 g**	**19 g**	**70 g**	**327 g**	**948 g**	**1445 g**	**4092 g**
$c = -1.1$	3.16	5.29	10.01	53.90	128.68	230.85	322.83
$c = -1.05$	3.15	5.29	10.01	53.94	129.01	230.60	323.59
$c = -1.0$	3.18	5.30	10.02	53.90	129.06	229.24	322.21
$c = -0.95$	3.18	5.30	10.01	53.89	129.02	229.55	321.14
$c = -0.9$	3.17	5.28	10.01	54.09	129.36	228.61	322.99
ME	3.51	5.74	11.60	55.10	133.30	234.95	332.34
Mid. Trap.	5.13	7.63	13.27	58.09	138.76	239.24	348.67

with time step of 1 ms is taken as the reference. From the figure, it can be observed that there is no appreciable difference between the methods in the first swing. The system becomes unstable after 8.5 seconds. The maximum errors upto 2s and 8.5s in rotor angle are given in Table 6.5. The total execution time for 10s simulation is also reported in Table 6.5. From the Table 6.4, it can be observed that as the c value changes from -1.1 to -0.8 the maximum time step increases and the execution time decreases. Since the errors are not much significant upto 2 seconds MHAM can be used to speed up the transient stability analysis for first swing instability detection. It can be explored as a coarse solver with different values of c as it allows larger time step and faster than ME and Midpoint Trapezoidal.

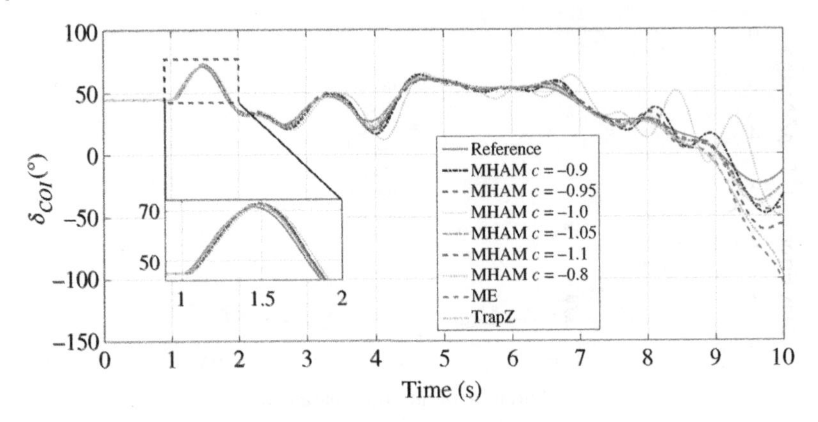

Figure 6.16 δ Response of Gen-7 for the Respective Maximum Time Steps, 3ϕ Self-Cleared Fault at Bus-5, 4 Cycles, 10 Generator 39 Bus System. Source: [11] (©[2019] IEEE).

Table 6.5 Maximum time step, execution time and maximum errors [11].

Method	Maximum Time step (ms)	Execution Time (s)	Max error Till 8.5 s (°)	Max error Till 2 s (°)
c = −1.1	16.4	0.2261	4.1727	0.5872
c = −1.05	18.1	0.2214	3.3992	1.6347
c = −1.0	20.0	0.2031	4.4228	2.0163
c = −0.95	22.1	0.1926	5.3924	3.2768
c = −0.9	24.4	0.1863	5.5912	1.0779
c = −0.8	30.0	0.1653	14.6528	4.8058
ME	20.0	0.2057	6.5071	2.3996
Mid. Trap.	20.1	0.2885	5.8514	1.5151

6.5 Application of Homotopy for Error Estimation

From Sections 6.2–6.4, it is seen that MHAM is

1. Simple to implement albeit it needs some offline effort to compute its components.
2. User can choose how many terms should be employed in the MHAM solution.

3. 3 term MHAM is sufficient for power system simulations. Also, its faster than Modified Euler and higher order methods.
4. MHAM gives least deviations when the auxiliary term $c = -1.0$ and converges to Multi-stage Homotopy Perturbation Method (MHPM) and MADM.
5. An issue was the deviation with respect to conventional solver, it grows as simulation duration increases.

Therefore, if MHAM is made to work with a smaller simulation duration then the deviation issue would not occur. In [11], a suggestion to use MHAM in Parareal Coarse solver is made, since the Coarse solver is used to provide initial conditions and corrections in the Parareal method. Another utilization can be to use MHAM as an in-between entity for a conventional ODE solver. One such application is adaptive adjustment of simulation step size, h.

6.5.1 MHAM-Assisted Adaptive Step Size Adjustment for Modified Euler Method

The Modified Euler method contains two numerical values for x_{n+1} as seen in (6.19). The first estimate is got by a prediction step, the second is the final value from the correction step of ME method. Hence there exists two sources of computing error. These errors cause numerical instabilities especially when discontinuities such as faults are involved [2]. Lets consider the initial value of x_n as a true solution of (6.14) denoted by $\hat{x}_n$, Hence, an error called Local Truncation Error (LTE) can be defined as the error in the solution when starting from an exact solution [20, 22]. Hence considering that x^P_{n+1} and x^C_{n+1} are the numerical solution of both the steps

$$LTE^{(1)} = \hat{x}_{n+1} - x^P_{n+1} \tag{6.40}$$

$$LTE^{(2)} = \hat{x}_{n+1} - x^C_{n+1} \tag{6.41}$$

Subtracting (6.41) from (6.40),

$$LTE^{(1)} - LTE^{(2)} = | - x^P_{n+1} + x^C_{n+1}|$$

neglecting $LTE^{(2)}$ since its of a higher order method [20]

$$LTE = \frac{h}{2}|f(x_n, t) - f(x^P_{n+1}, t + h)| \tag{6.42}$$

Since the exact solution $\hat{x}_n$ is not known, this estimate is that of the local error but LTE is considered to be the same as the local error [20, 22]. Given an estimate of the LTE, the new step size can be computed [20, 22] which increases the simulation accuracy

$$h_{new} = h_{old}\sqrt{\frac{k\tau}{LTE}} \tag{6.43}$$

where, τ is the required LTE and k is safety factor [20]. Hence at the end of a ME step, the LTE can be estimated and a decision to redo the step or go forward with a new step as per (6.43) can be decided, which results in adaptive ME algorithm. This step size adaptation increases the numerical stability of the simulations during discontinuities. The flowchart for single time step of an adaptive ME method in partitioned transient stability simulation is given in Figure 6.17. In the figure one can observe that two network solutions are needed, one at the start and other at the end of computation of x^P in side the LTE estimation loop. Under the condition that LTE is sufficient there are two function evaluations of both the network solution and $f(\cdot)$. But for each time LTE is beyond the limit, one set of network and $f(\cdot)$ need to be repeated, this increases the computation cost.

6.5.2 Non-iterative Adaptive Step Size Adjustment

The main disadvantage with the conventional approach of Figure 6.17 is that the LTE is computed at the end of the integration step. Hence if the LTE, which is an approximate quantity, can be computed before x^P then number of function evaluations of network solution and $f(\cdot)$ can be kept at two. However, the estimation of LTE should not be expensive.

As seen in Section 6.2.2, a 3-term MHAM has the same formulation and error as ME method, and [4, 11] have shown that semi-analytical techniques are good enough for power system dynamic computation for 10seconds windows. Here the duration of interest is between h_{min} *to* h_{max}, which would be much smaller than a overall simulation window, hence the deviation from the actual solution should be minimal. As shown in Figure 6.10, there are two function evaluations in MHAM. Hence the possibility of easily computing x^P_{n+1} from MHAM output, so that LTE estimation can aid ME method, is investigated.

Consider $x^M_{n+1} = u_0 + u_1 + u_2$ as per (6.24), the computation of u_1 for (6.23) using steps of Section 6.2.2 are as follows

$$\begin{aligned}\mathcal{L}[u_1] &= c\mathcal{R}_1\\ &= c\left(\dot{u}_0 + \dot{u}_1 q + \dot{u}_2 q^2 - \frac{\partial^0 f(u_0 + u_1 q + u_2 q^2, t)}{\partial q^0}\right)\Bigg|_{q=0}\\ &= c\left(\dot{u}_0 + \dot{u}_1 q + \dot{u}_2 q^2 - f(u_0 + u_1 q + u_2 q^2, t)\right)\Big|_{q=0}\end{aligned}$$

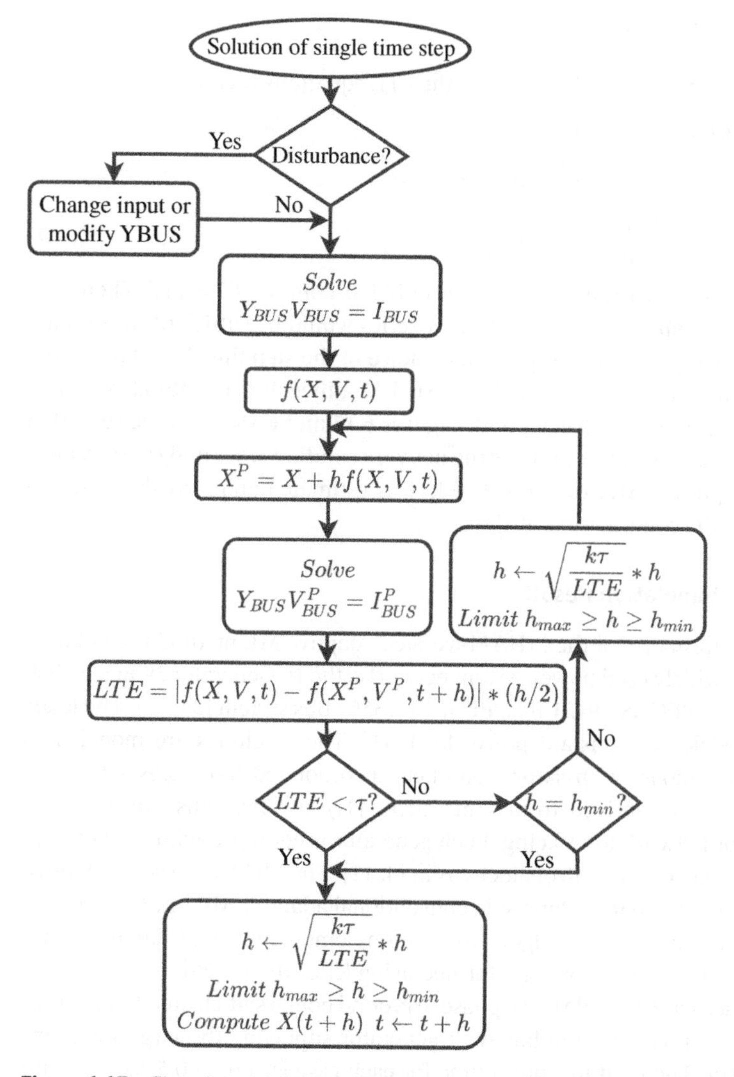

Figure 6.17 Flowchart of Adaptive Modified Euler. Source: [23] (©[2022] IEEE).

$$\begin{aligned} &= c\left(\dot{u}_0 - f(u_0, t)\right) \\ &= -c\left(f(u_0, t)\right) \end{aligned} \tag{6.44}$$

Integrating (6.44) and using $u_0 = x_n$

$$\begin{aligned} u_1 &= -chf(x_n, t) \\ f(x_n, t) &= \frac{-u_1}{ch} \end{aligned} \tag{6.45}$$

$$x_{n+1}^{P} = u_0 + \frac{-u_1}{c} \tag{6.46}$$

Hence using x_{n+1}^{M} in place of x_{n+1}^{C}, the LTE equations become

$$\begin{aligned} LTE &= |x_{n+1}^{P} - x_{n+1}^{M}| \\ &= |u_0 + \frac{-u_1}{c} - u_0 - u_1 - u_2| \\ &= \left|\left(\frac{1}{c} + 1\right) u_1 + u_2\right| \end{aligned} \tag{6.47}$$

Setting $c = -1.0$ the estimated value of LTE becomes, $LTE = |u_2|$. Therefore, a single computation of the MHAM series would be sufficient to compute LTE and x_{n+1}^{P} after the appropriate choice of the step time based on (6.43). The flowchart for the MHAM-assisted Modified Euler method is shown in Figure 6.18. Comparing both Figures 6.17 and 6.18, it can be seen that MHAM-assisted ME evaluates the network solution twice and $f(\cdot)$ only once apart from the MHAM step. The LTE is estimated and step size adjusted prior to the computation of x_{n+1}^{P}.

6.5.3 Simulation Results

The performance of the MHAM-assisted adaptive ME method is compared using 4 widely used power system networks, the 10 Generator 39 Bus, Polish 2383 Bus, PEGASE 9421 Bus, PEGASE 13569 Bus system [24, 25]. The loads are modeled as constant power loads [1]. The machines are modeled as IEEE 2.2 model with saturation effect incorporated (both axis saturation constants are derived from same curve [1]) and fictitious impedance is used for network interfacing. Each generator is equipped with an AVR and turbine–governor control blocks as in [4, 11]. The MHAM-assisted adaptive method is compared along side conventional adaptive ME implementation and a normal ME. The algorithms are implemented using C language and the network solution was performed using KLU library [26].

In each case, a single three phase to ground bolted self-cleared fault of four cycles is applied on the bus with generator supplying the largest amount of power. The fault inception time for each case is kept at 0.5 seconds and simulation is run for 5 seconds, the min and max step size for each simulation is kept at 1 and 100 ms, respectively. The conventional Modified Euler uses a constant 1 ms step time. Three variations of desired LTE value $k\tau = 0.1, 0.01, 0.001$ *where* $k = 0.8$ are used for each network.

6.5.4 Tracking of LTE

The first validation is to check the tracking capability of the LTE by the MHAM method. For this purpose, the estimated LTE by the MHAM

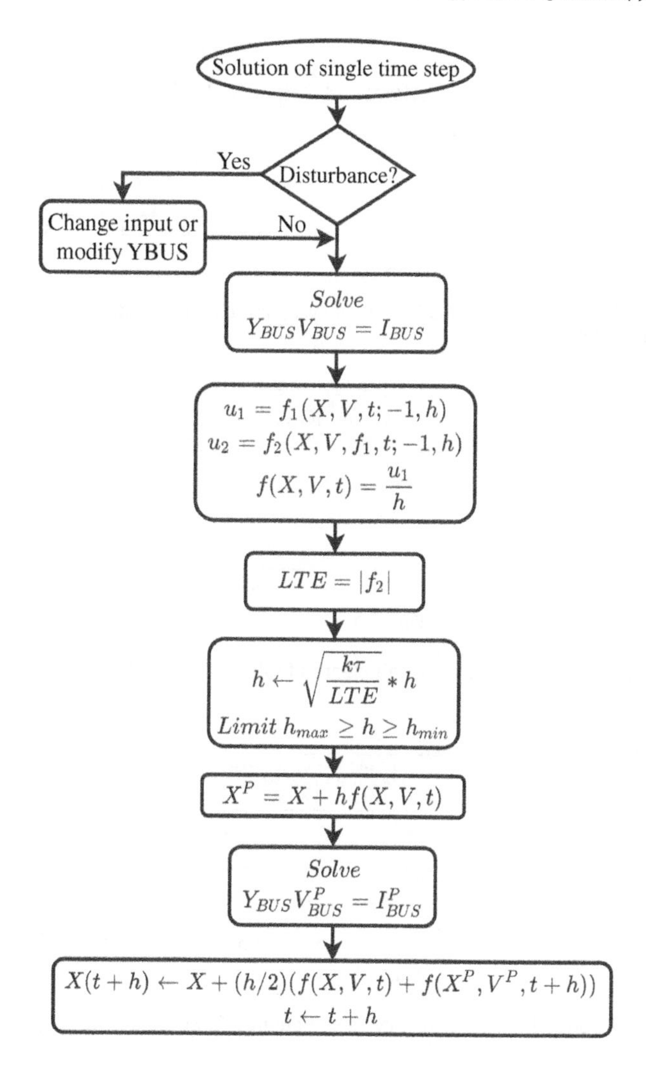

Figure 6.18 Flowchart of Adaptive Modified Euler with MHAM. Source: [23] (©[2022] IEEE).

formulation and corrected LTE after the ME integration are compared. The Figures 6.19–6.22 show LTE for each case. In all the cases, the corrected value of LTE follows the desired LTE value post the disturbance removal. During the time of fault and its removal, the LTE estimated is very high and the corrected step size reduces to the minimum step time specified. While the step size is floored to the lower limit, reduction of LTE is not possible. From the time 0–0.5 seconds (fault inception), the LTE is 0 since there is no

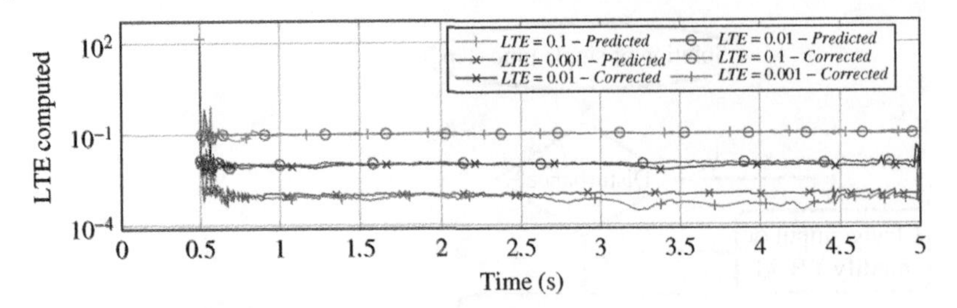

Figure 6.19 LTE of 10 Generator 39 Bus System, 4 cycle self cleared fault. Source: [23] (©[2022] IEEE).

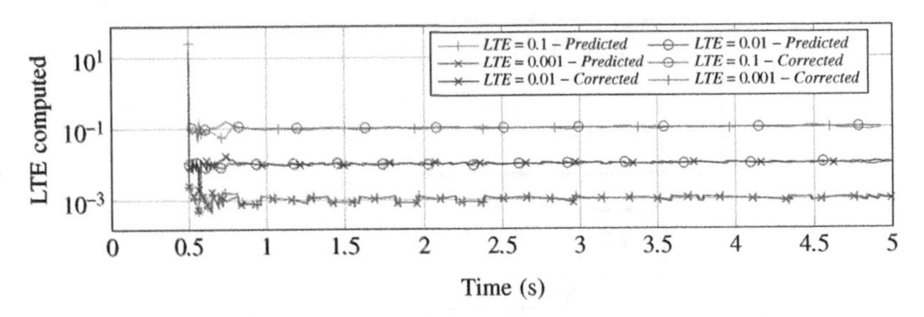

Figure 6.20 LTE of Polish System, 4 cycle self cleared fault. Source: [23] (©[2022] IEEE).

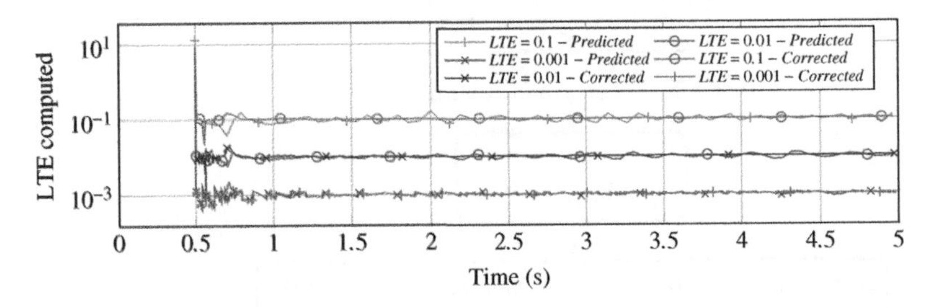

Figure 6.21 LTE of PEGASE9241 System, 4 cycle self cleared fault. Source: [23] (©[2022] IEEE).

change in the states, hence not represented in the graphs. Therefore, the MHAM method is able to track and maintain the desired level of LTE while adapting the step size.

6.5.5 Accuracy with Variation of Desired LTE

The desired value of LTE allows the users to dictate the allowable accuracy of the solution. It can be observed in Figure 6.23 for the 10 generator 39 bus system, that even under instability occurring when the δ_{COI} exceeds 100°

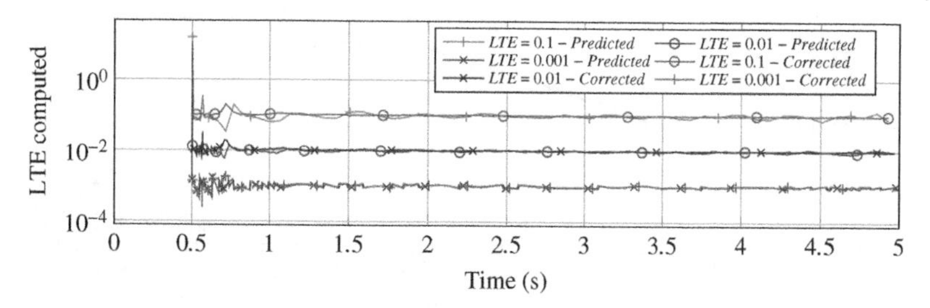

Figure 6.22 LTE of PEGASE13659 System, 4 cycle self cleared fault. Source: [23] (©[2022] IEEE).

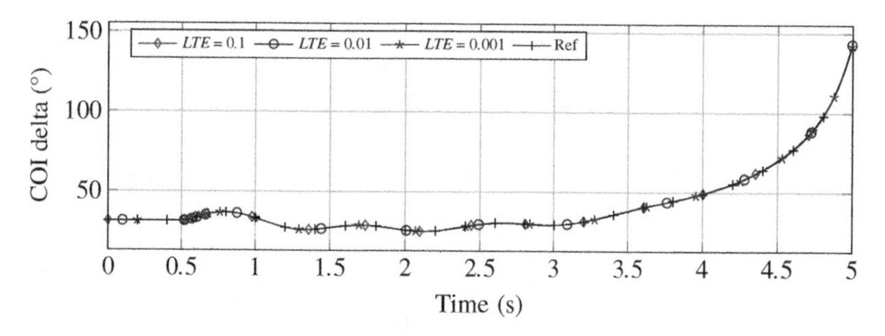

Figure 6.23 δ_{COI} of Generator-1, 10 Generator 39 Bus System, 4 cycles self cleared fault at Bus 2. Source: [23] (©[2022] IEEE).

[11] the method is able to track the solution. The impact of LTE selection is not easily seen since the system is unstable. However for the larger networks shown in Figures 6.24–6.26 show stable power swing for the duration of simulation. It can be seen in the case of the larger systems, the solution obtained at different LTE deviates from the reference solution given by the constant step ME solution. It is evident in the Polish system, Figure 6.24,

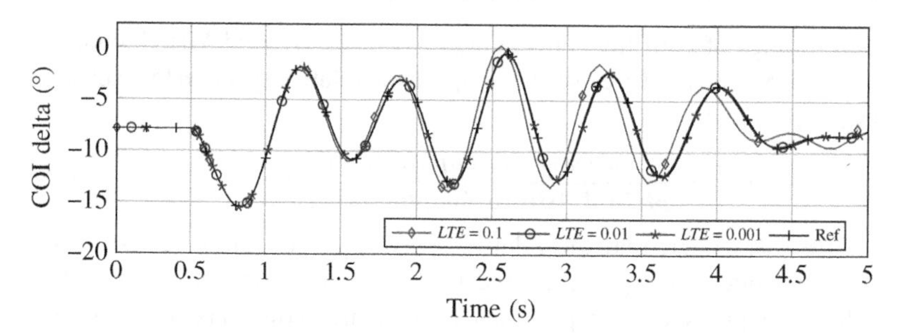

Figure 6.24 δ_{COI} of Generator-10, Polish System, 4 cycles self-cleared fault at Bus 18. Source: [23] (©[2022] IEEE).

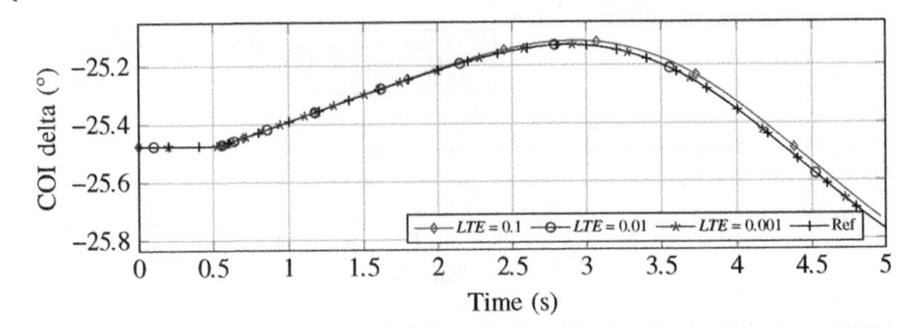

Figure 6.25 δ_{COI} of Generator-2, PEGASE9241 System, 4 cycles self-cleared fault at Bus 5490. Source: [23] (©[2022] IEEE).

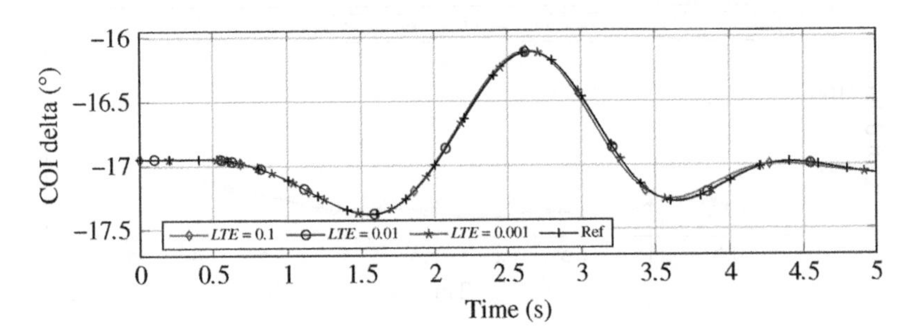

Figure 6.26 δ_{COI} of Generator-1, PEGASE13659 System, 4 cycles self-cleared fault at Bus 2067. Source: [23] (©[2022] IEEE).

that lower value of desired LTE ($k\tau = 0.001$) tracks the reference solution well whereas the highest LTE value solution deviates after the first power swing. Hence one can select the required LTE based on the simulation time and the acceptable deviation from the actual solution.

6.5.6 Computational Time and Speedup

The timings of both the conventional Adaptive ME and MHAM-assisted adaptive ME method are used to compute the relative speedup for each case as given by (6.48)

$$Speedup = \frac{Conventional\ Adaptive\ ME\ Time}{MHAM\ Assisted\ Adaptive\ ME\ Time} \tag{6.48}$$

Figure 6.27 shows that the relative speedup for all the cases is greater than 1. Under the highest desired LTE value, the conventional adaptive solution requires fewer step size corrections due to the relaxed error limits. At lower LTE values these step size corrections increase as shown in the

LTE adjustment loop in Figure 6.17. For each step size adjustment, the conventional adaptive solution requires a network solution, which is an expensive step compared to the state computation [4]. Hence the MHAM based method that can adjust the step size prior to the network solution will perform it only once per compute stage. It can be seen in Figure 6.27, at lower LTE values the MHAM method outperforms the conventional method by over 20%.

The execution time of the cases are provided in Table 6.6. It can be observed that the adaptive methods are indeed faster than the constant step ME by at least three times. The execution time of the PEGASE9241 for a 5 seconds simulation using the MHAM method is 5.70 seconds, but with the conventional ME its 7.15 seconds. If LTE is selected as 0.01 seconds the simulation time is 3.99 seconds with the MHAM method and hence faster-than-real-time simulation could be achieved with the MHAM method.

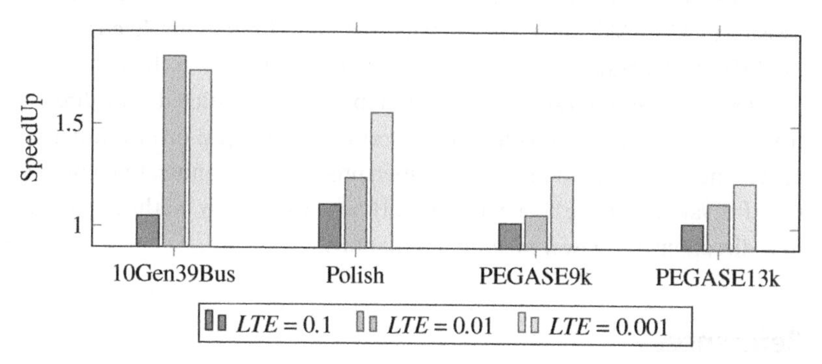

Figure 6.27 Relative Speedup of MHAM-assisted Adaptive ME over Adaptive ME. Source: [23] (©[2022] IEEE).

Table 6.6 Execution time (s) for each network [23].

Method	MHAM assisted adaptive ME			ME	Adaptive ME		
Case	LTE = 0.1	LTE = 0.01	LTE = 0.001		LTE = 0.1	LTE = 0.01	LTE = 0.001
10Gen39Bus	0.01	0.01	0.03	0.16	0.01	0.03	0.05
Polish	0.43	0.86	1.52	7.50	0.48	1.06	2.38
PEGASE9241	3.38	3.99	5.70	26.14	3.44	4.24	7.15
PEGASE13659	6.07	7.46	11.65	50.17	6.20	8.35	14.16

6.6 Summary

The application of Homotopy methods for solution of differential equations and power system time domain simulation is demonstrated. An approach for developing approximate semi-analytical models using MHAM has been presented for a linear second order system and synchronous machine models of varying complexities. A family of solutions has been created using MHAM and the most accurate solution has been found with $c = -1.0$. The impact of number of terms to be taken in the MHAM solution is investigated. 3-terms are found to be sufficient for stability simulations. An exhaustive study on the effect of variation of the auxiliary parameter, c, has been carried out. It has been shown that for the range of $c = [-1.05, -0.95]$, the error is within $4°$–$5°$ for a wide range of operating conditions in all the test systems considered. An alternate application of Homotopy method as an Error Estimator is presented. An adaptive Modified Euler method assisted by MHAM is shown for transient stability simulations. Even though the MHAM formulation requires a new set of equations to be derived, it enables easy computation of LTE, new step size and predicted state directly from the components of its individual terms. The MHAM error estimation makes the step adaptation a non iterative process and significant computational advantages have been witnessed. Hence, Homotopy methods provide new directions for power systems simulations.

References

1 Padiyar, K.R. (2002). *Power System Stability and Control*. Hyderabad: BS Publications.

2 Stott, B. (1979). Power system dynamic response calculations. *Proceedings of the IEEE* 67 (2): 219241.

3 Liao, S. (2004). *Beyond Perturbation Introduction to the Homotopy Analysis Method*. Chapman & Hall, USA.

4 Gurrala, G., Dinesha, D.L., Dimitrovski, A. et al. (2017). Large multi-machine power system simulations using multi-stage adomian decomposition. *IEEE Transactions on Power Systems* 32 (5): 35943606. https://doi.org/10.1109/TPWRS.2017.2655300.

5 Liao, S. (2009). Notes on the homotopy analysis method: some definitions and theorems. *Communications in Nonlinear Science and Numerical Simulation* 14 (4): 983–997.

6 Guo, S.X. and Salam, F.M.A. (1992). The real homotopy-based method for computing solutions of electric power systems. *IEEE International Symposium on Circuits and Systems.*

7 Mehta, D., Nguyen, H.D., and Turitsyn, K. (2016). Numerical polynomial homotopy continuation method to locate all the power flow solutions. *IET Generation, Transmission and Distribution* 10 (12): 2972–2980.

8 Chandra, S., Mehta, D., and Chakrabortty, A. (2015). Equilibria analysis of power systems using a numerical homotopy method. *IEEE PES General Meeting.*

9 Mehta, D., Molzahn, D.K., and Turitsyn, K. (2016). Recent advances in computational methods for the power flow equations. *2016 American Control Conference (ACC).*

10 Wang, S., Du, P., and Zhou, N. (2014). Power system transient stability analysis through a homotopy analysis method. *Nonlinear Dynamics* 76 (2): 10791086. https://doi.org/10.1007/s11071-013-1191-2.

11 Dinesha, D.L. and Gurrala, G. (2019). Application of multi-stage homotopy analysis method for power system dynamic simulations. *IEEE Transactions on Power Systems* 34 (3): 22512260. https://doi.org/10.1109/TPWRS.2018.2880605.

12 Park, B., Sun, K., Dimitrovski, A. et al. (2021). Examination of semi-analytical solution methods in the coarse operator of parareal algorithm for power system simulation. *IEEE Transactions on Power Systems* 36 (6): 50685080. https://doi.org/10.1109/TPWRS.2021.3069136.

13 Liao, S. (2004). *Beyond Perturbation Introduction to the Homotopoy Analysis Method.* Chapman & Hall, USA.

14 Liao, S.-J. (1992). The proposed homotopy analysis technique for the solution of nonlinear problems. PhD thesis. Shanghai Jiao Tong University.

15 He, J.-H. (1998). Homotopy perturbation technique. *Computer Methods in Applied Mechanics and Engineering* 178 (3–4): 257–262.

16 Li, J.-L. (2009). Adomian's decomposition method and homotopy perturbation method in solving nonlinear equations. *Journal of Computational and Applied Mathematics* 228 (1): 168–173.

17 He, J.H. (2004). Comparison of homotopy perturbation method and homotopy analysis method. *Applied Mathematics and Computation* 156 (2): 527–539.

18 Liao, S. (2005). Comparison between the homotopy analysis method and homotopy perturbation method. *Applied Mathematics and Computation* 169 (2): 1186–1194.

19 Liang, S. and Jeffrey, D.J. (2009). Comparison of homotopy analysis method and homotopy perturbation method through an evolution

equation. *Communications in Nonlinear Science and Numerical Simulation* 14 (12): 4057–4064.

20 Patil, M.B., Ramanarayanan, V., and Ranganathan, V.T. (2009). *Simulation of Power Electronic Circuits*. New Delhi: Narosa.

21 Padiyar, K.R. (2002). *Power System Dynamics Stability and Control*. B.S.Publications.

22 Shampine, L.F. (1994). *Numerical Solution of Ordinary Differential Equations*. Taylor and Francis.

23 Joseph, F.C. and Gurrala, G. (2022). Homotopy based error estimator for adaptive Modified Euler in transient stability simulations. In *ICECET*, 16.

24 Zimmerman, R.D. and Murillo-Sanchez, C.E. MATPOWER. https://matpower.org (accessed 18 May 2023).

25 Fliscounakis, S., Panciatici, P., Capitanescu, F., and Wehenkel, L. (2013). Contingency ranking with respect to overloads in very large power systems taking into account uncertainty, preventive, and corrective actions. *IEEE Transactions on Power Systems* 28 (4): 49094917. https://doi.org/10.1109/TPWRS.2013.2251015.

26 Davis, T.A. and Palamadai Natarajan, E. (2010). Algorithm 907: KLU, a direct sparse solver for circuit simulation problems. *ACM Transactions on Mathematical Software* 37 (3): 1–17. https://doi.org/10.1145/1824801.1824814.

7

Utilizing Semi-Analytical Methods in Parallel-in-Time Power System Simulations

Byungkwon Park

School of Electrical Engineering, Soongsil University, Dongjak-Gu, Seoul, South Korea

7.1 Introduction to the Parallel-in-Time (Parareal Algorithm) Simulation

In order to improve the computational performance of solving large-scale nonlinear differential algebraic equations, parallelization techniques can serve as powerful tools. Among many, the "Parallel-in-Time (Parareal)" algorithm, first introduced [1], has demonstrated its potential to significantly improve the computational performance of time-domain simulation by reducing the wall-clock time of the simulations and thus ultimately achieve faster than real-time (hence algorithm's name) simulations. The Parareal algorithm has enjoyed the wide-spread adoption due to its applicability and flexibility for a wide range of problems (e.g. [2, 3]). In particular, the Parareal algorithm has been applied in recent years to the research area of power system dynamic simulations [4–7].

7.1.1 Overview of Parareal Algorithm

The main idea of the Parareal algorithm is to decompose the time interval into smaller subintervals. Then, the evolution of these subintervals is carried out in parallel and converges to a true solution after a number of iterations between a coarse approximate solution and a fine true solution over the whole simulation period. In a simplified way, the Parareal algorithm computes the numerical solution for multiple subintervals simultaneously.

To solve multiple subintervals in parallel, the Parareal algorithm requires initial states for all subintervals, which are provided by the sequential in time, coarse operator (less accurate but computationally cheap

Power System Simulation Using Semi-Analytical Methods, First Edition. Edited by Kai Sun.

Published 2024 by John Wiley & Sons, Inc.

numerical integration method). Then, the fine operator (more accurate but computationally expensive numerical integration method) is implemented to find the true trajectory and correct the evolution of each independent subinterval in parallel. The Parareal algorithm iterates this process until all coarse solution converges to the true solution. As an example, consider an initial value problem:

$$\dot{x} = f(x, t) \tag{7.1}$$

$$x(t_0) = x_0 \quad \text{with } t_0 \le t \le T \tag{7.2}$$

where x represents the vector of state variables with the initial condition x_0.

In this original form, the Parareal algorithm divides the time interval $[t_0, T]$ into N subintervals $[t_j, t_{j+1}]$ such that $[t_0, T] = [t_0, t_1] \cup [t_1, t_2] \cup \ldots \cup [t_{N-1}, t_N]$. For simplicity, assume that the size of $t_{n+1} - t_n = \Delta T_n$ is equivalent to each other for all $0 \le n < N$ (i.e., $\Delta T = \Delta T_n$). We denote x^F and x^C as the system states obtained from the fine operator and the coarse operator, respectively. The corrected coarse solution is denoted as x^*. The following steps roughly summarize the Parareal implementation:

1. Design the coarse ($C_{\Delta t}$) and fine ($F_{\delta t}$) operator. In general, the coarse operator uses a larger time step (Δt), and the fine operator uses a smaller time step (δt). These operators obtain the solution to (7.1) at time t_n with an initial state x_{n-1} at time t_{n-1}

$$Fine : x_n^F = F_{\delta t}(t_{n-1}, x_{n-1}) \tag{7.3}$$

$$Coarse : x_n^C = C_{\Delta t}(t_{n-1}, x_{n-1}) \tag{7.4}$$

2. An initial coarse solution using the coarse operator ($C_{\Delta t}$) is generated in serial as:

$$x_n^{*,0} = x_n^{C,0} = C_{\Delta t}(x_{n-1}^{*,0}), \quad n = [1, \ldots, N] \tag{7.5}$$

$$\text{Set} \quad x_0^{*,1} = x_0^{*,0} \tag{7.6}$$

where the superscript denotes the iteration count and $x_0^{*,0}$ is the given initial point at t_0. Its graphical description is shown below.

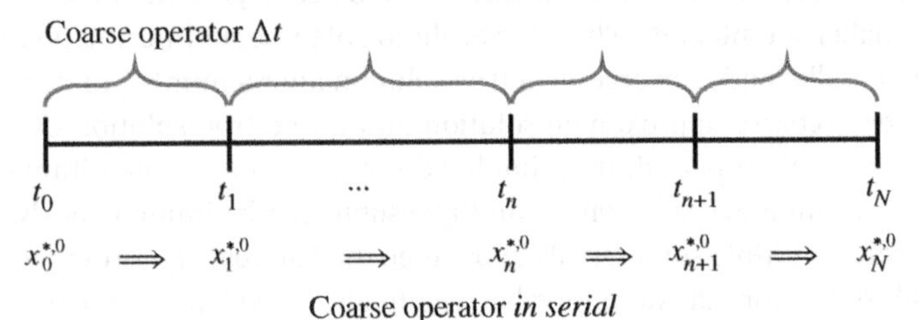

3. Iterations start $k = 1$. Using the initial condition obtained from the coarse operator for each subinterval, propagate the fine solution in parallel over each independent subinterval using the fine operator

$$x_n^{F,k} = F_{\delta t}\left(x_{n-1}^{*,k-1}\right), \quad n = [1, \dots, N] \tag{7.7}$$

where $x_n^{F,k}$ denotes the solution at t_n. Its graphical description is shown below.

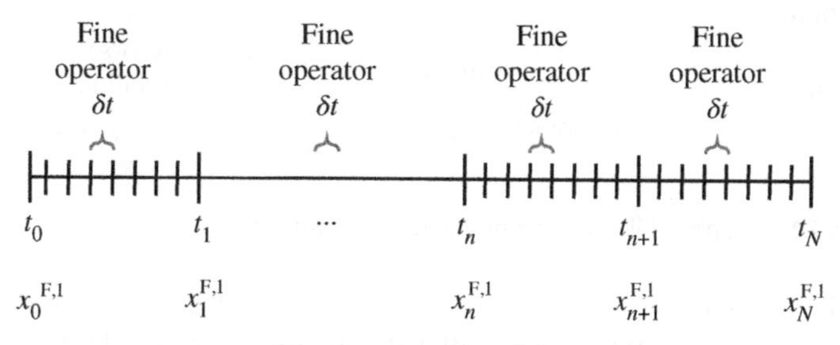

Fine operator *in parallel*

4. Using the fine solution obtained from the fine operator, update the coarse solution in serial

for $\quad n = k : N$

$$x_n^{C,k} = C_{\Delta t}(x_{n-1}^{*,k}) \tag{7.8}$$
$$x_n^{*,k} = x_n^{C,k} + x_n^{F,k} - x_n^{C,k-1} \tag{7.9}$$

end

Its graphical description is shown below.

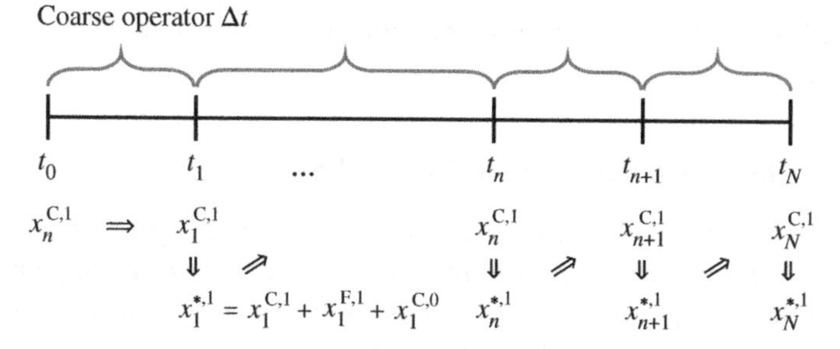

Coarse solution update *in serial*

5. Go to Step 3 and update the coarse solution iteratively until $x_n^{*,k} - x_n^{*,k-1} \leq tol$ for $n = [1, \dots, N]$.

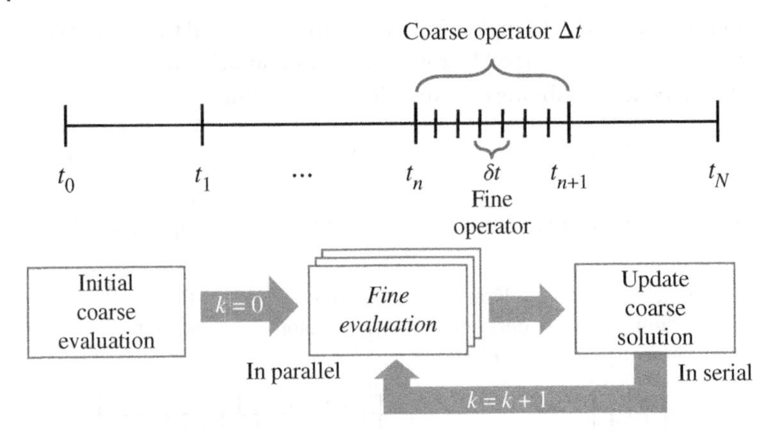

Figure 7.1 Graphical illustration of the Parareal algorithm.

To better understand the behavior of the Parareal algorithm, consider the updated coarse solution at $n = 1$ after the first iteration ($k = 1$). Then, one has $x_1^{*,1} = x_1^{C,1} + x_1^{F,1} - x_1^{C,0}$. Note that this updated coarse solution at t_1 is corrected to the fine solution as $x_1^{*,1} = x_1^{F,1}$ since $x_1^{C,1} = x_1^{C,0}$. This suggests that all the coarse solutions should theoretically be corrected to the fine solution (i.e., true solution) in N iterations, which is equivalent to implement the fine operator sequentially until it covers all N subintervals. Thus, one can see the speedup only when the Parareal algorithm converges to the true solution in kth iteration, which is less than N. This means that the theoretical speedup of the Parareal algorithm is $\frac{N}{k}$ under the assumption that the solving time for the coarse operator and other factors (e.g. communication time) related to the parallelization are negligible. In addition, note that Step 4 updates the coarse solution at t_n from k to N since the updated coarse solutions $x_1^*, x_2^*, \dots, x_{k-1}^*$ have been corrected to the true solutions in k iterations. The graphical illustration of the Parareal algorithm is also represented in Figure 7.1.

7.1.2 The Derivation of Parareal Algorithm

An alternative perspective to understand the Parareal algorithm is based on the observation that the Parareal algorithm is indeed the approximation of the multiple-shooting method using the Newton–Raphson method. Consider the following nonlinear problem:

$$f(X) = \begin{cases} X_0 - x^0 = 0 \\ X_1 - F(X_0) = 0 \\ \vdots \\ \vdots \\ X_N - F(X_{N-1}) = 0 \end{cases} = \begin{cases} f_0(X) = 0 \\ f_1(X) = 0 \\ \vdots \\ \vdots \\ f_N(X) = 0 \end{cases} \tag{7.10}$$

where $F()$ is the true solution with the initial condition X_0; variables are $X = [X_0, \dots, X_N]$. Notice that the subscript denotes the time interval. Then, the update rule can be constructed as:

$$f(X^{k+1}) = f(X^k) + f'(X^k)(X^{k+1} - X^k) + h(X^{k+1} - X^k) \tag{7.11}$$

Notice that each function f is a function of only two variables. Ignoring the higher-order term h, the derivative can be obtained as:

$$f'(X^k) = \begin{bmatrix} 1 & 0 & \cdots & 0 \\ -\frac{\partial F(X_0^k)}{\partial X_0^k} & 1 & \cdots & 0 \\ \vdots & \ddots & \ddots & 0 \\ 0 & \cdots & -\frac{\partial F(X_{N-1}^k)}{\partial X_{N-1}^k} & 1 \end{bmatrix} \tag{7.12}$$

Then, the following relation can be obtained using (7.11):

$$f_n(X^{k+1}) = f_n(X^k) - \frac{\partial F(X_{n-1}^k)}{\partial X_{n-1}^k}(X_{n-1}^{k+1} - X_{n-1}^k) + (X_n^{k+1} - X_n^k)$$

$$0 = f_n(X^k) - \frac{\partial F(X_{n-1}^k)}{\partial X_{n-1}^k}(X_{n-1}^{k+1} - X_{n-1}^k) + (X_n^{k+1} - X_n^k)$$

$$-f_n(X^k) = -\frac{\partial F(X_{n-1}^k)}{\partial X_{n-1}^k}(X_{n-1}^{k+1} - X_{n-1}^k) + (X_n^{k+1} - X_n^k)$$

$$-(X_n^k - F(X_{n-1}^k)) = -\frac{\partial F(X_{n-1}^k)}{\partial X_{n-1}^k}(X_{n-1}^{k+1} - X_{n-1}^k) + (X_n^{k+1} - X_n^k)$$

Therefore,

$$X_n^{k+1} = F(X_{n-1}^k) + \frac{\partial F(X_{n-1}^k)}{\partial X_{n-1}^k}(X_{n-1}^{k+1} - X_{n-1}^k) \tag{7.13}$$

Suppose that $F(X_{n-1}^{k})$ is the solution from the fine operator (*FINE*) and that

$$\frac{\partial F(X_{n-1}^{k})}{\partial X_{n-1}^{k}} \approx \frac{F(X_{n-1}^{k+1}) - F(X_{n-1}^{k})}{(X_{n-1}^{k+1} - X_{n-1}^{k})}$$

Then, approximate these solution as the solution from the coarse operator

$$F(X_{n-1}^{k+1}) \approx COARSE(X_{n-1}^{k+1})$$

$$F(X_{n-1}^{k}) \approx COARSE(X_{n-1}^{k})$$

Finally,

$$X_{n}^{k+1} = FINE(X_{n-1}^{k}) + COARSE(X_{n-1}^{k+1}) - COARSE(X_{n-1}^{k}) \tag{7.14}$$

Note that the Parareal update scheme (7.14) derived here with $k = k - 1$ is equivalent to the Parareal update scheme (7.9).

7.1.3 Implementation of Parareal Algorithm

To implement the Parareal algorithm, we focus on the partitioned-explicit method, which separately solves the succession of the ordinary differential equations and the algebraic equations. The partitioned-explicit method with explicit integration is commonly used in production-grade power system simulation packages.

7.1.3.1 Standard Coarse Operator

For benchmarking purpose, the following Midpoint-Trapezoidal predictor-corrector (Trap) method is used as the standard numerical predictor–corrector method in the coarse operator of the Parareal algorithm.

Midpoint Predictor :

$$x_{n+1}^{j} = x_{n}^{j} + \Delta t f\left(t_n + \frac{\Delta t}{2}, x_n + \frac{1}{2} f(t_n, x_n)\right)$$

Trapezoidal Corrector :

$$x_{n+1}^{j+1} = x_{n}^{j} + \frac{\Delta t}{2}\left[f(t_n, x_n) + f(t_{n+1}, x_{n+1}^{j})\right]$$

where only one iteration ($j = 1$) is used to obtain an approximate solution in the simulations. This Trap method serves as the standard coarse operator and is compared with two semi-analytical solution methods that will be discussed in Section 7.3.

7.1.3.2 Fine Operator

For the fine operator, the Runge–Kutta 4th order (RK-4) method, widely used in power system dynamic simulations, is employed. This fine operator has remained unchanged for all simulation results.

$$
\begin{aligned}
k_1 &= f(t_n, x_n),\ k_2 = f\left(t_n + \frac{\delta t}{2}, x_n + \frac{\delta t}{2} k_1\right) \\
k_3 &= f\left(t_n + \frac{\delta t}{2}, x_n + \frac{\delta t}{2} k_2\right) \\
k_4 &= f\left(t_n + \delta t, x_n + \delta t k_3\right) \\
x_{n+1} &= x_n + \frac{1}{6}[k_1 + 2k_2 + 2k_3 + k_4]
\end{aligned}
$$

Network equations are solved within each time step for the both coarse and fine operators in order to mitigate errors due to alternating solutions of the ordinary differential equations and the algebraic equations.

7.2 Examination of Semi-Analytical Solution Methods in the Parareal Algorithm

As illustrated, the computational performance of the Parareal algorithm is heavily dependent on the number of iterations for convergence. Hence, the design of coarse operators, that is reasonably accurate and fast, is crucial to improve its stability and reduce the number of iterations for fast convergence. To this end, this chapter describes the semi-analytical solution methods, discussed in Chapters 2, 5, 6 in detail, to construct the coarse operators of the Parareal algorithm. In particular, this chapter focuses on the two time-power series-based semi-analytical solution methods: Adomian decomposition method and Homotopy analysis method. Here, we briefly review these two methods.

In the semi-analytical solution methods, the basic idea is that the exact analytical solution $x(t)$ to the initial value problem of (7.1) can be represented as an infinite series of simple analytical terms (i.e. an approximate closed-form analytical solution) [8]:

$$
x(t) = \sum_{i=0}^{\infty} a_i (t - t_0)^i = a_0 + a_1(t - t_0) + \cdots \tag{7.15}
$$

where t_0 represents the initial time; a_0 indicates the initial state x_0; and a_i for $i \geq 1$ depends on a_0 and system parameters. Then, the semi-analytical solution method approximates the solution $x(t)$ by truncating higher order terms of the true solution (7.15) as follows:

$$x(t) \approx x^m_{SAS}(t) = \sum_{i=0}^{m} a_i (t - t_0)^i \tag{7.16}$$

where m is the semi-analytical solution order.

This solution strategy is to shift the computational burden of deriving an approximate but analytical solution, in a certain time interval, to the offline stage that mathematically derives unknown coefficients $a_1, a_2, \dots, a_m$. In the online stage, values are simply substituted into these symbolic semi-analytical solution terms, which are already derived offline, over consecutive time intervals until the end of the whole simulation period. With this, one can avoid the numerical integration of the ordinary differential equation and thus enjoy a fast online simulation. In addition, the accuracy of solution can be improved to a certain degree by increasing the semi-analytical solution order. There can be multiple ways to derive such unknown coefficients $a_1, a_2, \dots, a_m$. Sections 7.2.1 and 7.2.2 review the Adomian decomposition method and Homotopy analysis method to obtain these terms for ordinary differential equations of power systems.

7.2.1 Adomian Decomposition Method

This section briefly reviews the basic concept of Adomian decomposition method presented in Chapter 5 and also closely follows [6] for the description. Consider the initial value problem for a nonlinear ordinary differential equation problem in the following form:

$$Lx + Rx + N(x) = g \tag{7.17}$$

where x is the state variable of the system; g is the constant term; L is the linear operator to be inverted (typically highest order differential operator); R is the linear remainder operator; and N is the nonlinear operator. Then, solve for x as follows:

$$x = x_0 + L^{-1}g - L^{-1}Rx - L^{-1}N(x) \tag{7.18}$$

where the inverse operator can be regarded as $L^{-1} = \int_0^t dt$, and x_0 is the given initial condition. Now, assume that the solution $x(t)$ can be presented as an infinite series of the form:

$$x(t) = \sum_{i=0}^{\infty} x_i(t) \tag{7.19}$$

Furthermore, the nonlinear operator $N(x)$ is written as infinite series:

$$N(x) = \sum_{i=0}^{\infty} A_i \tag{7.20}$$

where $A_i(x_0, x_1, \dots, x_i)$ denote the Adomian polynomials that can be obtained for $N(x)$ using the following formula:

$$A_i = \frac{1}{i!}\frac{\partial^i}{\partial \lambda^i}\left[N\left(\sum_{k=0}^{\infty} x_k \lambda^k\right)\right]_{\lambda=0}, i = 0, 1, 2, \dots, \tag{7.21}$$

where λ is a grouping parameter. Then, substituting (7.19) and (7.20) in (7.18) gives

$$x(t) = \sum_{i=0}^{\infty} x_i = x_0 + L^{-1}g - L^{-1}R\sum_{i=0}^{\infty} x_i - L^{-1}\sum_{i=0}^{\infty} A_i$$

Thus, the Adomian decomposition method terms in power series forms can be obtained as:

$$\begin{aligned}
x_0 &= x_0 \\
x_1 &= L^{-1}g - L^{-1}Rx_0 - L^{-1}A_0 \\
x_2 &= -L^{-1}Rx_1 - L^{-1}A_1 \\
&\vdots \\
x_{i+1} &= -L^{-1}Rx_i - L^{-1}A_i
\end{aligned} \tag{7.22}$$

The Adomian decomposition method provides a fast converging series, and thus an approximate solution by the truncated series $\sum_{i=0}^{m} x_i = x_{SAS}^m(t)$ can serve as a good practical solution. Here, the coefficients $a_0, a_1, \dots, a_m$ in (7.16) correspond to the terms $x_0, x_1, \dots, x_m$ in (7.22).

The multistage Adomian decomposition method (MADM) uses the Adomian decomposition method to approximate dynamical response in a sequence of time intervals $[0, t_1], [t_1, t_2], \dots, [t_{N-1}, t_N]$. Note that the solution at t_n becomes an initial condition in interval $[t_n, t_{n+1}]$. This chapter utilizes the MADM as one of the coarse operators to obtain an approximation solution $x(t)$ with the equal time step Δt for all intervals, which is the step size of integration for the coarse operator.

7.2.2 Homotopy Analysis Method

This section briefly reviews the basic concept of the Homotopy analysis method presented in Chapter 6 and also closely follows [6] for the description. Consider a nonlinear differential equation in a general form:

$$N[x(t)] = 0 \tag{7.23}$$

where N is a nonlinear operator, t denotes time, and $x(t)$ is an unknown variable. Let $x_0(t)$ denote an initial guess of the exact solution $x(t)$, and L

denote an auxiliary linear operator with the property

$$L[f(t)] = 0 \quad \text{when } f(t) = 0 \tag{7.24}$$

Then, using $q \in [0, 1]$ as an embedding parameter, c as an auxiliary parameter (referred to as the convergence-control parameter), and $H(t)$ as a nonzero auxiliary function, one can construct a homotopy as follows:

$$\begin{aligned}&\mathcal{H}[\phi(t;q); x_0(t), H(t), c, q] = \\ &(1-q)L[\phi(t;q) - x_0(t)] - qcH(t)N[(\phi(t;q)]\end{aligned} \tag{7.25}$$

By enforcing (7.25) to be zero, one may obtain a family of equations, the so-called zero-order deformation equation

$$(1-q)L[\phi(t;q) - x_0(t)] = qcH(t)N[\phi(t;q)] \tag{7.26}$$

where $\phi(t;q)$ is the solution which depends on the initial guess $x_0(t)$, the auxiliary function $H(t)$, the auxiliary parameter c, and the embedding parameter $q \in [0, 1]$. Due to these parameters (e.g. the parameter c), the Homotopy analysis method is more general and flexible than other traditional methods. So, when $q = 0$, (7.26) becomes $L[\phi(t; 0) - x_0(t)] = 0$. Using (7.24), this gives

$$f(t) = 0 \Rightarrow \phi(t;0) = x_0(t) \tag{7.27}$$

When $q = 1$, since $c \neq 0$ and $H(t) \neq 0$, (7.26) is equivalent to $N[\phi(t; 1)] = 0$ which exactly corresponds to the original equation (7.23) if $\phi(t; 1) = x(t)$.

Therefore, the solution $\phi(t, q)$ varies continuously from the initial condition $x_0(t)$ to the exact solution $x(t)$ of the original equation (7.23) as the embedding parameter q increases from 0 to 1. Thus, one can obtain a family of solution curves by changing the values of c and $H(t)$, which provides a simple way to control and adjust the convergence of the approximate solution series. Here, the function $\phi(t, q)$ can be approximated by many different base functions (e.g., polynomial, fractional, exponential function). By Taylor's theorem, we expand $\phi(t, q)$ in a power series of the embedding parameter q as follows:

$$\phi(t;q) = x_0(t) + \sum_{i=1}^{\infty} x_i q^i \tag{7.28}$$

assuming that x_i exists and is defined as

$$x_i = \frac{1}{i!} \frac{\partial^i \phi(t;q)}{\partial q^i}\bigg|_{q=0} \tag{7.29}$$

Suppose that the auxiliary linear operator L, parameter c, and function $H(t)$ are properly chosen so that the power series (7.29) of $\phi(t; q)$ converges at $q = 1$. Then, one can obtain the solution series

$$\phi(t;1) = x(t) = x_0(t) + \sum_{i=1}^{\infty} x_i(t) \tag{7.30}$$

Next, the terms $x_i(t)$ are determined by the so-called high-order deformation equations. By differentiating the zero-order deformation equation (7.26) i times with respect to q, and then dividing it by $i!$ and setting $q = 0$, one can construct the ith-order deformation equation

$$L[x_i(t) - \mathcal{X}_i x_{i-1}(t)] = cR_i(t) \tag{7.31}$$

where $\mathcal{X}_i$ is defined by

$$\begin{cases} 0, & i \leq 1 \\ 1, & i > 1 \end{cases}$$

and $R_i(t)$ is defined as

$$R_i(t) = \frac{1}{(i-1)!} \frac{\partial^{i-1} N[\phi(t;q)]}{\partial q^{i-1}} \bigg|_{q=0} \tag{7.32}$$

Hence, one can obtain $x_i(t)$ by sequentially solving (7.31) i times. Notice that we select the polynomial as the base function, and thus $H(t)$ is uniquely defined as $H(t) = 1$ based on [9]. Interestingly, it has been demonstrated in [10] that the Adomian decomposition method described in Section 7.2.1, if using polynomial as the base function, is a special case of the Homotopy analysis method with $c = -1$. Likewise, the approximate solution of the Homotopy analysis method can be obtained by the truncated series $\sum_{i=0}^{m} x_i = x_{SAS}^{m}(t)$, and the coefficients $a_0, a_1, \ldots, a_m$ in (7.16) correspond to the terms $x_0, x_1, \ldots, x_m$ in (7.30). Similar to the multistage Adomian analysis method, the multistage Homotopy analysis method (MHAM) uses the Homotopy analysis method over multiple intervals of time.

7.2.3 Summary

This section provides a summary for each method and the detailed procedure to derive corresponding semi-analytical solution terms of each method. It also closely follows [6] for the description with the example of a first order governor equation:

$$\dot{P}_{sv} = \frac{1}{T_{sv}} \left(-P_{sv} + P_c - \frac{\omega}{R_D} \right) \tag{7.33}$$

where P_{sv} is the valve position of generators; ω is the slip speed; $\frac{1}{T_{sv}}$ is the governor time constant; P_c is the power command setting; and R_D is the governor droop constant. The following steps summarize the development of the Adomian decomposition method:

1. Recognize linear, nonlinear, and constant terms of (7.33) according to (7.18) as

$$P_{sv} = P_{sv,0} + \frac{1}{T_{sv}} \int_0^{\Delta t} (-P_{sv} + P_c - \frac{\omega}{R_D})dt$$

   where L^{-1} is replaced with $\int_0^{\Delta t} dt$.
2. Find nonlinear terms, and approximate them using the Adomian polynomials (7.21). Since there is no nonlinear term in (7.33), this step is not needed for the governor equation.
3. Obtain the Adomian decomposition method terms based on (7.22) and integrate each term analytically as

$$\begin{aligned}
P_{sv,0} &= \boldsymbol{P_{sv,0}} \\
P_{sv,1} &= \frac{1}{T_{sv}} \int_0^{\Delta t} \left(-P_{sv,0} + P_c - \frac{\omega_0}{R_D}\right) dt \\
&= \frac{\boldsymbol{1}}{\boldsymbol{T_{sv}}} \left(-\boldsymbol{P_{sv,0}} + \boldsymbol{P_c} - \frac{\boldsymbol{\omega_0}}{\boldsymbol{R_D}}\right) \Delta t \\
P_{sv,2} &= \frac{1}{T_{sv}} \int_0^{\Delta t} \left(-P_{sv,1} t - \frac{\omega_1}{R_D} t\right) dt \\
&= \frac{\boldsymbol{1}}{\boldsymbol{T_{sv}}} \frac{\boldsymbol{1}}{\boldsymbol{2}} \left(-\boldsymbol{P_{sv,1}} - \frac{\boldsymbol{\omega_1}}{\boldsymbol{R_D}}\right) \Delta t^2 \\
P_{sv,i+1} &= \frac{1}{T_{sv}} \int_0^{\Delta t} \left(-P_{sv,i} t^i - \frac{\omega_i}{R_D} t^i\right) dt
\end{aligned}$$

   where letters in bold indicate the corresponding constant of each term for $i = \{0, 1, \ldots\}$, and ω_0 is an initial condition and ω_1 is the first term, which needs to be computed within synchronous machine differential equations simultaneously, of slip speed, respectively.
4. Obtain the closed form approximate solution for the desired number of terms m.

$$P_{sv}(\Delta t) = P_{sv,0} + P_{sv,1}\Delta t + \cdots + P_{sv,m}\Delta t^m$$

In the Adomian decomposition method, the integration is straightforward since each term is constant and is only a function of the initial values. The desired number of terms can be efficiently derived offline. Now, let us focus on the development of the Homotopy analysis method that can be summarized in the following steps:

1. Rearrange (7.33) according to (7.23) as

$$N[P_{sv}(t)] = \dot{P}_{sv} - \frac{1}{T_{sv}}\left(-P_{sv} + P_c - \frac{\omega}{R_D}\right)$$

2. Select the linear operator L for the time $[t_n, t_{n+1}]$ as

$$L[P_{sv}(t)] = \dot{P}_{sv}(t) = \frac{dP_{sv}(t)}{dt} \quad 0 \le t \le t_{n+1} - t_n$$

3. Build the m-term approximate solution using (7.28) as

$$\phi(t;q) = P_{sv}(t) = P_{sv,0} + P_{sv,1}q + \cdots + P_{sv,m}q^m$$

4. Form the mth order deformation using (7.31) and substitute the m-term approximate solution for the variables. For example, a two-term approximate can be obtained as:

$$L[P_{sv,1}] = cR_1, \quad L[P_{sv,2} - P_{sv,1}] = cR_2$$

Here, the two-term approximation for P_{sv} is substituted for R_i using (7.32) as

$$L[P_{sv,1}] = cR_1 = c\left[\left.\frac{\partial(\dot{P}_{sv,0} + \dot{P}_{sv,1}q + \dot{P}_{sv,2}q^2)}{\partial q^0}\right|_{q=0} + \partial\left.\frac{\frac{1}{T_{sv}}\left((P_{sv,0} + P_{sv,1}q + P_{sv,2}q^2) - P_c\right)}{\partial q^0}\right|_{q=0} + \partial\left.\frac{\frac{1}{T_{sv}R_D}\left(\omega_0 + \omega_1 q + \omega_2 q^2\right)}{\partial q^0}\right|_{q=0}\right]$$

Therefore, one can obtain

$$L[P_{sv,1}] = c\left[\dot{P}_{sv,0} + \frac{1}{T_{sv}}\left(P_{sv,0} - P_c + \frac{\omega_0}{R_D}\right)\right]$$

Then, by integrating it and substituting $\dot{P}_{sv,0} = 0$,

$$\frac{d}{dt}P_{sv,1} = c\left[\frac{1}{T_{sv}}\left(P_{sv,0} - P_c + \frac{\omega_0}{R_D}\right)\right]$$

$$P_{sv,1} = \int_0^{\Delta t} c\left[\frac{1}{T_{sv}}\left(P_{sv,0} - P_c + \frac{\omega_0}{R_D}\right)\right] dt$$

$$P_{sv,1} = c\left[\frac{1}{T_{sv}}\left(P_{sv,0} - P_c + \frac{\omega_0}{R_D}\right)\right]\Delta t$$

Table 7.1 Comparison of the multistage Adomian decomposition method (MADM) and multistage Homotopy decomposition method (MHAM) for (7.33).

	Method	
Terms	**MADM**	**MHAM**
$P_{sv,0}$	$P_{sv}(t_n)$	$P_{sv}(t_n)$
$P_{sv,1}$	$\frac{1}{T_{sv}}\left(-P_{sv,0}+P_c-\frac{\omega_0}{R_D}\right)\Delta t$	$c\frac{1}{T_{sv}}\left(P_{sv,0}-P_c+\frac{\omega_0}{R_D}\right)\Delta t$
$P_{sv,2}$	$\frac{1}{T_{sv}}\frac{1}{2}\left(-P_{sv,1}-\frac{\omega_1}{R_D}\right)\Delta t$	$(1+c)P_{sv,1}+c\frac{1}{T_{sv}}\frac{1}{2}(P_{sv,1}+\frac{\omega_1}{R_D})\Delta t$

Following the similar steps, $P_{sv,2}$ can be obtained as

$$P_{sv,2} = (1+c)P_{sv,1} + c\frac{1}{T_{sv}}\frac{1}{2}\left(P_{sv,1}+\frac{\omega_1}{R_D}\right)\Delta t$$

where ω_1 is the first term of the slip speed, which also needs to be computed within synchronous machine differential equations when one computes other first terms simultaneously.

5. Obtain the closed form approximate solution for the desired number of terms m by setting $q = 1$

$$P_{sv}(\Delta t) = P_{sv,0} + P_{sv,1} + \cdots + P_{sv,m}$$

Notice that both the MADM and the MHAM, described so far, are the time-power series-based semi-analytical solution method such that each term is a function of time, initial conditions, and system parameters. In the MADM, we separated each term and time as a constant and a variable, respectively (e.g., $P_{sv,2}$ and Δt^2) since the ith-order term is only multiplied by the ith power of Δt. However, in the MHAM, this convenient separation is not possible, and each term is a function of Δt.

Table 7.1 summarizes the derived terms of each method with $m = 2$, 2-term approximate solution for the time interval $[t_n, t_{n+1}]$. One can recognize that terms between the MADM and the MHAM are equivalent when $c = -1$.

7.3 Numerical Case Study

This section presents numerical examples on the MADM and MHAM, briefly described in this chapter, as the coarse operators of the Parareal algorithm. In addition to these two methods, this section also includes comparison results with the differential transformation method (please refer to Chapter 3 and [11, 12] for more details). Two test networks are considered; the New England 10-generator 39-bus system [13] as a small test network, and the Polish 327-generator 2383-bus system [14] as a large test network. The following disturbances are considered: three-phase faults on buses with four cycles fault duration; three-phase faults on branches with four cycles fault duration.

For the MADM and MHAM, it incorporates the **dummy coil model [13]** for (7.1). For the differential transformation method, it incorporates the 6th **order generator model [15]**. Both are based on IEEE Model 2.2 [16] with two damper windings on the q-axis and one damper winding on the d-axis along with the field winding for the synchronous generators, IEEE Type 1 excitation system, and first order turbine-governor models. Loads are modelled as aggregate static loads employing polynomial representation (ZIP load models, which consist of constant-impedance, constant-current, and constant-power loads).

For the Parareal algorithm, the number of processors for parallel computing of the fine solver is chosen as 50. The 10 s simulation is conducted and thus divided into 50 subintervals. For each processor, 100 subintervals are used for the fine solver, indicating $\delta t = \frac{10}{50\times100} = 0.002\,\text{s}$. This setup for the Parareal algorithm has remained unchanged for all simulation results.

7.3.1 Validation of Parareal Algorithm

We first illustrate the convergence of the Parareal algorithm. This section only describes the representative result of the three-phase fault at the bus 1 with four cycles. Here, 10 subintervals are used for the coarse operators in each processor (thus $\Delta t = 0.02\,\text{s}$), and the true solution is obtained using the standard sequential RK4 method with the time step of 0.002 s. The $m = 3$ is used for the number of semi-analytical solution terms for both the MADM and MHAM. For the MHAM, $c = -0.9$ is used.

Figures 7.2 and 7.3 show simulation results of the Parareal algorithm for the New England system and Polish system, respectively. As depicted, rotor angle and slip speed at bus 1 of the Parareal solution converged to the true solution with all coarse operators considered. Note that our extensive numerical experiments showed this convergence for all other variables and disturbances, which demonstrates the convergence of the Parareal algorithm.

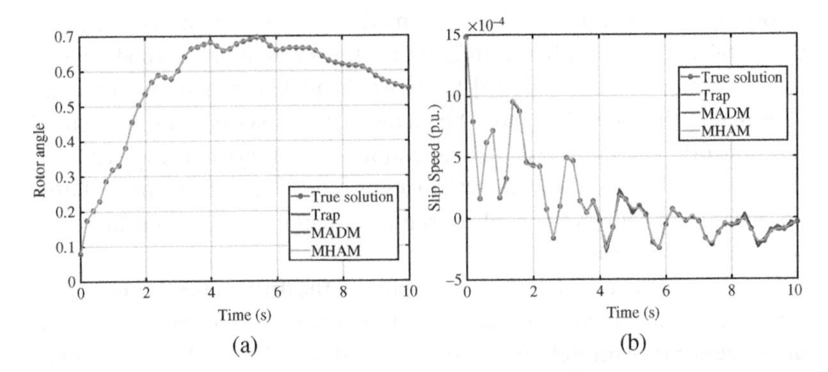

Figure 7.2 MADM and MHAM results followed by the three-phase fault at bus 1 with four cycles for 10 s simulation using the New England system. (a) Rotor angle of generator 1 and (b) Slip speed of generator 1.

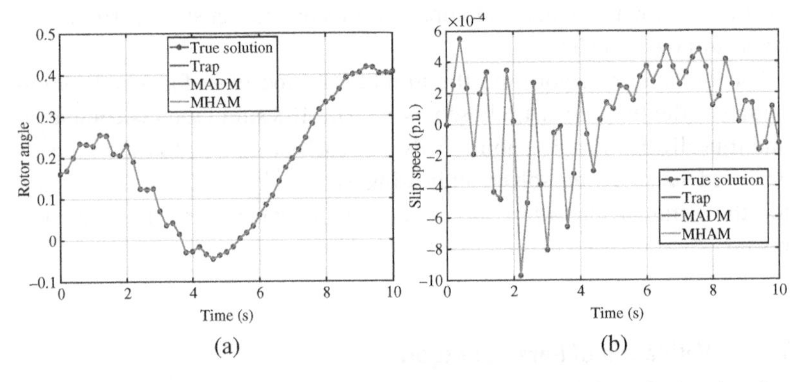

Figure 7.3 MADM and MHAM results followed by the three-phase fault at bus 1 with four cycles for 10 s simulation using the Polish system. (a) Rotor angle of generator 1 and (b) Slip speed of generator 1.

To illustrate the comparison with the differential transformation method, this section has considered two other coarse operators [11, 12]: the simultaneous-implicit method based on the trapezoidal rule (SI-Trap); the modified Euler (ME) method. Figure 7.4 shows simulation results for the differential transformation method as the coarse operator of the Parareal algorithm using the Polish system. Similarly, the Parareal solution converged to the true solution with all coarse operators considered, and this convergence is also checked with all other variables and disturbances.

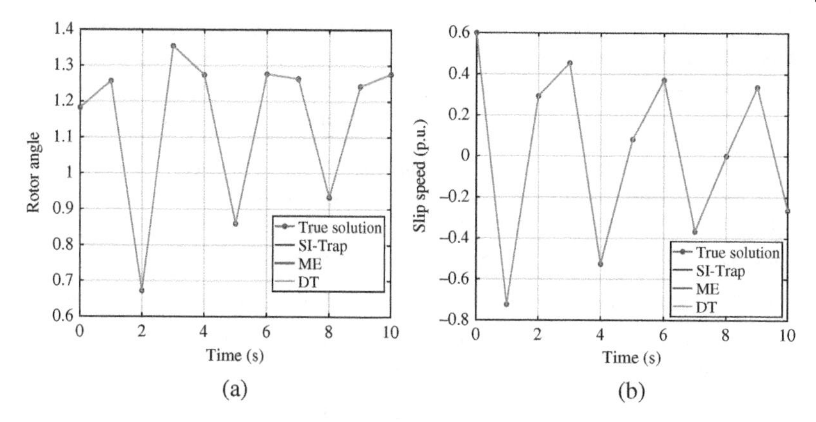

Figure 7.4 Differential transformation results followed by the 3-phase fault at bus 10 with 4 cycles for 10 s simulation using the Polish system. (a) Rotor angle of generator 1 and (b) Slip speed of generator 1.

7.3.2 Benefits of Semi-Analytical Solution Methods

This section illustrates the benefits of semi-analytical solution methods. To this end, we consider the number of Parareal iterations for the convergence of all coarse solutions in each processor to the true solution. For this analysis, the number of the coarse intervals in each processor is increased from 4 to 45. For the MADM, we have increased the number of semi-analytical solution terms from $m = 3$ to $m = 10$. For the MHAM, while the number of semi-analytical solution terms is set to $m = 3$, the value of c is decreased from $c = -0.6$ to -1.2.

The values in Tables 7.2 and 7.3 show the number of Parareal iterations to converge to the true solution using the MADM and MHAM for the New England system and Polish system, respectively. In the table, "fail" denotes cases that do not converge to a true solution. From this, we obtain the following observations: (i) the Trap method tends to require fewer iterations with a small number of coarse intervals, but it can cause the divergence of the Parareal algorithm; (ii) the MADM and MHAM can improve the stability of the Parareal algorithm (can employ a larger time step for the coarse operator) with more semi-analytical solution terms and larger values of c, respectively, but a smaller number of coarse intervals requires more iterations; (iii) While it is not shown here, both the MADM and MHAM were slightly faster than the Trap method since it does not need the corrector step

Table 7.2 Number of Parareal iterations for the New England system with different numbers of coarse intervals using the MADM and MHAM.

# of Coarse intervals	Trap	MADM			
		$m = 3$	$m = 4$	$m = 5$	$m = 10$
4 = 0.05 s	Fail	Fail	Fail	Fail	Fail
5 = 0.04 s	Fail	Fail	Fail	Fail	23
6 = 0.033 s	Fail	Fail	Fail	Fail	19
7 = 0.028 s	Fail	Fail	Fail	16	16
8 = 0.025 s	7	13	13	13	13
9 = 0.022 s	6	11	12	12	11
10 = 0.020 s	6	10	10	10	10
15 = 0.013 s	5	7	7	7	7
20 = 0.010 s	4	5	5	5	5
25 = 0.008 s	4	4	4	4	4
30 = 0.007 s	4	4	4	4	4
35 = 0.006 s	4	4	4	4	4
40 = 0.005 s	4	4	4	4	4

# of Coarse intervals	MHAM ($m = 3$)					
	$c = -0.6$	$c = -0.7$	$c = -0.8$	$c = -0.9$	$c = -1.1$	$c = -1.2$
4	31	Fail	Fail	Fail	Fail	Fail
5	27	24	Fail	Fail	fail	Fail
6	23	22	20	Fail	fail	Fail
7	22	19	17	16	fail	Fail
8	20	17	15	14	Fail	Fail
9	18	15	13	12	fail	Fail
10	16	13	12	11	10	Fail
15	14	9	8	7	7	7
20	12	7	5	5	5	5
25	11	6	4	4	4	4
30	10	5	4	4	4	4
35	10	4	4	4	4	4
40	10	4	4	4	4	4

Table 7.3 Number of Parareal iterations for the Polish system with different numbers of coarse intervals using the MADM and MHAM.

# of Coarse intervals	Trap	MADM $m=3$	MADM $m=4$	MADM $m=5$	MADM $m=10$
4 = 0.05 s	Fail	Fail	Fail	Fail	Fail
5 = 0.04 s	Fail	Fail	Fail	Fail	34
6 = 0.033 s	Fail	Fail	Fail	Fail	31
7 = 0.028 s	Fail	Fail	Fail	Fail	27
8 = 0.025 s	Fail	Fail	Fail	25	25
9 = 0.022 s	12	Fail	23	23	23
10 = 0.020 s	11	20	21	21	21
15 = 0.013 s	8	13	13	13	13
20 = 0.010 s	6	10	10	10	10
25 = 0.008 s	5	8	8	7	7
30 = 0.007 s	5	6	6	6	6
35 = 0.006 s	4	5	5	5	5
40 = 0.005 s	4	5	5	5	5
45 = 0.004 s	4	4	4	4	4

# of Coarse intervals	MHAM ($m = 3$) $c=-0.6$	$c=-0.7$	$c=-0.8$	$c=-0.9$	$c=-1.1$	$c=-1.2$
4	44	Fail	Fail	Fail	Fail	Fail
5	39	37	Fail	Fail	fail	Fail
6	36	34	Fail	Fail	fail	Fail
7	33	31	29	Fail	fail	Fail
8	31	28	26	25	Fail	Fail
9	29	25	24	23	Fail	Fail
10	27	23	22	21	Fail	Fail
15	21	17	14	14	14	16
20	19	13	11	10	10	11
25	18	11	8	8	8	9
30	17	10	7	6	6	7
35	17	9	6	5	5	6
40	16	8	5	5	4	5
45	15	7	4	4	4	5

Table 7.4 Number of Parareal iterations for the Polish system with different numbers of coarse intervals using the differential transformation method.

# of Coarse intervals	SI-Trap	ME	Differential transformation			
			$m = 5$	$m = 10$	$m = 15$	$m = 20$
$2 = 0.1$ s	10	Fail	10	8	4	4
$4 = 0.05$ s	8	Fail	1	1	1	1
$8 = 0.025$ s	2	7	1	1	1	1
$10 = 0.020$ s	1	6	1	1	1	1

used in the standard numerical integration method; (iv) the Parareal algorithm converges to the true solution in a reasonable number of iterations, which is significantly less than the number of subintervals (i.e. 50 in this example).

Similarly, the comparison result with the differential transformation for the Polish system is shown in Table 7.4. For this comparison, we have increased the number of semi-analytical solution terms for the differential transformation from $m = 10$ to $m = 20$. As illustrated, the differential transformation method also improves the stability of the Parareal algorithm and requires fewer Parareal iterations.

Overall, the semi-analytical solution methods can improve the stability of the Parareal algorithm, reduce the simulation time spent on the coarse operator as compared to the standard numerical integration-based method, require fewer Parareal iterations. Also, for this particular example, one can observe that the theoretical speedup (assuming ideal parallelization and negligible communication time) of the Parareal algorithm can be more than 10, which is an encouraging result for the semi-analytical solution methods to be a good candidate as the coarse operator of the Parareal algorithm.

To further improve the computational performance of the Parareal algorithm, one can employ the windowing approach [6], which divides the 10 s simulation window into smaller windows and run the Parareal algorithm for each smaller window sequentially. For example, it solves 1 s simulation window using the Parareal algorithm 10 times sequentially to obtain 10 s simulation trajectories. The reason is that the simulation length can have a significant impact on the convergence of the Parareal algorithm and the number of Parareal iterations; the Parareal algorithm typically shows good convergence results (converged in 1–2 iterations) using a smaller simulation window (e.g., 1 s simulation window). In this scenario, the Parareal algorithm can benefit from the enhanced stability facilitated by the semi-analytical solution methods to reduce the time spent on the coarse operator without increasing the number of Parareal iterations.

7.3.3 Results with the High Performance Computing Platform

To further illustrate the speedup described in Section 7.3.2 and practical values of the Parareal algorithm, this section describes the Parareal algorithm results which have been deployed on Compute and Data Environment for Science (CADES) high performance computing platform [17]. Using the CADES, the simulations are conducted and profiled from serial to 100's of processors. The CADES condo machine consists of 160 nodes with 32 core per node with 124 GB RAM.

In particular, this section considers the Eastern Interconnection, 5617-generator and 70 285-bus system [18], which has been officially validated by the Eastern Interconnection reliability assessment group. Note that this section employs the same dynamic model as used in Sections 7.3.1 and 7.3.2 without any approximations for this extra-large test network. With the different number of processors, the resulting computational time (seconds) of the Parareal algorithm using the high performance computing platform is shown in Table 7.5.

As the number of processors increases, the computational performance is significantly improved; the Parareal algorithm with 500 processors is completed about six times faster than one with 10 processors. This is because as we scale the problem to large processors, the algorithm allows one to have more (smaller) subintervals and thus more parallel computations in the fine and coarse operators, which leads to reducing the computation time in each operator on each processor significantly. While the simulation with more processors increases the communication time, the reduction from a larger number of parallel computations is more significant and thus reduces the total simulation time.

Table 7.5 The full Eastern Interconnection simulation result (s) with the high performance computing platform.

Processors	Last processor	Fine	Coarse	Communication	Total time
10	857	774	40.3	42.7	857
100	243	125	7.24	56.4	243
250	178	73.6	4.67	93.5	179
500	145	46.1	3.33	85.3	145

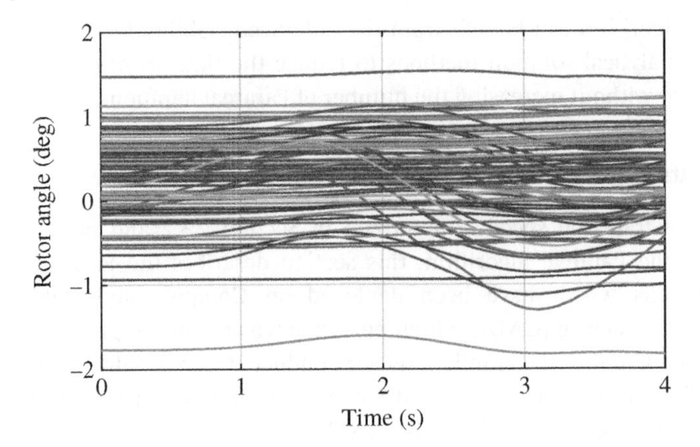

Figure 7.5 High performance computing result for trajectories of state variables in the Eastern Interconnection system using the Parareal algorithm. It only shows the rotor angle of randomly selected 100 generators until 4 s.

Figure 7.5 shows the rotor angle of randomly selected 100 generators in the Eastern Interconnection system. Note that all other solutions for the Eastern Interconnection system from the Parareal algorithm have converged to the true solution. These results illustrate the great potential of the Parareal algorithm equipped with semi-analytical solution methods to achieve an ambitious goal of facilitating "faster than real-time simulation" for predicting power system dynamic behaviors.

7.3.4 Results with Variable Order Variable Step Adaptive Parareal Algorithm

As discussed in Chapter 3, the differential transformation method has the flexibility to balance the accuracy and time performance in simulation by adjusting its semi-analytical solution orders and the time step length. Such flexibility has facilitated the development of the variable-order variable-step differential transformation method and the adaptive Parareal method. This section briefly illustrates the variable-order variable-step differential transformation method and corresponding results as the coarse operator of the Parareal algorithm (readers may refer to Chapter 3 for more details).

Algorithm 7.1 shows the variable-order variable-step differential transformation method, where the outputs include not only the trajectories of state variables but also the variable time step lengths $\mathbf{h}(t_n)$ and variable orders $\mathbf{K}(t_n)$ during the simulation; lines 1 to 7 are similar to the differential

Algorithm 7.1 Variable-order variable-step differential transformation strategy

Input: $t_0, t_{end}, x_0, h_0, K_0$
Output: $x(t_n), \mathbf{h}(t_{end}), \mathbf{K}(t_n), t_0 \leq t_n \leq t_{end}$
Offline stage: Derive the recursive equation $X(k+1) = F(X(0:k))$

1: $x(t_0) \leftarrow x_0, \mathbf{h}(t_0) \leftarrow h_0, \mathbf{K}(t_0) \leftarrow K_0$
2: $t \leftarrow t_0, h \leftarrow h_0, K \leftarrow K_0, X(0) \leftarrow x_0, n \leftarrow 0$
3: **While** $t + h < t_{end}$
4:    **for** $k = 0 : K$ // compute power series coefficients
5:       $K(k+1) = F(X(0:k))$
6:    **end**
7: $x(t_{n+1}) \leftarrow \sum_{k\in\{0,1,\dots,K\}} X(k)h^k$
8: err $\leftarrow ||X(K+1)h^{K+1}||_\infty$
9: **if** err $< \varepsilon_1; h \leftarrow min(q_1, h, h_{max})$ **end**
10: **if** err $< \varepsilon_2; h \leftarrow max(q_2, h, h_{min})$ **end**
11: **if** err $< \varepsilon_3; K \leftarrow max(K - \Delta K, K_{min})$ **end**
12: **if** err $< \varepsilon_4; K \leftarrow min(K + \Delta K, K_{max})$ **end**
13: $\mathbf{h}(t_{n+1}) \leftarrow h, \mathbf{K}(t_{n+1}) \leftarrow K, t \leftarrow t + h, n \leftarrow n + 1$
14: **end**

transformation method, except for the initialization of additional variables associated with the time step length and the order of the differential transformation method; line 8 is error estimation using the $(K+1)$th power series term; lines 9 to 12 are the adaptive change of time step length and the order of the differential transformation method based on the error estimation. In Algorithm 8.1, $\varepsilon_1, \varepsilon_2, \varepsilon_3, \varepsilon_4$ are predefined error thresholds; q_1, q_2 are the factors to adjust the time step length; ΔK is an incremental adjustment of the order of the differential transformation method; h_{max}, h_{min} are the predefined maximum and minimum time step length; K_{max}, K_{min} are the maximum and minimum order.

The variable-order variable-step differential transformation strategy has the following compelling advantages when used as the coarse operator in the Parareal algorithm: (i) it increases the order and reduces the time step length of the differential transformation-based coarse operator when the error is larger than a threshold during the simulation, which could reduce the needed number of iterations for the Parareal algorithm to converge; (ii) it decreases the order and increases the time step length of the differential transformation-based coarse operator when the error is smaller than a threshold during the simulation, which might slightly increase the number of iterations but can avoid unnecessary computation burden and

Table 7.6 The number of iterations and approximate CPU time in the 2383-bus system test.

Method	# of iterations	CPU time (sec)
M1	Divergent	N/A
M2	1	158.46
M3	1	117.60

improve the overall efficiency. Note that these advantages are not easily attainable from other existing methods.

Here, we illustrate the differential transformation-based adaptive Parareal algorithm with the Polish system. The simulated contingency is a permanent three-phase fault at bus 9 applied at $t = 1$ second and cleared after 0.4 s by tripping the line between bus 6 and bus 9. The following three methods are compared: (i) M1: the traditional Parareal method, (ii) M2: the differential transformation-based Parareal method, and (iii) M3: the differential transformation-based adaptive Parareal algorithm. Note that M3 integrates the adaptive strategies compared with M2. Table 7.6 gives the number of iterations and approximate CPU time of the three methods when the Parareal window size is 0.75 s for 10 s simulation. The traditional Parareal method diverges while both the differential transformation-based Parareal method and the differential transformation-based adaptive Parareal method converge and need only one iteration in each window. Since the differential transformation-based adaptive Parareal method adjusts the window size, the order, and the time step length adaptively during the simulation, the computation time (note that this is the approximate time without using the high performance computing platform) is reduced by 25% compared with the differential transformation-based Parareal method without adaptive strategies.

Figures 7.6–7.8, respectively, give the trajectories of rotor angles of all generators; the window length and the number of iterations of each Parareal window; and the time step length and order of the differential transformation-based coarse operator. From Figures 7.7 and 7.8, it can be seen that the number of iterations is 12 in the first window but it is reduced in the subsequent windows, benefiting from the flexibility of adaptively

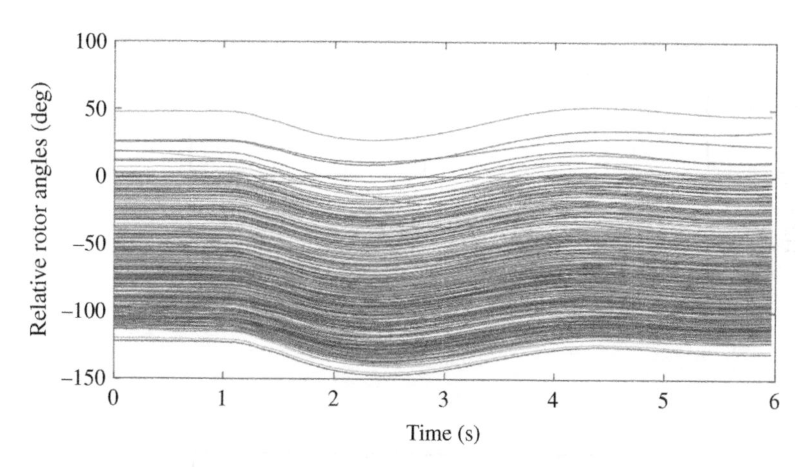

Figure 7.6 Rotor angle trajectories in the Polish system using the differential transformation-based adaptive Parareal method.

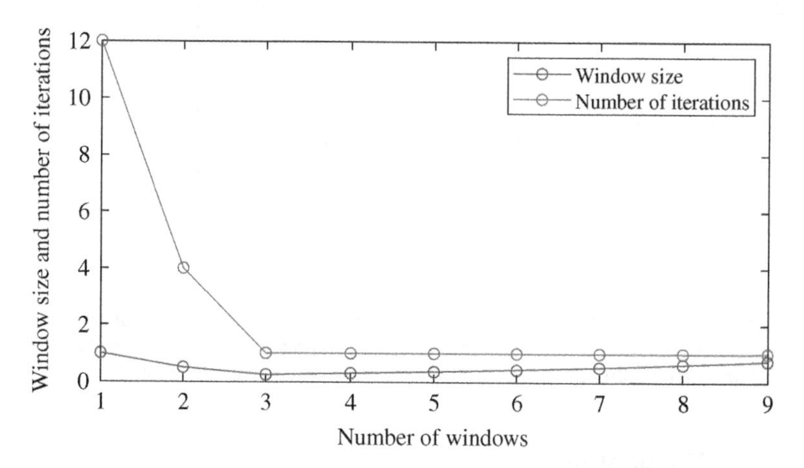

Figure 7.7 The window size and the number of iterations of each window in the Polish system using the differential transformation-based adaptive Parareal method.

adjusting the window size, the order, and the time step length of the coarse operator during the simulation. These results illustrate the better convergence performance of the differential transformation-based adaptive Parareal method in a large power system with detailed dynamic models.

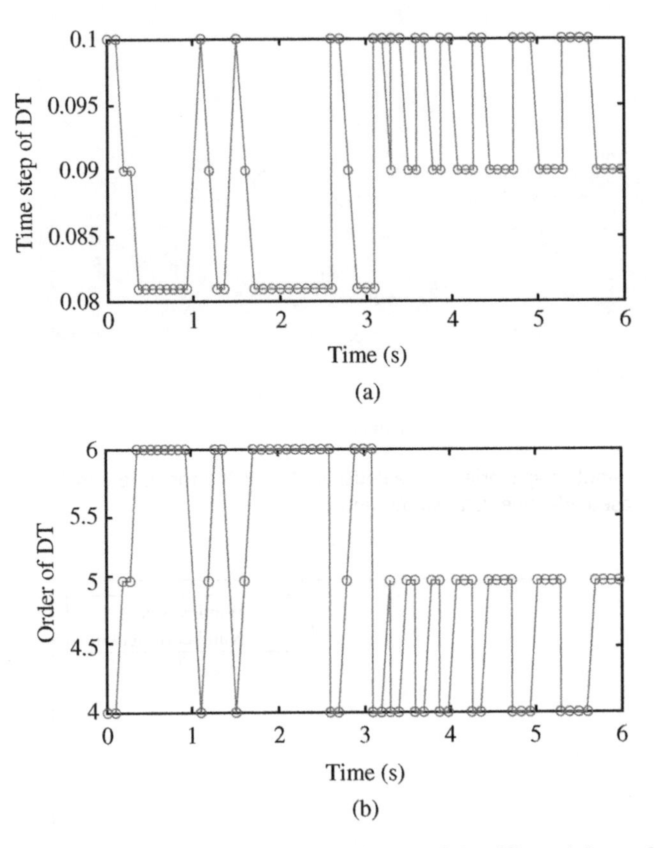

Figure 7.8 The time step length and order of the differential transformation-based coarse operator in the Polish system using the differential transformation-based adaptive Parareal method: (a) the time step length and (b) the order.

7.4 Conclusions

This chapter has described the benefits of semi-analytical solution methods as the coarse operator of the Parareal algorithm for power system dynamic simulations. Using three different semi-analytical solution methods: Adomian decomposition method, Homotopy analysis method, and differential transformation method, the great potential of semi-analytical solution methods to design a novel Parareal algorithm is illustrated in detail. In particular, semi-analytical solution methods improve the stability of the Parareal algorithm, decrease the time spent on the coarse operator, and reduce the number of Parareal iterations. All these benefits from semi-analytical solution methods can enhance the performance of the

Parareal algorithm significantly and thus allow of developing an advanced simulation framework toward high-fidelity and high-speed simulations of large-scale electric grids.

As the grid complexity is increasing exponentially with the ongoing grid modernization efforts, an improved simulation of dynamic behavior is becoming more important. In this scenario, it is imperative to improve the computational performance of time-domain simulation, which solves a large number of nonlinear differential algebraic equations. As discussed in this chapter, one key approach is to utilize parallel algorithms, and one can leverage the benefits of semi-analytical solution methods to design more innovative parallel algorithms. These emerging solution techniques (i.e. semi-analytical solution methods and parallel algorithms) in the dynamic modeling and simulation will greatly help to address a wide range of challenges associated with future electric grids whose components are governed by faster and more complex dynamics.

References

1. Lions, J., Maday, Y., and Turinici, G. (2001). A "parareal" in time discretization of PDE's. *Comptes Rendus de l Académie des Sciences – Series I – Mathematics* 332 (7): 661–668.
2. Nielsen, A.S. (2012). Feasibility study of the parareal algorithm. Master's thesis. Universities in Denmark.
3. Baffico, L., Bernard, S., Maday, Y. et al. (2002). Parallel in time molecular dynamics simulations. *Physical Review E* 66 (5): 057701.
4. Gurrala, G., Dimitrovski, A., Pannala, S. et al. (2016). Parareal in time for fast power system dynamic simulations. *IEEE Transactions on Power Systems* 31 (3): 1820–1830.
5. Cheng, T., Duan, T., and Dinavahi, V. (2020). Parallel-in-time object-oriented electromagnetic transient simulation of power systems. *IEEE Open Access Journal of Power and Energy* 7: 296–306.
6. Park, B., Sun, K., Dimitrovski, A. et al. (2021). Examination of semi-analytical solution methods in the coarse operator of parareal algorithm for power system simulation. *IEEE Transactions on Power Systems* 36 (6): 5068–5080.
7. Park, B., Allu, S., Sun, K. et al. (2022). Resilient adaptive parallel sImulator for griD (RAPID): an open source power system simulation toolbox. *IEEE Open Access Journal of Power and Energy* 9: 361–373.

8 Wang, B., Duan, N., and Sun, K. (2019). A time-power series-based semi-analytical approach for power system simulation. *IEEE Transactions on Power Systems* 34 (2): 841–851.

9 Liao, S. (2009). Notes on the homotopy analysis method: some definitions and theorems. *Communications in Nonlinear Science and Numerical Simulation* 14 (4): 983–997.

10 Liao, S. (2003). *Beyond Perturbation: Introduction to the Homotopy Analysis Method*. Chapman and Hall/CRC.

11 Liu, Y., Sun, K., Yao, R., and Wang, B. (2019). Power system time domain simulation using a differential transformation method. *IEEE Transactions on Power Systems* 34 (5): 3739–3748.

12 Liu, Y. and Sun, K. (2020). Solving power system differential algebraic equations using differential transformation. *IEEE Transactions on Power Systems* 35 (3): 2289–2299.

13 Padiyar, K.R. (2002). *Power System Dynamics Stability and Control*. Hyderabad: BS Publications.

14 Zimmerman, R.D., Murillo-Sanchez, C.E., and Thomas, R.J. (2011). MATPOWER: steady-state operations, planning and analysis tools for power systems research and education. *IEEE Transactions on Power Systems* 26 (1): 12–19.

15 Milano, F. (2010). *Power System Modeling and Scripting*. Berlin, Heidelberg: Springer-Verlag.

16 Dandeno, P.L. (2003). *IEEE Std. 1110-2002 – IEEE guide for synchronous generator modeling practices and applications in power system stability analyses*, 1–72.

17 CADES (2022). *Oak Ridge National Laboratory*. https://docs.cades.ornl.gov/ (accessed 18 May 2023).

18 Multiregional Modeling Working Group (2021). A library of power flow models and associated dynamics simulation models of the eastern interconnection. https://rfirst.org/ProgramAreas/ESP/ERAG/MMWG.

8

Power System Simulation Using Holomorphic Embedding Methods

Rui Yao[1], *Kai Sun*[2] *and Feng Qiu*[1]

[1] *Division of Energy Systems, Argonne National Laboratory, Lemont, IL, USA*
[2] *Department of Electrical Engineering & Computer Science, University of Tennessee, Knoxville, Knoxville, TN, USA*

- Learn how the Holomorphic Embedding (HE) approach is extended from a pure steady-state analysis method to dynamic modeling and analysis in power systems.
- Learn the general holomorphic embedding approach for dynamic security analysis.
- Learn how holomorphic embedding can be used for extended-term simulation for enhanced efficiency and without losing accuracy by using a hybrid steady-state and dynamic simulation approach.
- Learn how simulation can be further accelerated without losing numerical robustness with parallel or distributed holomorphic embedding.

8.1 Holomorphic Embedding from Steady State to Dynamics

In recent years, the holomorphic embedding (HE) method has been utilized to solve algebraic equations (AE) of power system steady-state problems and achieved considerable success in improving the convergence and computational efficiency [1–7]. The holomorphic embedding method represents the solution of the equations as a power series of an embedding variable in the complex domain. At the original point of the power series, the solution is already known or easily acquired. Thus, the computation

Power System Simulation Using Semi-Analytical Methods, First Edition. Edited by Kai Sun.

Published 2024 by John Wiley & Sons, Inc.

of holomorphic embedding is to derive the coefficients in the power series. The Padé approximation can also be used to further increase the convergence range of the series [8]. Moreover, the multistage holomorphic embedding method proposed in [9] can improve precision, extend effective range and significantly reduce the number of terms in Padé approximants.

With the development of holomorphic embedding method in steady-state analysis of power systems, it is natural to extend its use for dynamic modeling. The voltage stability analysis is a typical use case. Steady-state methods have been used for voltage stability analysis, but essentially a realistic system involves lots of dynamics. For example, induction motor loads have significant influence on voltage stability of power systems [10–12]. The self-restorative characteristic of induction motors contributes to voltage instability. Moreover, the low voltage caused by insufficient voltage support or faults can cause motors to stall. The stalling motors cause delayed voltage recovery and thus exacerbate voltage instability [13]. The voltage stability analysis (VSA) with induction motor loads has been studied or discussed in some literature over the past years. Strictly speaking, the VSA with induction motors or other dynamic load models is a dynamic problem [10, 14–16], but the traditional dynamic simulation methods based on numerical integration are computationally expensive [17, 18]. And in the meantime, the traditional numerical methods (e.g. Newton–Raphson (NR) method and its variants) for solving differential-algebraic equations (DAE) also suffer from non-convergence problems.

As commented in [2] and [5], the holomorphic embedding is a high-order analytic continuation method and thus has a much larger effective range than the traditional homotopy method which uses a limited order for continuation. Considering the essence of the popular traditional numerical integration methods for dynamic simulation (e.g. the Euler method and Runge–Kutta method), they can only limit the error to a relatively low order, and thus the time step is constrained to a very short length. Ideally, using a higher order would help extend the effective range of approximation. Thus, we generalize the use of holomorphic embedding from the space of an auxiliary variable in the complex domain in steady-state analysis, to the time domain in dynamic analysis. Analogously, the idea of holomorphic embedding can also be utilized in dynamic simulation to achieve much larger time steps of computation, thus significantly reducing the number of steps and improving the speed of simulation.

In this section, we will use the voltage stability analysis with induction motors as an example to show the generalization of holomorphic embedding from steady-state analysis to dynamic modeling and analysis. We will derive the steady-state and dynamic induction motor load models with holmorphic embedding, and use these models to perform quasi-steady-state (QSS), dynamic or hybrid voltage analysis.

8.1.1 Holomorphic Embedding Formulations

8.1.1.1 Classic Formulation from Trivial Germ Solution

From the grid side, the power flow equation on each bus is

$$S_i^* W_i^* - \sum_l Y_{il}^{tr} V_l - Y_i^{sh} V_i - I_{Li} = 0 \tag{8.1}$$

where "*" means conjugation, and V_i is the voltage phasor on bus i, whose reciprocal is W_i. For PQ buses, $S_i = P_i + jQ_i$ is given. For PV buses, P_i is given, and the voltage should satisfy magnitude constraint $V_i V_i^* = |V_i^{sp}|^2$. The V_i^{sp} means a specified voltage phasor. For PV buses, only its magnitude $|V_i^{sp}|$ is needed, so the angle of phasor V_i^{sp} can be simply set as 0, and the magnitude is set as $|V_i^{sp}|$. For the swing bus, the complete phasor V_i^{sp} is known. Y_{il}^{tr} is the serial branch admittance between buses i and l, and Y_i^{sh} is the shunt admittance on bus i. I_{Li} is the total load current (except const-PQ and const-impedance loads that have been included in S_i and Y_i^{sh}) on bus i. The holomorphic embedding formulation for solving power flow is constructed as follows:

$$S_i(\alpha)^* W_i^*(\alpha) - \sum_l Y_{il}^{tr} V_l(\alpha) - \alpha Y_i^{sh} V_i(\alpha) - I_{Li}(\alpha) = 0 \tag{8.2}$$

where the embedding variable $\alpha \in \mathbb{C}$. For PQ buses, $S_i(\alpha) = \alpha(P_i + jQ_i)$ is known. For PV buses $S_i(\alpha) = \alpha P_i + jQ_i(\alpha)$ and reactive power $Q_i(\alpha)$ needs to be calculated. And the holomorphic embedding formulations of the voltages on PV and slack buses are

$$\begin{aligned} V_i(\alpha)V_i^*(\alpha) - 1 - \alpha(|V_i^{sp}|^2 - 1) = 0, i \in S_{PV} \\ V_i(\alpha) - 1 - \alpha(V_i^{sp} - 1) = 0, i \in S_{SL} \end{aligned} \tag{8.3}$$

where S_{PV} and S_{SL} represent the sets of PV and slack buses, respectively. The term "trivial germ solution" is the state $V_i[0] = 1$, $Q_i[0] = 0$ (for PV buses) and $I_{Li}[0] = 0$. The trivial germ solution satisfies (8.2) and (8.3) when $\alpha = 0$. The variables in holomorphic embedding are represented as power series of α, such as

$$V_i(\alpha) = V_i[0] + V_i[1]\alpha + V_i[2]\alpha^2 + V_i[3]\alpha^3 + \cdots \tag{8.4}$$

In this paper, each coefficient of the power series is denoted by the variable name followed by a square-bracketed index, e.g. $V_i[k]$ for $V_i(\alpha)$. holomorphic embedding solves the coefficients like $V_i[n]$. Similar to [3], $V_i[n]$ and $Q_i[n]$

are obtained by solving (8.5):

$$\begin{bmatrix} -B_{11} & G_{11} & \cdots & -B_{1i} & 0 & \cdots & -B_{1N} & G_{1N} \\ G_{11} & B_{11} & \cdots & G_{1i} & 0 & \cdots & G_{1N} & B_{1N} \\ \vdots & \vdots & \vdots & \vdots & \vdots & \vdots & \vdots & \vdots \\ -B_{i1} & G_{i1} & \cdots & -B_{ii} & 0 & \cdots & -B_{iN} & G_{iN} \\ G_{i1} & B_{i1} & \cdots & G_{ii} & 1 & \cdots & G_{iN} & B_{iN} \\ \vdots & \vdots & \vdots & \vdots & \vdots & \vdots & \vdots & \vdots \\ -B_{N1} & G_{N1} & \cdots & -B_{Ni} & 0 & \cdots & -B_{NN} & G_{NN} \\ G_{N1} & B_{N1} & \cdots & G_{Ni} & 0 & \cdots & G_{NN} & B_{NN} \end{bmatrix} \begin{bmatrix} D_1[n] \\ C_1[n] \\ \vdots \\ D_i[n] \\ Q_i[n] \\ \vdots \\ D_N[n] \\ C_N[n] \end{bmatrix}$$
$$= \begin{bmatrix} \Re(\Gamma_1[n]) \\ \Im(\Gamma_1[n]) \\ \vdots \\ \Re(\Gamma_i[n]) \\ \Im(\Gamma_i[n]) \\ \vdots \\ \Re(\Gamma_N[n]) \\ \Im(\Gamma_N[n]) \end{bmatrix} - \sum_{j \in S_{PV} \cup S_{SL}} \begin{bmatrix} G_{1j} \\ B_{1j} \\ \vdots \\ G_{ij} \\ B_{ij} \\ \vdots \\ G_{Nj} \\ B_{Nj} \end{bmatrix} C_j[n] - \sum_{j \in S_{SL}} \begin{bmatrix} -B_{1j} \\ G_{1j} \\ \vdots \\ -B_{ij} \\ G_{ij} \\ \vdots \\ -B_{Nj} \\ G_{Nj} \end{bmatrix} D_j[n] \tag{8.5}$$

where G_{il} and B_{il} are real and imaginary parts of the admittance Y_{il}^{tr}. $C_i[n]$ and $D_i[n]$ are real and imaginary parts of $V_i[n]$. $\Re(\cdot)$ and $\Im(\cdot)$ are real and imaginary operators. In (8.5), for PQ buses,

$$\Gamma_i[n] = S_i^* W_i^*[n-1] - Y_i^{sh} V_i[n-1] - I_{Li}[n] \tag{8.6}$$

and for PV buses,

$$\Gamma_i[n] = P_i W_i^*[n-1] - j\sum_{k=0}^{n-1} Q_i[k] W_i^*[n-k] - Y_i^{sh} V_i[n-1] - I_{Li}[n] \tag{8.7}$$

From (8.3) the $C[n]$ terms on the RHS of (8.5) are

$$C_i[n] = \frac{1}{2C_i[0]} \left(\delta_{n,1}(|V_i^{sp}|^2 - 1) - \sum_{k=1}^{n-1} V_i[k] V_i^*[n-k] \right) \tag{8.8}$$

where $\delta_{n,1}$ is 1 for $n = 1$ and 0 for $n \neq 1$. To solve (8.5), the load current $I_{Li}[n]$ should be determined first. $I_{Li}[n]$ can be viewed as a load-grid interface, which decomposes the computation of the grid and loads. Here we only discuss $I_{Li}[n]$ as the current of induction motors. If there are multiple induction motors (labeled as m) on bus i, then

$$I_{Li}[n] = \sum_{m} I_{Lim}[n] \tag{8.9}$$

Next, the holomorphic embedding formulation of an induction motor will be established. As a general guideline, the holomorphic embedding formulation should satisfy the following requirements:

(1) The correctness of the mathematical model. At $\alpha = 1$, the holomorphic embedding formulation must be equivalent to the original mathematical model of the studied element or system.
(2) The germ solution should be easy to obtain. At $\alpha = 0$, the solution of the holomorphic embedding formulation is either trivial or known.

Take the most commonly used model of induction motors for analysis, whose equivalent circuit is shown in Figure 8.1. The circuit has an internal node whose voltage is denoted as $V_{Mim}(\alpha)$. Following the abovementioned rules, the construction of the trivial germ solution should satisfy: (i) the voltages on all buses are identical. (ii) The current on each branch is 0. Thus consequently, the admittance of every shunt branch should be 0. So in the holomorphic embedding formulation of induction motors, the impedance of the excitation branch is constructed as Z_{eim}/α to make 0 admittance. While for the rotor side, the manipulation can be achieved by modifying the mechanical torque as $\alpha T_{im}(s_{im})$. When $\alpha = 0$, the zero mechanical torque will guarantee $I_{Rim} = 0$. The following equations are listed for holomorphic embedding:

$$\begin{aligned}
V_{Mim}(\alpha) &= V_i(\alpha) - I_{Lim}(\alpha)Z_{1im} \\
I_{Lim}(\alpha) &= I_{Rim}(\alpha) + \frac{\alpha V_{Mim}(\alpha)}{Z_{eim}} \\
I_{Rim}(\alpha)I^*_{Rim}(\alpha) &= \frac{\alpha T_{im}(s_{im}(\alpha))s_{im}(\alpha)}{R_{2im}} \\
I_{Rim}(\alpha)\left(R_{2im} + jX_{2im}s_{im}(\alpha)\right) &= V_{Mim}(\alpha)s_{im}(\alpha)
\end{aligned} \tag{8.10}$$

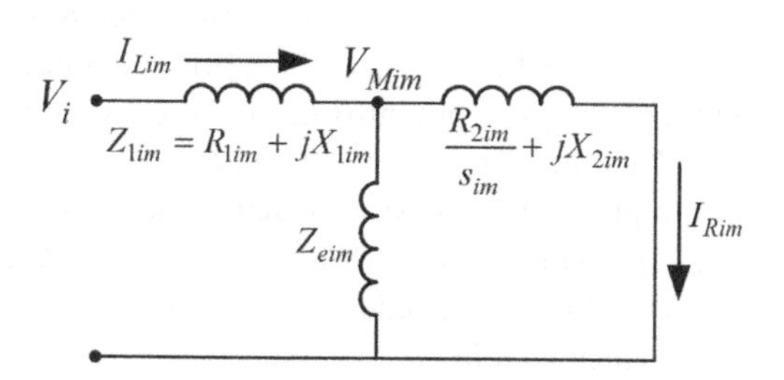

Figure 8.1 Equivalent circuit of induction motor.

where the mechanical torque T_{im} depends on the slip s_{im}, and usually it is assumed as a quadratic function of s_{im} [19]

$$T_{im}(s_{im}) = T_{0im} + T_{1im}s_{im} + T_{2im}s_{im}^2 \tag{8.11}$$

The germ solution of the embedding form (8.10) is $V_{Mim}[0] = 1$, $I_{Lim}[0] = 0$, $I_{Rim}[0] = 0$, $s_{im}[0] = 0$. From (8.10), the equations of the coefficients (8.12)–(8.15) are obtained ((8.14) and (8.15) are at the bottom of the page):

$$I_{Lim}[n] = I_{Rim}[n] + \frac{V_{Mim}[n-1]}{Z_{eim}} \tag{8.12}$$

$$V_{Mim}[n] = V_i[n] - I_{Lim}[n]Z_{1im} \tag{8.13}$$

$$\begin{aligned}\sum_{k=0}^{n+1} I_{Rim}[k]I_{Rim}^*[n+1-k] &= \frac{1}{R_{2im}}\left(T_{0im}s_{im}[n] + T_{1im}\sum_{k=0}^{n} s_{im}[k]s_{im}[n-k]\right.\\ &\left.\quad + T_{2im}\sum_{k=0}^{n}\sum_{l=0}^{n-k} s_{im}[k]s_{im}[l]s_{im}[n-k-l]\right)\end{aligned} \tag{8.14}$$

$$I_{Rim}[n]R_{2im} + jX_{2im}\sum_{k=0}^{n} I_{Rim}[k]s_{im}[n-k] = \sum_{k=0}^{n} V_{Mim}[k]s_{im}[n-k] \tag{8.15}$$

When $n = 1$, (8.14) and (8.15) become:

$$I_{Rim}[1]I_{Rim}^*[1] = \frac{T_{0im}s_{im}[1]}{R_{2im}} \tag{8.16}$$

$$I_{Rim}[1]R_{2im} = V_{Mim}[0]s_{im}[1] \tag{8.17}$$

From (8.16) and (8.17),

$$I_{Rim}[1] = T_{0im},\ s_{im}[1] = T_{0im}R_{2im} \tag{8.18}$$

When $n > 1$, note that $I_{Rim}[0] = 0$, $s_{im}[0] = 0$, (8.14) and (8.15) become linear equations of $I_{Rim}[n]$ and $s_{im}[n]$. Therefore, $I_{Rim}[n]$ and $s_{im}[n]$ can be solved from (8.14)–(8.15), and $I_{Lim}[n]$ is obtained from (8.12). The load current term $I_{Lim}[n]$ is put into (8.5) and the states in the grid $V_i[n]$, $W_i[n]$, $Q_i[n]$ are solved. Finally, with $V_i[n]$ and $I_{Lim}[n]$, $V_{Mim}[n]$ is obtained from (8.13). Such a procedure is clearly structured as the load-grid two-stage computation. Also, since solving (8.12)–(8.15) does not involve the variables other than the studied induction motor, here the states of all induction motors can be calculated in parallel.

8.1.1.2 Continuation from Practical States

The HE-I formulation based on trivial, nonrealistic germ solution is for the computation of a single power flow state. In voltage stability analysis, it is desirable to study the trace of system states when the system configuration changes, e.g. load increases in a direction. This is a quite typical QSS simulation scenario. Deriving and solving such a problem (denoted as HE-II) is similar, i.e. describing the variables as a power series of α and then solving the coefficients of the series. Denoting the embedding variable α as the loading factor, the holomorphic embedding formulation for bus i is

$$(P_i(\alpha) - jQ_i(\alpha))W_i^*(\alpha) - \sum_l Y_{il}V_l(\alpha) - I_{Li}(\alpha) = 0 \tag{8.19}$$

For PQ buses, the $P_i(\alpha)$ and $Q_i(\alpha)$ are known, while for PV buses, only $P_i(\alpha)$ is known and $Q_i(\alpha)$ is to be calculated. The voltage equations for PV and slack buses are:

$$\begin{aligned} V_i(\alpha)V_i^*(\alpha) &= |V_i^{sp}|^2, i \in S_{PV} \\ V_i(\alpha) &= V_i^{sp}, i \in S_{SL} \end{aligned} \tag{8.20}$$

From (8.19)–(8.20) the coefficients of the holomorphic embedding satisfy

$$\sum_{k=0}^{n} \left(P_i[k] - jQ_i[k]\right) W_i^*[n-k] - \sum_l Y_{il}V_l[n] - I_{Li}[n] = 0 \tag{8.21}$$

$$\sum_{k=0}^{n} V_i[k]V_i^*[n-k] = \begin{cases} |V_i^{sp}|^2, & n = 0 \\ 0, & n > 0 \end{cases} \tag{8.22}$$

For the induction motor loads, the equations are:

$$\begin{aligned} V_{Mim}(\alpha) &= V_i(\alpha) - I_{Lim}(\alpha)Z_{1im} \\ I_{Lim}(\alpha) &= I_{Rim}(\alpha) + \frac{V_{Mim}(\alpha)}{Z_{eim}} \\ \Re\{V_{Mim}(\alpha)I_{Rim}^*(\alpha)\} &= (r_{0im} + r_{1im}\alpha)T_{im}(s_{im}(\alpha)) \\ I_{Rim}(\alpha)\left(R_{2im} + jX_{2im}s_{im}(\alpha)\right) &= V_{Mim}(\alpha)s_{im}(\alpha) \end{aligned} \tag{8.23}$$

where r_{0im} is the base mechanical torque level, and r_{1im} represents the growth direction of mechanical torque.

It should be noted that although the formulations of HE-I and HE-II look similar, the underlying studied problems and the physical meanings are different. For HE-I, the purpose is to obtain a single steady-state power flow solution corresponding to $\alpha = 1$. In HE-I, α is only an embedding variable connecting between the trivial germ solution ($\alpha = 0$) and the desired solution ($\alpha = 1$), and it does not have a physical meaning. While the purpose of HE-II is to study the trace of system state when the load increases. In HE-II, α represents the loading factor. Any $\alpha \in \mathbb{R}$ can represent a physically existing state of the system. So in (8.23) the induction motor should keep the excitation impedance as Z_{eim} rather than Z_{eim}/α in (8.10). In steady-state VSA, it is usually assumed that the load grows linearly with the loading factor, so in (8.19), the active and reactive powers are expressed as $P_i(\alpha) = P_{0i} + \alpha P_{1i}$ and $Q_i(\alpha) = Q_{0i} + \alpha Q_{1i}$, respectively. Similarly in (8.23), the term $(r_{0im} + r_{1im}\alpha)T_{im}(s_{im}(\alpha))$ represents the linear growth of mechanical torque with loading factor α. To solve HE-II, it is assumed that the germ solution (which is a steady-state physical solution) is known. The germ solution can be given by HE-I/II or other methods for solving power flows. Generally, given the analytic algebraic equations, a proper holomorphic embedding formulation can be established and the problem can be solved. So the application can also be further extended to include generation control [20], outage analysis [21, 22], etc.

From (8.20)–(8.22), and renumbering the PQ buses before the PV buses, the holomorphic embedding coefficients can be derived by solving the following equations:

$$
\left[\begin{array}{c:c:c:c:c}
-\mathbf{G} & \mathbf{B} & \mathcal{D}(\mathbf{P}[0]) & -\mathcal{D}(\mathbf{Q}[0]) & \begin{matrix}\mathbf{0}\\ -\mathcal{D}(\mathbf{F}_{PV}[0])\end{matrix} \\ \hdashline
-\mathbf{B} & -\mathbf{G} & -\mathcal{D}(\mathbf{Q}[0]) & -\mathcal{D}(\mathbf{P}[0]) & \begin{matrix}\mathbf{0}\\ -\mathcal{D}(\mathbf{E}_{PV}[0])\end{matrix} \\ \hdashline
\mathbf{0}\ \mathcal{D}(\mathbf{C}_{PV}[0]) & \mathbf{0}\ \mathcal{D}(\mathbf{D}_{PV}[0]) & \mathbf{0} & \mathbf{0} & \mathbf{0} \\ \hdashline
\mathcal{D}(\mathbf{E}[0]) & -\mathcal{D}(\mathbf{F}[0]) & \mathcal{D}(\mathbf{C}[0]) & -\mathcal{D}(\mathbf{D}[0]) & \mathbf{0} \\ \hdashline
\mathcal{D}(\mathbf{F}[0]) & \mathcal{D}(\mathbf{E}[0]) & \mathcal{D}(\mathbf{D}[0]) & \mathcal{D}(\mathbf{C}[0]) & \mathbf{0}
\end{array}\right]
\begin{bmatrix}\mathbf{C}[n]\\ \mathbf{D}[n]\\ \mathbf{E}[n]\\ \mathbf{F}[n]\\ \mathbf{Q}_{PV}[n]\end{bmatrix}
$$

$$
= \left[\begin{array}{c}
\Re\left(-\sum_{k=1}^{n}\mathbf{P}_{PQ}[k]\circ\mathbf{W}^*_{PQ}[n-k] + j\sum_{k=1}^{n}\mathbf{Q}_{PQ}[k]\circ\mathbf{W}^*_{PQ}[n-k] + \mathbf{I}_{LPQ}[n]\right) \\
\Re\left(-\sum_{k=1}^{n}\mathbf{P}_{PV}[k]\circ\mathbf{W}^*_{PV}[n-k] + j\sum_{k=1}^{n-1}\mathbf{Q}_{PV}[k]\circ\mathbf{W}^*_{PV}[n-k] + \mathbf{I}_{LPV}[n]\right) \\ \hdashline
\Im\left(-\sum_{k=1}^{n}\mathbf{P}_{PQ}[k]\circ\mathbf{W}^*_{PQ}[n-k] + j\sum_{k=1}^{n}\mathbf{Q}_{PQ}[k]\circ\mathbf{W}^*_{PQ}[n-k] + \mathbf{I}_{LPQ}[n]\right) \\
\Im\left(-\sum_{k=1}^{n}\mathbf{P}_{PV}[k]\circ\mathbf{W}^*_{PV}[n-k] + j\sum_{k=1}^{n-1}\mathbf{Q}_{PV}[k]\circ\mathbf{W}^*_{PV}[n-k] + \mathbf{I}_{LPV}[n]\right) \\ \hdashline
-\frac{1}{2}\sum_{k=1}^{n-1}\mathbf{V}_{PV}[k]\circ\mathbf{V}^*_{PV}[n-k] \\ \hdashline
\Re\left(-\sum_{k=1}^{n-1}\mathbf{W}[k]\circ\mathbf{V}[n-k]\right) \\ \hdashline
\Im\left(-\sum_{k=1}^{n-1}\mathbf{W}[k]\circ\mathbf{V}[n-k]\right)
\end{array}\right] \tag{8.24}
$$

where E_i and F_i are real and imaginary parts of W_i, $\mathcal{D}(\cdot)$ forms a diagonal matrix from a vector, "o" produces the element-wise multiplication.

Similar to HE-I, Eq. (8.24) contains terms of induction motor currents $\mathbf{I}_{LPQ}[n]$ and $\mathbf{I}_{LPV}[n]$ (denoting $I_{Lim}[n]$ on PQ and PV buses, respectively). From (8.23), the holomorphic embedding formulation of each induction motor has Eq. (8.25). The terms a_{im}, b_{im}, and c_{im} in (8.25) are listed separately as follows:

$$\left[\begin{array}{c:c:c:c:c:c:c} R_{2im} & -X_{2im}s_{im}[0] & R_{1im}s_{im}[0] & -X_{1im}s_{im}[0] & a_{im} & -s_{im}[0] & 0 \\ \hdashline X_{2im}s_{im}[0] & R_{2im} & X_{1im}s_{im}[0] & R_{1im}s_{im}[0] & b_{im} & 0 & -s_{im}[0] \\ \hdashline \Re\{V_{Mim}[0]\} & \Im\{V_{Mim}[0]\} & 0 & 0 & c_{im} & 0 & 0 \\ \hdashline -1 & 0 & 1+\Re\left(\frac{Z_{1im}}{Z_{eim}}\right) & -\Im\left(\frac{Z_{1im}}{Z_{eim}}\right) & 0 & -\Re\left(\frac{1}{Z_{eim}}\right) & \Im\left(\frac{1}{Z_{eim}}\right) \\ \hdashline 0 & -1 & \Im\left(\frac{Z_{1im}}{Z_{eim}}\right) & 1+\Re\left(\frac{Z_{1im}}{Z_{eim}}\right) & 0 & -\Im\left(\frac{1}{Z_{eim}}\right) & -\Re\left(\frac{1}{Z_{eim}}\right) \end{array}\right]$$

$$\begin{bmatrix} I_{Rim\mathrm{Re}}[n] \\ I_{Rim\mathrm{Im}}[n] \\ I_{Lim\mathrm{Re}}[n] \\ I_{Lim\mathrm{Im}}[n] \\ s_{im}[n] \\ C_i[n] \\ D_i[n] \end{bmatrix}$$

$$= \left[\begin{array}{c} \Re\left(\sum_{k=1}^{n-1} V_{Mim}[k]s_{im}[n-k] - jX_{2im}\sum_{k=1}^{n-1} I_{Rim}[k]s_{im}[n-k]\right) \\ \hdashline \Im\left(\sum_{k=1}^{n-1} V_{Mim}[k]s_{im}[n-k] - jX_{2im}\sum_{k=1}^{n-1} I_{Rim}[k]s_{im}[n-k]\right) \\ \hdashline -\Re\{\sum_{k=1}^{n} V_{Mim}[k]I_{Rim}^{*}[n-k]\} + r_{0im}\left(T_{2im}\sum_{k=1}^{n-1} s_{im}[k]s_{im}[n-k]\right) \\ +r_{1im}\left(T_{0im}\delta(n,1) + T_{1im}s_{im}[n-1] + T_{2im}\sum_{k=0}^{n-1} s_{im}[k]s_{im}[n-1-k]\right) \\ \hdashline 0 \\ \hdashline 0 \end{array}\right] \tag{8.25}$$

$$a_{im} = -I_{Rim\mathrm{Im}}[0]X_{2im} - C_i[0] + I_{Lim\mathrm{Re}}[0]R_{1im} - I_{Lim\mathrm{Im}}[n]X_{1im} \tag{8.26}$$

$$b_{im} = I_{Rim\mathrm{Re}}[0]X_{2im} - D_i[0] + I_{Lim\mathrm{Re}}[0]X_{1im} + I_{Lim\mathrm{Im}}[n]R_{1im} \tag{8.27}$$

$$c_{im} = -r_{0im}\left(T_{1im} + 2T_{2im}s_{im}[0]\right) \tag{8.28}$$

Equation (8.25) has seven unknowns and five equations. Eliminating $I_{Rim\mathrm{Re}}[n]$, $I_{Rim\mathrm{Im}}[n]$ and $s_{im}[n]$, the load currents $I_{Lim\mathrm{Re}}[n]$, $I_{Lim\mathrm{Im}}[n]$ change linearly with $C_i[n]$ and $D_i[n]$:

$$\begin{bmatrix} I_{Lim\mathrm{Re}}[n] \\ I_{Lim\mathrm{Im}}[n] \end{bmatrix} = A_{im}\begin{bmatrix} C_i[n] \\ D_i[n] \end{bmatrix} + B_{im} \tag{8.29}$$

Aggregating all the currents of motor loads on each bus

$$\begin{bmatrix} I_{Li\mathrm{Re}}[n] \\ I_{Li\mathrm{Im}}[n] \end{bmatrix} = \left(\sum_m A_{im}\right) \begin{bmatrix} C_i[n] \\ D_i[n] \end{bmatrix} + \sum_m B_{im} \tag{8.30}$$

and substituting (8.30) into (8.24) and canceling $\mathbf{I}_L$ terms, the grid terms $\mathbf{C}[n]$, $\mathbf{D}[n]$, $\mathbf{E}[n]$, $\mathbf{F}[n]$, and $\mathbf{Q}_{PV}[n]$ are solved. Placing $\mathbf{C}[n]$, $\mathbf{D}[n]$ back to (8.25) will solve the states of induction motors $I_{Rim\mathrm{Re}}[n]$, $I_{Rim\mathrm{Im}}[n]$, $I_{Lim\mathrm{Re}}[n]$, $I_{Lim\mathrm{Im}}[n]$, and $s_{im}[n]$.

As for the computational complexity of HE-II, the major computation burden is on solving the linear equations in (8.24) and (8.25). Since the matrices are constant, the factorization of matrices is needed only once and the computation can be significantly accelerated.

8.1.1.3 Enabling Dynamic Modeling

HE-II is based on the QSS assumption without dynamic modeling. However, when analyzing motor stalling, dynamic modeling and simulation in time domain is necessary. In the formulation of HE-II, with a common assumption in steady-state VSA that the loading factor increases linearly with time, the loading factor α also implicitly represents the time elapse. If t is explicitly used instead of α as the embedding variable, the solution naturally gives the evolution of states in time domain. In this regard, we can take a further step to apply the holomorphic embedding into dynamic simulation which is denoted as HE-III.

In VSA, assume the system does not have any angle stability problem, so the dynamic models of synchronous generators are not considered, and the generator buses are still modeled as constant-voltage sources connected to PV or slack buses. Therefore, in HE-III, the power flow equations of the system except for induction motors are the same as those in HE-II. In long-term VSA, the electro-magnetic transients of motors can also be neglected, and differential equations only include rotor acceleration equations:

$$2H_{im}\frac{ds_{im}}{dt} = (r_{0im} + r_{1im}t)T_{im}(s_{im}) - \Re\{V_{Mim}I^*_{Rim}\} \tag{8.31}$$

In HE-III, the embedding variable becomes time t, and the solution represents the evolution of states in time domain. Take (8.31) as an example. Assume the slip $s_{im}(t)$ as

$$s_{im}(t) = \sum_{k=0}^{\infty} s_{im}[k]t^k \tag{8.32}$$

then the derivative of s_{im} with respect to t is

$$\frac{ds_{im}}{dt} = \sum_{k=1}^{\infty} ks_{im}[k]t^{k-1} \tag{8.33}$$

From (8.31) the equations for the coefficients of the terms of the same order can be obtained:

$$
\begin{aligned}
&2(n+1)H_{im}s_{im}[n+1] = \\
&r_{0im}\left(T_{0im}\delta_{n,0} + T_{1im}s_{im}[n] + T_{2im}\sum_{i+j=n} s_{im}[i]s_{im}[j]\right) + \\
&r_{1im}\left(T_{0im}\delta_{n,1} + T_{1im}s_{im}[n-1] + T_{2im}\sum_{i+j=n} s_{im}[i]s_{im}[j]\right) - \\
&\Re\left\{\sum_{i+j=n} V_{Mim}[i]I^{*}_{Rim}[j]\right\}
\end{aligned}
\tag{8.34}
$$

It shows that $s_{im}[n]$ can be derived from all the terms of orders up to $n-1$. The equations of induction motors in HE-III are similar to (8.25). Remove the third equation of (8.25), substitute $s_{im}[n]$ into other Eq. (8.25), and eliminate $I_{RimRe}[n]$, $I_{RimIm}[n]$ terms. Then one can get a similar relationship between $I_{LimRe}[n]$, $I_{LimIm}[n]$, and $C_i[n]$, $D_i[n]$ like (8.29), and then the following procedure is the same with HE-II.

HE-III flexibly allows dynamic modeling on only part of the motors while maintaining steady-state modeling on the other motors. The motors modeled in steady-state have the same formulation as that in HE-II.

8.1.2 VSA Using Holomorphic Embedding

8.1.2.1 Extend Effective Range by Using Padé Approximation

A power series usually has a limited radius of convergence, so the series derived by holomorphic embedding become inaccurate beyond the radius. Moreover, the truncated power series tends to have an even smaller effective range under given accuracy tolerance. To overcome such a disadvantage of power series approximation, the Padé approximants can be used to extend the range of satisfactory accuracy [2, 3]. Although some extra computation is needed for deriving the Padé approximants from power series, the holomorphic embedding using Padé approximants still has advantage in VSA because of its much better convergence than power series. So the Padé approximants are used for the approximation of solutions.

8.1.2.2 Multistage Holomorphic Embedding

Even with Padé approximants, the effective range of approximation is still limited [9]. Increasing the terms of Padé approximants can help extend the effective range. But when the number of terms reaches a certain level, increasing it has little effect in extending the effective approximation range. Also, adding terms costs more time in deriving Padé approximants. The number of Padé approximants terms is limited by N_P, and in VSA, the effectiveness of approximants is checked by comparing the equation balance with a tolerance ε_E. The furthest point on the system state trace satisfying the equation imbalance below ε_E is designated as the end of the current stage as well as the starting point of the next stage. Such a multi-stage scheme guarantees the accuracy of simulation until the collapse point without requiring too many Padé terms, which saves computation time.

A remark is worth noticing on the modeling of the mechanical torque of induction motors. In this paper, the mechanical torque adopts the commonly used quadratic form (8.11), but holomorphic embedding can also handle more general mechanical torque models. For example, $T_{im}(s_{im})$ as even higher-order polynomials can also be solved with holomorphic embedding. Although the equations need modification, the methodology is the same. Even more generally, $T_{im}(s_{im})$ in other forms can also be solved with holomorphic embedding. If $T_{im}(s_{im})$ is not in a polynomial form, then it can be approximated with Taylor expansion as a truncated polynomial $T'_{im}(s_{im})$. In this case, the initial state s_{im0} can be selected as the original point of Taylor expansion, and since this involves approximation, the error can be tracked and controlled by using the multistage holomorphic embedding scheme. Also, the switching of segments can also be achieved with the multistage holomorphic embedding scheme.

8.1.2.3 Partial-QSS Voltage Stability Analysis Scheme

The procedure for VSA using the partial-QSS method [23] is shown in Figure 8.2. We assume the ZIP/motor load grows continuously at a certain pattern with time. Here we name the method as "partial-QSS" because the simulation starts with standard QSS method, but some motors switch to dynamic model during the simulation. The multistage holomorphic embedding is used in simulation, and when singularity occurs, the type of singularity is examined by using participation factors [23]. First, the Jacobian matrix **J** of all the algebraic equations (including the power flow equations on all buses and the equilibrium equations of rotor motion of motors that are under QSS assumption) is derived, then apply eigen decomposition to **J**:

$$\mathbf{J} = \mathbf{P}\boldsymbol{\Lambda}\mathbf{Q} \tag{8.35}$$

and select the left and right eigenvectors $\mathbf{p}_i$ and $\mathbf{q}_i$ corresponding to the smallest eigenvalue λ_i. Then the participation factors of the jth algebraic equation are:

$$\pi_j = p_{ij}q_{ij} \tag{8.36}$$

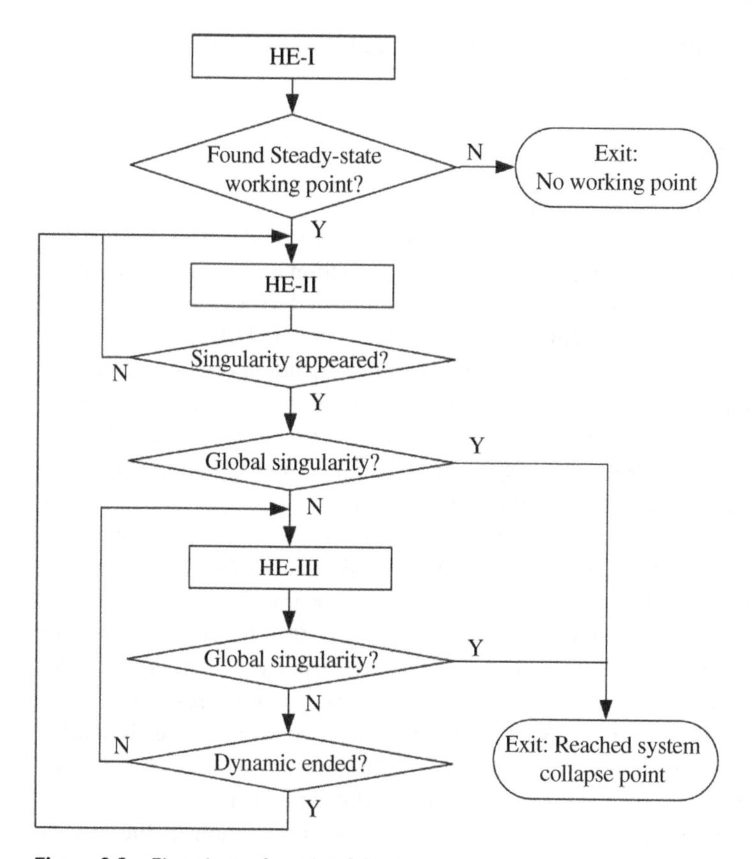

Figure 8.2 Flowchart of partial-QSS VSA.

If any motor has a participation factor much larger than the others (e.g. over 10 times that of any other), then it can be inferred that singularity is mostly contributed by the motor model. In this case we can switch the motor to dynamic model in HE-III, and continue simulation. Otherwise, if the participation factor of all motors are small, then the singularity indicates voltage stability issue on the whole system.

8.1.2.4 Full-Dynamic Simulation

With HE-III, the time-domain simulation with dynamic modeling of all motors can be implemented. The simulation starts from a steady-state starting point derived by HE-I, and then the multistage HE-III simulates system dynamics until meeting global singularity. The full-dynamic simulation with HE-III should be more accurate than the partial-QSS method, and a significant boost in computation speed is expected as compared with traditional numerical integration method.

8.1.3 Test Cases

8.1.3.1 IEEE 14-Bus System

IEEE 14-bus system has 11 PQ buses, 4 PV buses, and 1 slack bus. To test the holomorphic embedding in VSA with induction motors, the IEEE 14-bus system is modified so that each PQ bus is connected with an induction motor load. The average percentage of induction motor load is 42.2%.

First test the HE-I in solving power flow. In holomorphic embedding, different numbers of Padé approximant terms N_P are selected, and the results are checked with the maximum absolute imbalance of the equations. As shown in Table 8.1, as N_P grows, the maximum equation imbalance decreases significantly, but the computation time increases. The N_P can be chosen based on the actual demand of accuracy and computation speed.

Next, we study the system state evolution under load increase. The mechanical torque of induction motors is assumed as $T_{im} = T_0(1 - s_{im}^2)$. In the 14-bus system, increase the PQ load by 10% of base state value per second, and increase the mechanical load by 2% per second, then use HE-II and HE-III to do steady-state and dynamic simulations, respectively. For both HE-II and HE-III, $N_P = 20$. Under QSS assumptions, the simulation stops at $t = 6.286$s, as Figure 8.3 shows, the imbalance of equations under HE-II is below 10^{-5}, showing that the HE-II itself is accurate. As Figure 8.4 shows, the steady-state solution exists at the intersection of the electric torque and mechanical torque curves. At practical working points, the slip is close to 0 (the left intersection point in Figure 8.4). When mechanical load increases (mechanical torque rises) or voltage decreases (electric torque

Table 8.1 Accuracy and efficiency of HE-I in IEEE 14-bus system case.

N_P	Max. equation imbalance (pu)	Computation time (s)
5	9.6×10^{-3}	0.019
10	1.4×10^{-3}	0.030
15	4.9×10^{-5}	0.043
20	3.0×10^{-7}	0.058
30	6.3×10^{-10}	0.074
40	1.7×10^{-11}	0.091

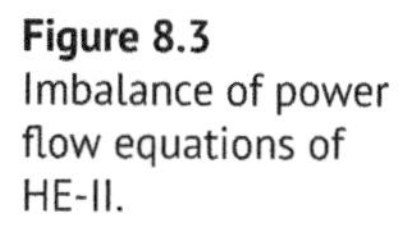

Figure 8.3 Imbalance of power flow equations of HE-II.

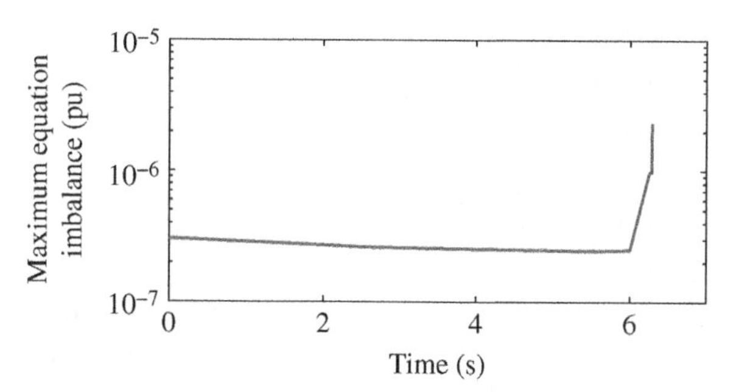

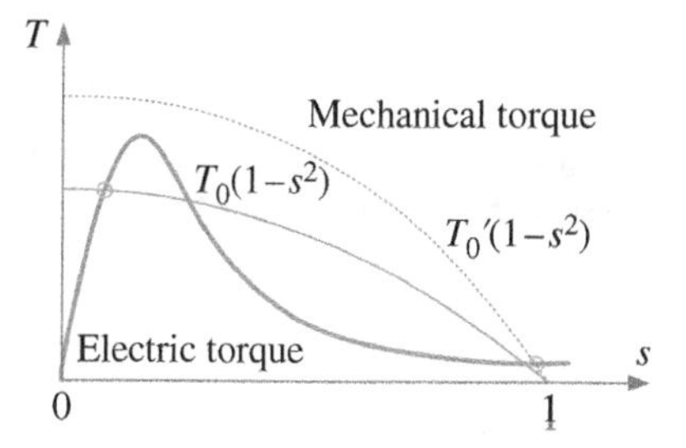

Figure 8.4 Illustration of induction motor steady-state solutions (denoted by the circles at the intersections of the curves).

drops), the intersection point will move toward the peak point of the electric torque with slip increase. As mechanical load increases or voltage drops to some extent, the intersection point near $s_{im} = 0$ disappears, which is a bifurcation, and the motor starts to stall. The state of a stalling motor then moves to the only steady-state solution near $s_{im} = 1$. As Figure 8.5 shows, the termination point of simulation approximately matches the maximum electric torque point of motor 4, which verifies the local bifurcation caused by the motor [23]. At the local bifurcation point, there is no steady-state solution in the vicinity, but it does not mean the collapse of the whole system. The system can still operate with increasing load, but the stalling motor should be simulated in dynamic model (as was treated in [23]), or alternatively, all the motors can be simulated with dynamic models from the very beginning. We use HE-III to simulate the dynamics of all the induction motor loads, and as Figure 8.5 shows, the HE-III successfully simulates the stalling of the motor. Figure 8.6 compares the results of HE-III with that of the modified Euler method (a commonly-used numerical integration method), it shows that the difference of bus voltages is below 10^{-4} for the entire time span, and for 99% of the time span, the difference is below 10^{-6}.

The results of holomorphic embedding is compared with those of the modified Euler method under different time steps. For the modified Euler method, time steps of $\Delta t = 0.01$ s, $\Delta t = 0.005$ s, and $\Delta t = 0.002$ s are selected. The metric selected for comparison is the maximum imbalance

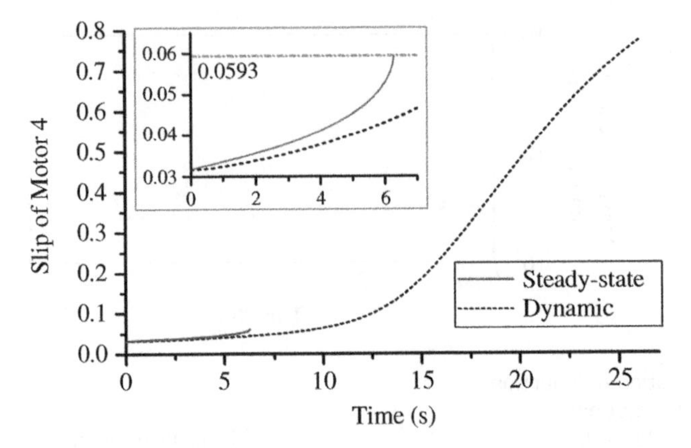

Figure 8.5 Slip of motor on bus 4 in IEEE 14-bus system. The horizontal line in the small figure corresponds to the slip at the maximum electric torque.

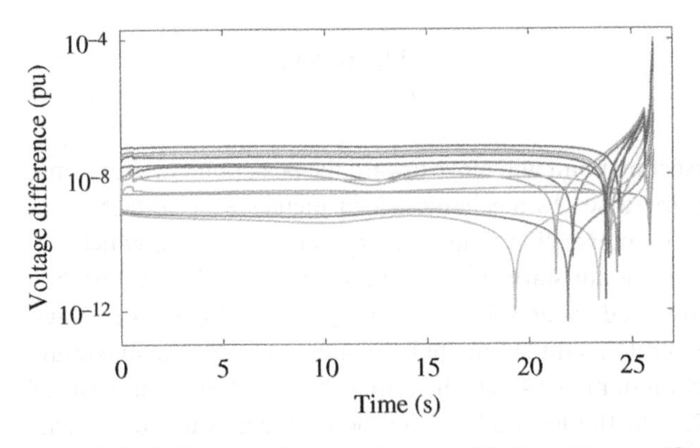

Figure 8.6 Difference of voltage between HE-III and the modified Euler methods.

of the DAEs. The results are shown in Figure 8.7. It can be seen that the equation imbalance of the modified Euler method is much higher than that of HE-III. Although reducing the time step achieves lower imbalance, to achieve the same level of accuracy as HE-III, the time step of the modified Euler method needs to be around $\Delta t = 10^{-4}$ s, which is impractically small. Therefore, HE-III has significant advantage in accuracy over the modified Euler method as a traditional numerical integration method.

8.1.3.2 NPCC 140-Bus System

The Northeast Power Coordinating Council (NPCC) system model [24] has 140 buses, including 45 PV buses, 94 PQ buses, and 1 slack bus. The system

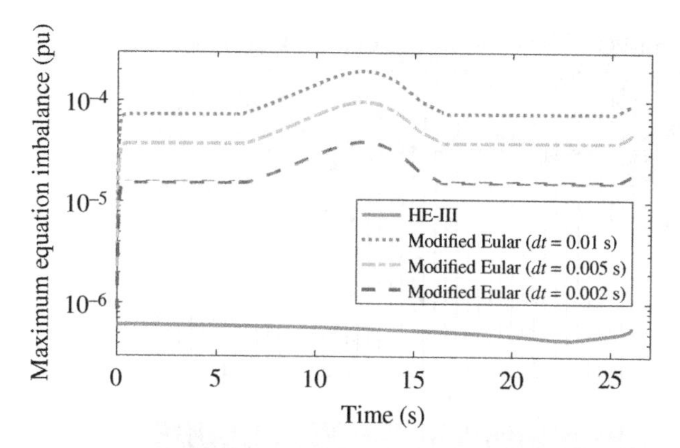

Figure 8.7 Imbalance of equations: HE-III and the modified Euler methods.

is also modified to add 54 induction motors to 54 PQ buses. Moreover, ZIP loads can also be added to these buses. To demonstrate the performance of the method and compare different load models, three system models are provided for analysis: (i) PQ modeling of load only, (ii) ZIP modeling of load, and (iii) ZIP and induction motors.

The parameters of the ZIP and motor loads are adjusted so that under the base condition, the system state is the same as the original PQ-load only system model. The percentage of each type of load on the 54 buses are shown in Figure 8.8. For the system model with ZIP load, the average percentage of the ZIP components are: 17.0% for "*Z*," 17.9% for "*I*" and 65.1% for "*P*." And for system model with ZIP and induction motor load, the average load components are: 8.0% for "*Z*," 6.4% for "*I*," 51.5% for "*P*," and 34.1% for motor.

The system models with only PQ loads or ZIP loads do not have dynamic models, so the long-term VSA can be studied with multistage HE-II. Assume the load increase on each bus is 5% of its base state value per second. The time consumption and the number of holomorphic embedding stages are listed in Table 8.2. For comparison, the continuation power flow (CPF) in PQ-load model case takes 57 steps and 2.69 s at best performance. This test verifies satisfactory efficiency of the holomorphic embedding method.

Next, test the dynamic simulation with induction motor loads. In the NPCC system model, the inertial constants of induction motors H_{im} range from 0.5 to 2 s, and the average value is 1.22 s. The mechanical loads of induction motors increase by 10% of base case values per second, and the other types of loads increase by 5% of the base case value per second. First use the partial-QSS method [23] implemented in holomorphic embedding for

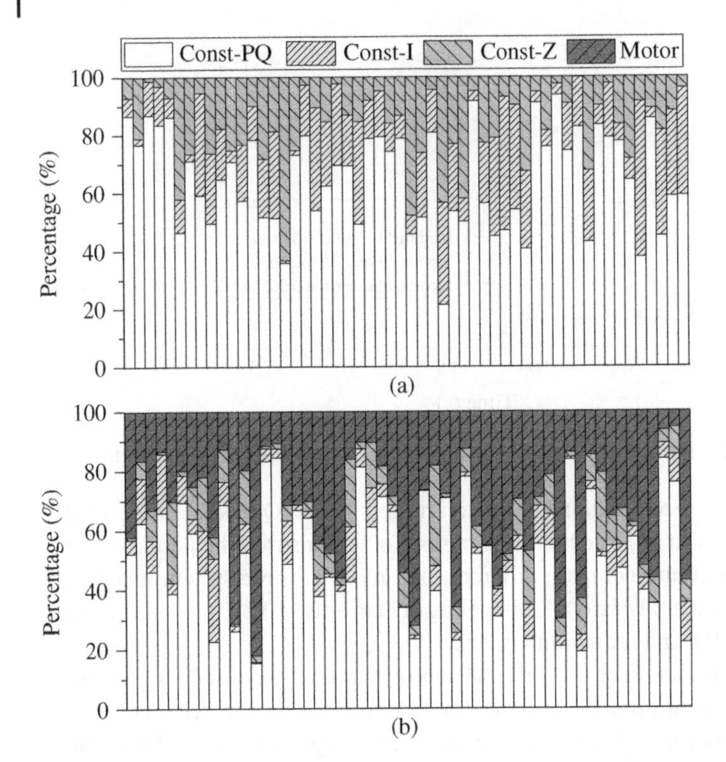

Figure 8.8 Percentage by load type on 54 load buses in NPCC system. (a) ZIP load model; (b) ZIP + Motor load model. The x-axis represents the load buses.

Table 8.2 Performance of holomorphic embedding in static VSA.

Load model	Stages	Time (s)
PQ	4	0.65
ZIP	7	1.41

simulation, and the motor slips are shown in Figure 8.9. Results show that the collapse point simulated by partial-QSS scheme is at 7.856 s.

Full-dynamic simulation with HE-III is also conducted in the same system. The voltage of buses and motor slips are shown in Figures 8.10 and 8.11. With full-dynamic simulation, the system collapse point is reached at 7.856 s. The system state traces are also benchmarked with traditional numerical integration methods, and the differences of bus voltages

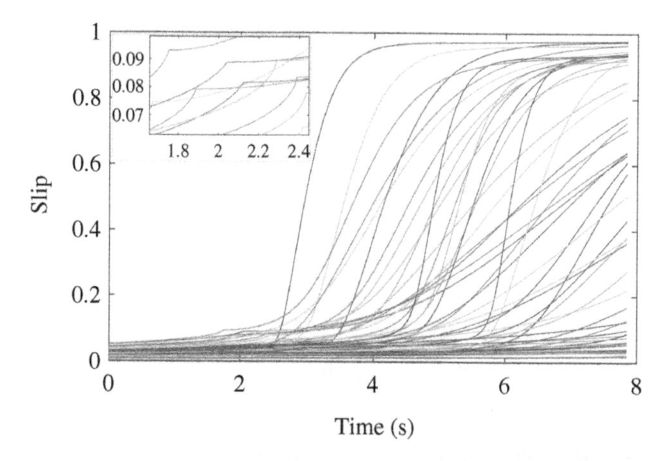

Figure 8.9 Motor slips of NPCC system obtained by partial-QSS with holomorphic embedding (ZIP+Motor loads). The small figure shows the switching points when induction motors start to stall.

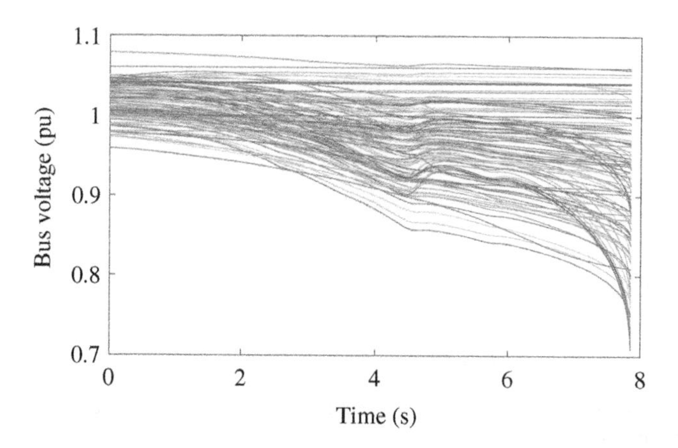

Figure 8.10 Bus voltage of NPCC system (ZIP+Motor load).

are shown in Figure 8.12. The result shows that the voltage difference is under 2×10^{-4} p.u. for over 99% of the time span, and the difference is under 1.5×10^{-3} p.u. for the entire process, which shows satisfactory accuracy of the holomorphic embedding method.

In this case, the partial-QSS simulation results are also quite close to those derived by full-dynamic simulation. But due to QSS assumption, the accuracy is compromised. Also, the holomorphic embedding using the partial-QSS scheme is not as efficient as full-dynamic holomorphic embedding. The partial-QSS scheme needs additional efforts for switching

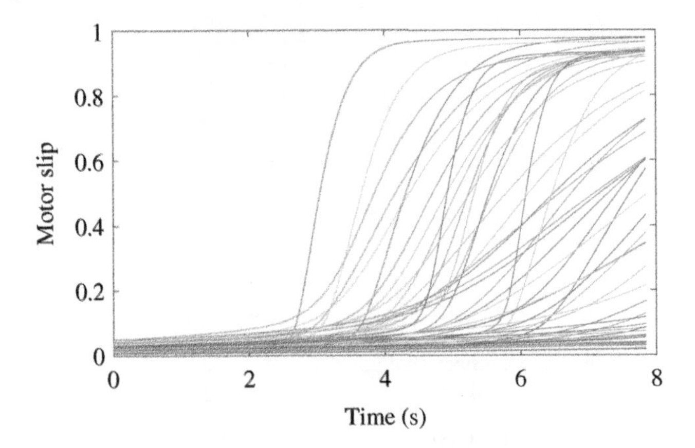

Figure 8.11 Motor slip in NPCC system (ZIP+Motor load).

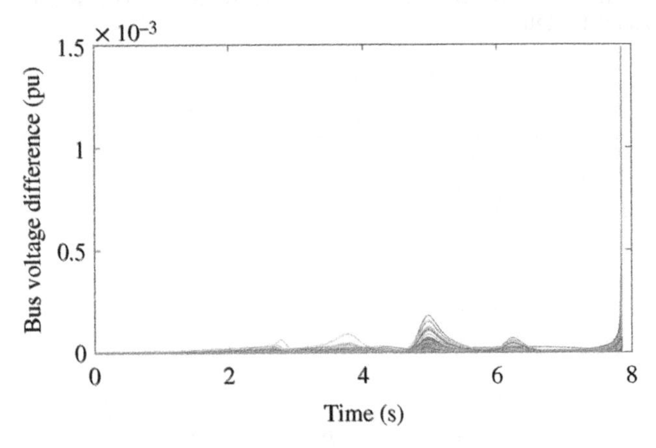

Figure 8.12 Comparison between holomorphic embedding (full-dynamic) and modified Euler method.

motors from static to dynamic model. Moreover, the switching of motor models causes errors with the equations, and thus shortens the length of each stage that holomorphic embedding can reach. Figure 8.13 shows that the equation imbalance of partial-QSS scheme grows significantly where many motors switch to dynamic model, and finally causing more than 10 times the equation imbalance that of full-dynamic simulation scheme. In contrast, the full-dynamic simulation scheme implemented in holomorphic embedding has nearly stable equation imbalance.

In terms of computation efficiency, Table 8.3 compares the modified Euler method, the holomorphic embedding with partial-QSS scheme, and

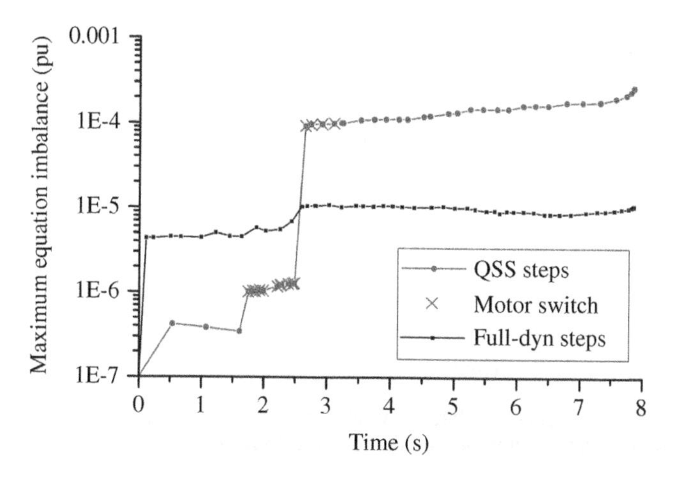

Figure 8.13 Comparison of steps and equation mismatch between partial-QSS and full-dynamic schemes.

Table 8.3 Computational efficiency comparison in NPCC system.

Simulation methods	Stages/Steps	Avgerage step len.(s)	Time cost(s)
Modified Euler	3928	0.002	1150.1
HE (partial-QSS)	105	0.0748	31.19
HE (full-dynamic)	55	0.143	13.95

holomorphic embedding full-dynamic simulation. The modified Euler method uses a fixed step length of 0.002 s, which is the largest available step length without causing a significant numerical simulation error. The partial-QSS simulation implemented with holomorphic embedding is much faster than the numerical integration method. Nevertheless, the full-dynamic simulation with holomorphic embedding is even faster, and it is also advantageous in accuracy. These test results exhibit the promising potential of holomorphic embedding in dynamic analysis of power systems.

8.1.3.3 Polish Test System

To further demonstrate the capability of holomorphic embedding, a large-scale system model modified from the Polish test system [25] is utilized in this case. The system has 2383 buses, including 2056 PQ buses and 326 PV buses. There are 1562 ZIP loads and 1542 induction motors in the system. In such a complex test, the NR method faces difficulty in solving the

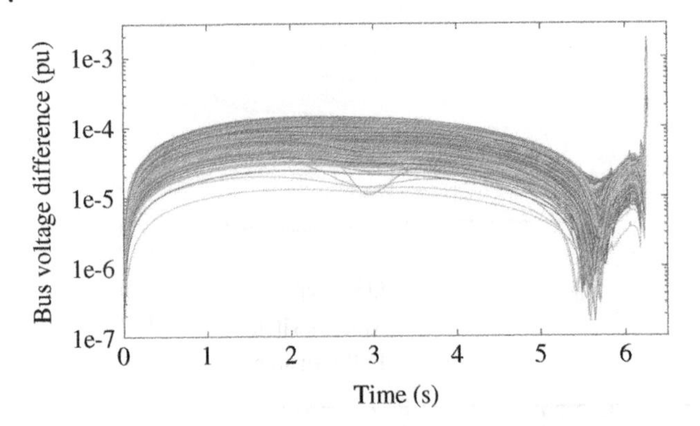

Figure 8.14 Comparison of results between holomorphic embedding (full-dynamic) and modified Euler method on Polish test system.

nonlinear algebraic equations of the system, so for the traditional numerical integration approaches, the time step is restricted to a very small value. By testing, the time step for the modified Euler method is set as 0.002 s. Assume the PQ/ZIP load increase on each bus is 10% of its base-state value per second, and the mechanical torque of each induction motor increases by 50% of its base-state value per second. The results of the two methods match well. As Figure 8.14 shows, the difference of bus voltage is below 2×10^{-4} p.u. for above 99% of the time span. But in terms of computational efficiency, the two approaches differ significantly. The holomorphic embedding is significantly faster than the modified Euler method. And the result shows that the VSA based on holomorphic embedding can be finished within a minute on the Polish system with many static and dynamic loads, so this holomorphic embedding-based approach is promising for accurate online VSA (Table 8.4).

8.1.4 Summary of the Section

In this section, three formulations of holomorphic are derived for solving a single steady state, QSS trace, and system dynamics in the context of

Table 8.4 Computational efficiency comparison in Polish system.

Simulation methods	Time cost(s)
Modified Euler	2840.4
HE (full-dynamic)	28.63

voltage stability analysis, respectively. The algorithms for solving the three holomorphic embedding formulations are also elucidated. Then, with Padé approximation and multistage continuation techniques to extend the effective range of holomorphic embedding, the partial-QSS simulation and full-dynamic simulation methods based on the holomorphic embedding formulations are implemented, respectively. The key takeaway of this section is the holomrphic embedding formulation for differential equations, which expands the use of holomorphic embedding from steady-state analysis (solving algebraic equations only) to dynamic anslysis (solving differential equations and DAE).

The voltage stability analysis based on holomorphic embedding are tested on the IEEE 14-bus test system, an NPCC 140-bus system and Polish 2383-bus system. By comparing with traditional methods such as CPF and dynamic simulation based on numerical integration, the advantages of the holomorphic embedding-based methods in accuracy is verified. Moreover, the tests on NPCC 140-bus system and Polish 2383-bus system show that holomorphic embedding is significantly faster than the traditional modified Euler method, which reveals very promising performance of holomorphic embedding methodology in dynamic analysis of power systems.

8.2 Generic Holomorphic Embedding for Dynamic Security Analysis

Power systems are large in scale and complex in composition (including transmission lines, transformers, generators, loads, and controllers). The complex structure and diverse element behaviors may bring various security issues. Of the approaches for studying system stability, in contrast to the specified analysis methods that are usually subject to simplifications and assumptions [26–28], dynamic simulation is a generic approach for dynamic security analysis (DSA) of power systems due to its flexibility in accommodating various elements, control measures, and disturbances.

Mathematically, dynamic simulation solves the initial-value problem (IVP) of DAE. Conventional approaches that include explicit and implicit methods [29–31] approximate the trace of state variables as numerical integration of differential equations with small time steps. Because numerical integration is a low-order approximation of system dynamics, the time step should be small enough to confine the error, which limits the speed of the simulation. To enhance performance and credibility of simulation, new methodologies that extend time steps and decrease errors is desired.

Moreover, complex nonlinear dynamic behaviors also pose challenges for simulation. To simplify computation, many dynamic simulation algorithms assume that loads use a constant-impedance model [32–34], which makes the DAE convertible to ordinary differential equations (ODE). However, the actual load behaviors may be quite different from those of constant impedance loads, and such an assumption may cause substantial errors in DSA [35, 36]. More accurate modeling of load behaviors leads to strong non-linearity in algebraic equations, and conventional approaches usually use the NR method to solve the equations. However, the NR method depends on the selection of the initial guess: it may fail to converge when the system is large, or the system is in stress, or the disturbance is relatively large.

This section will show how holomorphic embedding can significantly improve the performance and accuracy of simulation. The section first generalizes the holomorphic embedding methodology for power system dynamic simulation, including the algorithm for solving DAEs as well as solving instant system switches. First, rules for deriving holomorphic embedding formulation and solving holomorphic embedding approx-imate solutions of DAE. Then, by introducing the generator interface, holomorphic embedding formulations of synchronous generators and the coordinate transformation, and the holomorphic embedding formulation of power system dynamics are established. The instant switches in power systems caused by faults are also modeled and solved by holomorphic embedding. By putting together the holomorphic embedding formulations for solving system dynamics and instant switches, the overall dynamic simulation process based on holomorphic embedding is established.

8.2.1 General Holomorphic Embedding

8.2.1.1 Dynamic Simulation Formulation

The dynamic simulation of a power system typically solves a set of DAE:

$$\dot{x} = f(x, y, p(t)), \tag{8.37a}$$

$$0 = g(x, y, p(t)), \tag{8.37b}$$

where x denotes state variables, y stands for algebraic variables (e.g. voltage phasors on the buses), and $p(t)$ denotes system configuration (e.g. the con-nection status of branches, the extent of load increase). The disturbances introduced into a power system can be represented by changing $p(t)$.

In a numerical integration method for solving DAE, with a known system state at time t (i.e. $x(t)$ and $y(t)$), and given a time step Δt, the next-step state $x(t + \Delta t)$ is approximated as the following discrete form:

$$x(t + \Delta t) = x(t) + \Delta f_t \tag{8.38}$$

where Δf_t is an approximated increment of x based on (8.37a). Such a scheme introduces error at level $O(\Delta t^{k+1})$, and k is usually up to 5 (e.g. modified Euler method with $k = 2$ and Runge–Kutta method with $k = 4$). Methods with $k > 5$ also exist, but they are less frequently used because many more computation sub-steps are needed to increase the order k which affects the efficiency. Therefore, the commonly used numerical integration methods are low-order approximations, which confine the time step Δt. Moreover, solving (8.37) also includes solving algebraic equations (AE) (8.37b), but the Newton method may fail to converge. Therefore, to improve the efficiency and the robustness of dynamic simulation, a higher-order approximation of DAE with improved convergence is desirable. In Sections 8.2.1-8.2.3, the holomorphic embedding (HE) will be utilized to solve the DAE of power systems with arbitrarily higher-order approximations.

8.2.1.2 Approximation with Holomorphic Embedding

For an IVP in the complex domain:

$$0 = h\left(\frac{dz(\alpha)}{d\alpha}, z(\alpha), \alpha\right), \quad \alpha \in \mathbb{C}, z(0) = z_0 \tag{8.39}$$

the idea of holomorphic embedding is to solve z as a power series $\sum_{k=0}^{\infty} z[k]\alpha^k$, and the series converges to $z(\alpha)$ within a radius R_z:

$$z(\alpha) = \sum_{k=0}^{\infty} z[k]\alpha^k, \ |\alpha| < R_z \tag{8.40}$$

Therefore, within the convergence radius R_z, the truncated power series $\sum_{k=0}^{N_L} z[k]\alpha^k$ is a reasonable approximation of $z(\alpha)$. Moreover, the Padé approximant is a common post-processor of power series to further expand the usable region of approximation. For the truncated power series $\sum_{k=0}^{N_L} z[k]\alpha^k$, the corresponding Padé approximant has the following form:

$$[m/n]_z(\alpha) = \frac{a_{z0} + a_{z1}\alpha + \cdots + a_{zm}\alpha^m}{1 + b_{z1}\alpha + \cdots + b_{zn}\alpha^n} \tag{8.41}$$

where $m + n = N_L$. It is recommended that m and n are as close as possible to achieve a best convergence region [37].

As for solving problems in power systems, the embedded variable α is usually a real variable. Because $\mathbb{R} \subset \mathbb{C}$, the power series (8.40) and Padé approximant (8.41) still hold for $\alpha \in \mathbb{R}$. Holomorphic embedding has been used to solve power flows [3] and static voltage stability analysis [4]. In those steady-state studies, the embedded variable α either connects the desired solution from a trivial solution without physical meaning, or is an implicit representation of time. In dynamic studies, the time t can be explicitly used as the embedded variable. In power systems, the DAE (8.37) is holomorphic in segments, which means that in each segment without the occurrence of a singularity, the solutions $x(t)$ and $y(t)$ can be asymptotically approximated by their Taylor series. Thus, for the DAE of power systems (8.37) with initial values $x(0)$ and $y(0)$, in the time interval with continuous $p(t)$, we can use holomorphic embedding to obtain approximate solutions as truncated series:

$$x(t) \approx \sum_{k=0}^{N_L} x[k]t^k, \; y(t) \approx \sum_{k=0}^{N_L} y[k]t^k \tag{8.42}$$

or the corresponding Padé approximants.

8.2.1.3 General Computation Flow

For a general mathematical problem like

$$\mathbf{f}(\mathbf{x}, \dot{\mathbf{x}}, \mathbf{y}, t) = \mathbf{0}, \mathbf{x}(0) = \mathbf{x}_0, \mathbf{y}(0) = \mathbf{x}_0 \tag{8.43}$$

where $\mathbf{x}$ and $\mathbf{y}$ are the variables to be solved. The holomorphic embedding method derives the solutions as the following power series form:

$$\begin{aligned} \mathbf{x}(t) &= \sum_{n=0}^{\infty} \mathbf{x}[n]t^n \\ \mathbf{y}(t) &= \sum_{n=0}^{\infty} \mathbf{y}[n]t^n \end{aligned} \tag{8.44}$$

where $\mathbf{x}[n]$ and $\mathbf{y}[n]$ are the coefficients in the holomorphic embedding solution. Since infinite orders of power series cannot be obtained in practice, we usually use the truncated power series up to a certain order N.

$$\begin{aligned} \mathbf{x}(t) &= \sum_{n=0}^{N} \mathbf{x}[n]t^n \\ \mathbf{y}(t) &= \sum_{n=0}^{N} \mathbf{y}[n]t^n \end{aligned} \tag{8.45}$$

As we will show later, the holomorphic embedding approach can be used to solve algebraic and/or differential equations. Here we assume that in the studied system (8.43), the function $\mathbf{f}$ is (at least piece-wise) elementary function. Let us begin with a simple example on how the holomorphic embedding approach works.

Example 8.1. Try solving the equations below:

$$x^2 - xt - 1 = 0, t \geq 0 \tag{8.46}$$

Considering the formulation of holomorphic embedding solution (8.44), assume that

$$x(t) = \sum_{n=0}^{\infty} x[n]t^n \tag{8.47}$$

substitute (8.47) into (8.46), we can get:

$$\left(\sum_{n=0}^{\infty} x[n]t^n\right)\left(\sum_{n=0}^{\infty} x[n]t^n\right) - t\left(\sum_{n=0}^{\infty} x[n]t^k\right) - 1 = 0 \tag{8.48}$$

Expand all the terms and organize the terms with the same order:

$$\sum_{n=1}^{\infty}\left(\sum_{i=0}^{n} x[i]x[n-i] - x[n-1]\right)t^n + x[0]x[0] - 1 = 0 \tag{8.49}$$

(8.49) should hold for all $t \geq 0$, therefore for each order-n term on the left-hand side (including the constant term), the coefficient should be 0, and then we can get the following equations regarding $x[n]$:

$$x[0]x[0] - 1 = 0 \tag{8.50a}$$

$$\sum_{i=0}^{n} x[i]x[n-i] - x[n-1] = 0, n \geq 1 \tag{8.50b}$$

To solve (8.50), we can first solve (8.50a). (8.50a) has two solutions and here we select $x[0] = 1$. Then we can transform (8.50b) as

$$x[n]x[0] = -\sum_{i=1}^{n-1} x[i]x[n-i] + x[n-1] \tag{8.51}$$

And thus the $x[n]$ can be solved using $x[i], i = 0, \dots n-1$, which is an iterative process beginning from $k = 1$.

From the above procedure we can see that the basic idea of holomorphic embedding is to substitute power-series solutions into the equations, organizing the terms with the same order, and then derive and solve the coefficients of the equations. Solving the holomorphic embedding coefficients involves an iterative procedure that starts from the 0th order. And finally the power series form can approximate the true solution within a certain range.

We can see that a key step of the above procedures is the derivation of the equations of holomorphic embedding coefficients. When solving (8.46), the coefficients of terms x^2 and xt are manually derived, which is not convenient and does not fit for computerized derivation. Actually as long as $\mathbf{f}$ is elementary function, it comprises finite types of arithmetic operations and functions, and we can try to derive the rules for deriving the coefficients of each arithmetic operation or function.

8.2.1.4 Rules for Deriving Holomorphic Embedding Coefficients

For each kind of operation or function, the holomorphic embedding coefficient function can be generalized on both the left hand side (LHS) and right hand side (RHS). The LHS only involves the linear terms of the coefficients to be solved, and the RHS terms only involve the lower-order coefficients.

By using the rules in Table 8.5, the holomorphic embedding can be solved by gathering all the terms in the equations. To better illustrate the derivation of holomorphic embedding, a simple example DAE is shown:

$$\begin{aligned} \frac{dz_1(\alpha)}{d\alpha} &= 2z_1(\alpha)z_2(\alpha) + z_1(\alpha) \\ 0 &= z_2(\alpha)^2 + z_1(\alpha) - 1 \end{aligned} \tag{8.52}$$

Table 8.5 Holomorphic embedding coefficients for elementary operations and functions.

Operations	nth-order LHS	RHS (lower-order terms)
$x \pm y$	$x[n] \pm y[n]$	
xy	$y[0]x[n] + x[0]y[n]$	$-\sum_{k=1}^{n-1} x[k]y[n-k]$
$f_{div} \equiv x/y$	$\frac{1}{y[0]}x[n] - \frac{f_{div}[0]}{y[0]}y[n]$	$\frac{1}{y[0]}\sum_{k=1}^{n-1} f_{div}[k]y[n-k]$
ax	$ax[n]$	
α^k		$-\delta(n,k)$
$f_{der} \equiv \frac{dx}{d\alpha}$	$(n+1)x[n+1]$	
$f_{pow} \equiv x^b$	$b\frac{x[n]}{x[0]}f_{pow}[0]$	$-\frac{b}{nx[0]}\sum_{k=1}^{n-1} kx[k]f_{pow}[n-k] + \frac{1}{nx[0]}\sum_{k=1}^{n-1} kf_{pow}[k]x[n-k]$
$f_{\sin} \equiv \sin(x)$	$x[n]f_{\cos}[0]$	$-\frac{1}{n}\sum_{k=1}^{n-1} kx[k]f_{\cos}[n-k]$
$f_{\cos} \equiv \cos(x)$	$-x[n]f_{\sin}[0]$	$\frac{1}{n}\sum_{k=1}^{n-1} kx[k]f_{\sin}[n-k]$
$f_{\exp} \equiv \exp(x)$	$x[n]f_{\exp}[0]$	$-\frac{1}{n}\sum_{k=1}^{n-1} kx[k]f_{\exp}[n-k]$
$f_{expx} \equiv a^x$	$\ln ax[n]f_{expx}[0]$	$-\frac{1}{n}\sum_{k=1}^{n-1} k\ln ax[k]f_{expx}[n-k]$
$f_{sqrt} \equiv \sqrt{x}$	$\frac{1}{2f_{sqrt}[0]}x[n]$	$\frac{1}{nf_{sqrt}[0]}\sum_{k=1}^{n-1} kf_{sqrt}[k]f_{sqrt}[n-k]$
$f_{\ln} \equiv \ln(x)$	$\frac{1}{x[0]}x[n]$	$-\frac{1}{nx[0]}\sum_{k=1}^{n-1} kx[k]f_{\ln}[n-k]$

From Table 8.5, the corresponding equations of holomorphic embedding coefficients can be directly derived:

$$\begin{aligned}(k+1)z_1[k+1] &= 2\sum_{i=0}^{k} z_1[i]z_2[k-i] + z_1[k] \\ 0 &= \sum_{i=0}^{k} z_2[i]z_2[k-i] + z_1[k] - f_\delta(k,0)\end{aligned} \tag{8.53}$$

where $f_\delta(i,j) = 1$ only if $i = j$, and otherwise $f_\delta(i,j) = 0$.

8.2.1.5 Some Properties of Holomorphic Embedding

More generally, the holomorphic embedding approximates the solution of the following α-parameterized system

$$\mathbf{f}(\mathbf{x}(\alpha), \alpha) = \mathbf{0} \tag{8.54}$$

where $\mathbf{f}$ is an analytic function, $\mathbf{x}$ is $M \times 1$ vector, and $\mathbf{x}(\alpha)$ has the following power-series form:

$$\mathbf{x}(\alpha) = \mathbf{x}[0] + \mathbf{x}[1]\alpha + \mathbf{x}[2]\alpha^2 + \cdots \tag{8.55}$$

The core of holomorphic embedding is to derive the equations of the coefficients in (8.55), which transforms the operations in (8.54) to the relationship of the coefficients based on a set of rules, e.g.:

$$\begin{aligned} ax(\alpha) + b &\leftrightarrow ax[n] + b \\ x(\alpha)y(\alpha) &\leftrightarrow (x * y)[n] \end{aligned} \tag{8.56}$$

where $*$ is convolution: $(x * y)[n] = \sum_{k=0}^{n} x[k]y[n-k]$.

For classical power system steady-state analysis, the holomorphic embedding formulation can usually be generalized as the following form:

$$(\mathbf{A}_0 + \alpha\mathbf{A}_1)(\mathbf{x}(\alpha) \otimes \mathbf{x}(\alpha)) + (\mathbf{B}_0 + \alpha\mathbf{B}_1)\mathbf{x}(\alpha) + \mathbf{C}_0 + \alpha\mathbf{C}_1 = \mathbf{0} \tag{8.57}$$

where $\otimes$ is Kronecker product, $\mathbf{x} \otimes \mathbf{x} = (\mathbf{x} \otimes \mathbf{1}) \circ (\mathbf{1} \otimes \mathbf{x})$, the $\circ$ is Hardamard (element-wise) product. $\mathbf{A}_0$, $\mathbf{A}_1$, $\mathbf{B}_0$, $\mathbf{B}_1$, $\mathbf{C}_0$, and $\mathbf{C}_1$ are constant matrices/vectors. According to the rules for deriving the equations of holomorphic embedding coefficients, for nth-level terms, the equations are:

$$\begin{aligned} &\mathbf{A}_0\left((\mathbf{x} \otimes \mathbf{1}) * (\mathbf{1} \otimes \mathbf{x})\right)[n] + \mathbf{A}_1\left((\mathbf{x} \otimes \mathbf{1}) * (\mathbf{1} \otimes \mathbf{x})\right)[n-1] \\ &+ \mathbf{B}_0\mathbf{x}[n] + \mathbf{B}_1\mathbf{x}[n-1] + \delta_{n,0}\mathbf{C}_0 + \delta_{n,1}\mathbf{C}_1 = \mathbf{0} \end{aligned} \tag{8.58}$$

where $\delta_{n,m} = 1$ if $m = n$; otherwise $\delta_{n,m} = 0$.

Alternatively, note the following two formulas regarding high-order derivatives:

$$\begin{aligned} \left.\frac{d^n(\alpha f(\alpha))}{d\alpha^n}\right|_{\alpha=0} &= nf^{(n-1)}(0) \\ \frac{d^n(f(\alpha)g(\alpha))}{d\alpha^n} &= \sum_{k=0}^{n}\binom{n}{k} f^{(k)}(\alpha)g^{(n-k)}(\alpha) \end{aligned} \tag{8.59}$$

perform nth-order derivative on both sides of (8.57) to α at $\alpha = 0$:

$$\begin{aligned} &\mathbf{A}_0\sum_{k=0}^{n}\binom{n}{k}(\mathbf{x}\otimes\mathbf{1})^{(k)}(0)\circ(\mathbf{1}\otimes\mathbf{x})^{(n-k)}(0)+ \\ &n\mathbf{A}_1\sum_{k=0}^{n-1}\binom{n-1}{k}(\mathbf{x}\otimes\mathbf{1})^{(k)}(0)\circ(\mathbf{1}\otimes\mathbf{x})^{(n-1-k)}(0)+ \\ &\mathbf{B}_0\mathbf{x}^{(n)}(0) + n\mathbf{B}_1\mathbf{x}^{(n-1)}(0) + \delta_{n,0}\mathbf{C}_0 + \delta_{n,1}\mathbf{C}_1 = \mathbf{0} \end{aligned} \tag{8.60}$$

Dividing both sides of (8.60) with $n!$:

$$\begin{aligned} &\mathbf{A}_0\sum_{k=0}^{n}\frac{(\mathbf{x}\otimes\mathbf{1})^{(k)}(0)}{k!}\circ\frac{(\mathbf{1}\otimes\mathbf{x})^{(n-k)}(0)}{(n-k)!}+ \\ &\mathbf{A}_1\sum_{k=0}^{n-1}\frac{(\mathbf{x}\otimes\mathbf{1})^{(k)}(0)}{k!}\circ\frac{(\mathbf{1}\otimes\mathbf{x})^{(n-1-k)}(0)}{(n-1-k)!}+ \\ &\mathbf{B}_0\frac{\mathbf{x}^{(n)}(0)}{n!} + \mathbf{B}_1\frac{\mathbf{x}^{(n-1)}(0)}{(n-1)!} + \frac{\delta_{n,0}}{n!}\mathbf{C}_0 + \frac{\delta_{n,1}}{n!}\mathbf{C}_1 = \mathbf{0} \end{aligned} \tag{8.61}$$

By comparing (8.58) and (8.61) it is concluded that

$$\mathbf{x}[n] = \frac{\mathbf{x}^{(n)}(0)}{n!} \tag{8.62}$$

and the following remarks can be made:

Remark 8.1 The holomorphic embedding solution represents the Maclaurin series of $\mathbf{x}(\alpha)$ for problem (8.57).

Remark 8.2 If the nth-order truncated series of holomorphic embedding solution $\mathbf{x}_{ps,n}(\alpha) = \sum_{k=0}^{n}\mathbf{x}[k]\alpha^k$ is used to approximate $\mathbf{x}(\alpha)$, then the error is $o(\alpha^n)$.

Remark 8.3 It is shown that the derivation of the holomorphic embedding coefficient equations is equivalent to performing high-order derivatives on both sides of equations at $\alpha = 0$. For (8.54), calculate the derivative of α to both sides, and get

$$\frac{\partial\mathbf{f}}{\partial\mathbf{x}}\frac{d\mathbf{x}}{d\alpha} = -\frac{\partial\mathbf{f}}{\partial\alpha} \tag{8.63}$$

Perform $(n-1)$th-order derivative on (8.63) w.r.t. α:

$$\frac{\partial \mathbf{f}}{\partial \mathbf{x}}\frac{d^n\mathbf{x}}{d\alpha^n}+\sum_{k=1}^{n-1}\binom{n-1}{k}\frac{d^k\partial \mathbf{f}/\partial \mathbf{x}}{d\alpha^k}\frac{d^{n-k}\mathbf{x}}{d\alpha^{n-k}}=-\frac{\partial^n \mathbf{f}}{\partial \alpha^n}, \tag{8.64}$$

note that the second term on the left-hand side has an up-to-$(n-1)$th-order derivative of $\mathbf{x}$ to α. Once the derivatives of $(n-1)$th-order or lower are obtained, the nth-order derivative $\frac{d^n\mathbf{x}}{d\alpha^n}$ can be calculated. This also matches the recursive procedure of calculating holomorphic embedding coefficients. Moreover, (8.64) reveals that to solve the holomorphic embedding coefficient at any level, one needs to solve linear equations with the same coefficient matrix $\left.\frac{\partial \mathbf{f}}{\partial \mathbf{x}}\right|_{\alpha=0}$ [38–40]. This will facilitate the computation because the matrix only needs to be factorized once.

8.2.2 Solve State after Instant Switches

Instant disturbances may occur in power systems, such as faults, or opening or closing of switches. As (8.37) shows, at the instant, the state variable x does not change, and as system parameters change from $p(t_f^-)$ to $p(t_f^+)$, the algebraic variable y changes accordingly. Usually the algebraic variable before instant disturbance $y(t_f^-)$ is known, and the post-disturbance variable $y(t_f^+)$ needs to be solved. The most commonly used instant disturbance in dynamic simulation is a symmetric three-phase fault. In this case, the network admittance changes from Y^- to Y^+. In holomorphic embedding formulation, this denotes the embedded variable as α, the network algebraic equations are

$$\begin{aligned}(P_i-jQ_i)W_i^*(\alpha)-\sum_l\left(Y_{il}^-+\alpha(Y_{il}^+-Y_{il}^-)\right)V_l(\alpha)\\ -I_{Li}(\alpha)+I_{Gi}(\alpha)=0,\end{aligned} \tag{8.65}$$

then $\alpha=0$ corresponds to $y(t_f^-)$ and $\alpha=1$ corresponds to $y(t_f^+)$. The post-switch state can be obtained by solving the approximate solution of (8.65) and assigning $\alpha=1$. The equations of holomorphic embedding coefficients are shown in (8.66).

$$\left[\begin{array}{c|c|c|c}-\mathbf{G}^- & \mathbf{B}^- & \mathcal{D}(\mathbf{P}[0]) & -\mathcal{D}(\mathbf{Q}[0])\\ \hline -\mathbf{B}^- & -\mathbf{G}^- & -\mathcal{D}(\mathbf{Q}[0]) & -\mathcal{D}(\mathbf{P}[0])\\ \hline \mathcal{D}(\mathbf{W}_x[0]) & -\mathcal{D}(\mathbf{W}_y[0]) & \mathcal{D}(\mathbf{V}_x[0]) & -\mathcal{D}(\mathbf{V}_y[0])\\ \hline \mathcal{D}(\mathbf{W}_y[0]) & \mathcal{D}(\mathbf{W}_x[0]) & \mathcal{D}(\mathbf{V}_y[0]) & \mathcal{D}(\mathbf{V}_x[0])\end{array}\right]\begin{bmatrix}\mathbf{V}_x[n]\\ \mathbf{V}_y[n]\\ \mathbf{W}_x[n]\\ \mathbf{W}_y[n]\end{bmatrix}$$

$$
=\begin{bmatrix}
\Re\left(-\sum_{k=1}^{n}\mathbf{P}[k]\circ\mathbf{W}^*[n-k]+j\sum_{k=1}^{n}\mathbf{Q}[k]\circ\mathbf{W}^*[n-k]+\mathbf{I}_L[n]-\mathbf{I}_G[n]\right)\\
\Im\left(-\sum_{k=1}^{n}\mathbf{P}[k]\circ\mathbf{W}^*[n-k]+j\sum_{k=1}^{n}\mathbf{Q}[k]\circ\mathbf{W}^*[n-k]+\mathbf{I}_L[n]-\mathbf{I}_G[n]\right)\\
\Re\left(-\sum_{k=1}^{n-1}\mathbf{W}[k]\circ\mathbf{V}[n-k]\right)\\
\Im\left(-\sum_{k=1}^{n-1}\mathbf{W}[k]\circ\mathbf{V}[n-k]\right)
\end{bmatrix}
+\begin{bmatrix}
\mathbf{G}^+-\mathbf{G}^- & -\mathbf{B}^++\mathbf{B}^-\\
\mathbf{B}^+-\mathbf{B}^- & \mathbf{G}^+-\mathbf{G}^-\\
\mathbf{0} & \mathbf{0}\\
\mathbf{0} & \mathbf{0}\\
\mathbf{0} & \mathbf{0}
\end{bmatrix}
\begin{bmatrix}
\mathbf{V}_x[n-1]\\
\mathbf{V}_y[n-1]
\end{bmatrix}
\tag{8.66}
$$

At the instant of disturbance, the state variables are constant, and thus the equations of dynamic elements (e.g. synchronous machines and asynchronous machines) can be significantly simplified.

8.2.3 Overall Dynamic Simulation Process

With the methods introduced in Sections 8.2.1–8.2.2, the whole process of dynamic simulation can be realized. Figure 8.15 illustrates the key steps for transient stability analysis. The dynamic simulation starts with a certain state (labeled as 0 in the figure), and then the dynamic processes are simulated (labeled 2, 3, and 4, representing the pre-fault, fault-on, and post-fault stages, respectively) until a switch occurs in the system. The post-switch state needs to be solved as the starting point of the following simulation of dynamics. The starting state is usually given by the steady-state power flow. At the starting state, for the most part the system is not in stress and the solution does not deviate much from the normal state, so the conventional methods for solving power flow (e.g. NR method) are usually sufficient. While when solving the post-switch state, the system state may significantly deviate from the original one, and the conventional NR method may fail to converge. Thus, holomorphic embedding should be selected as a reliable method for solving the post-switch state in dynamic simulation.

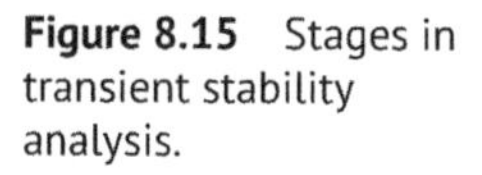

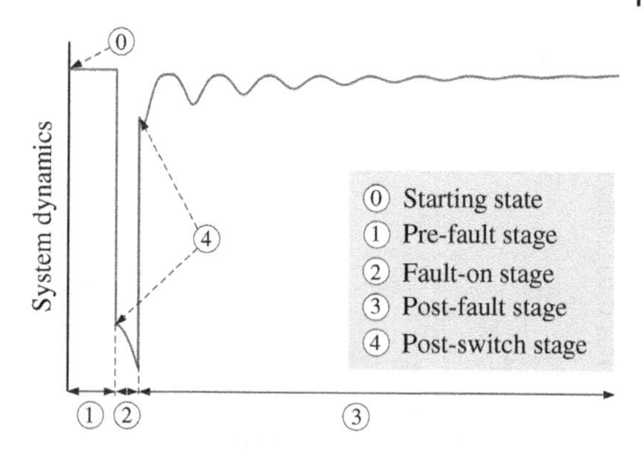

Figure 8.15 Stages in transient stability analysis.

8.2.4 Test Cases

8.2.4.1 Modified IEEE 39-Bus System

We show the performance of several simulation methods on a modified IEEE 39-bus system model. The system has 10 synchronous machines, 19 PQ loads, 18 ZIP loads, and 18 induction motors. The synchronous generators use 6th-order models. The traditional methods for benchmarking are modified Euler (ME) and 4th-order Runge–Kutta (RK4) methods, and the NR method is utilized for solving algebraic equations. To be concise, the traditional methods are abbreviated as ME-NR and RK4-NR, respectively. A three-phase fault is applied at bus 3 at 0.5 s and is then cleared at 0.75 s. The three-phase fault is applied with a larger fault impedance of $Z_f = j0.5$. System dynamics in 10 s are simulated. The methods are developed in Matlab and are tested on laptop computer with Intel Core™ i7-6600U CPU and 8GM RAM. Figure 8.16 shows the curves of voltage magnitude obtained by the simulation.

Table 8.6 compares the computation time of different methods, including ME-NR and RK4-NR with different time steps, and holomorphic embedding. Figure 8.17 also shows the mean error of state variables in each method for comparison. To calculate the error of each method, the result from RK4-NR with a very tiny time step $\Delta t = 1 \times 10^{-5}$ s is considered as the benchmark solution. The results show that holomorphic embedding is faster than ME-NR and RK4-NR and with lower errors. ME-NR and RK4-NR are expected to use very small time step ($\Delta t < 0.001$ s) to reach the same error level as holomorphic embedding and consequently the computation speed will be very slow.

The parameter N_L in holomorphic embedding influences the computational speed. The larger N_L is, the more computation is needed in each

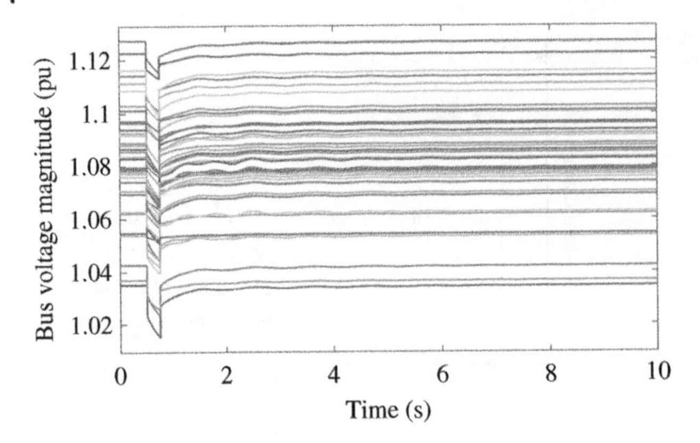

Figure 8.16 Voltage magnitude curves in IEEE 39-bus system.

Table 8.6 Computation time of different methods in IEEE 39-bus system.

Method	Computation time (s)	Steps
ME-NR ($\Delta t = 0.01$ s)	21.58	1000
ME-NR ($\Delta t = 0.005$ s)	34.44	2000
ME-NR ($\Delta t = 0.002$ s)	84.46	5000
RK4-NR ($\Delta t = 0.01$ s)	34.64	1000
RK4-NR ($\Delta t = 0.005$ s)	64.78	2000
RK4-NR ($\Delta t = 0.002$ s)	156.37	5000
HE ($N_L = 30$)	4.49	54

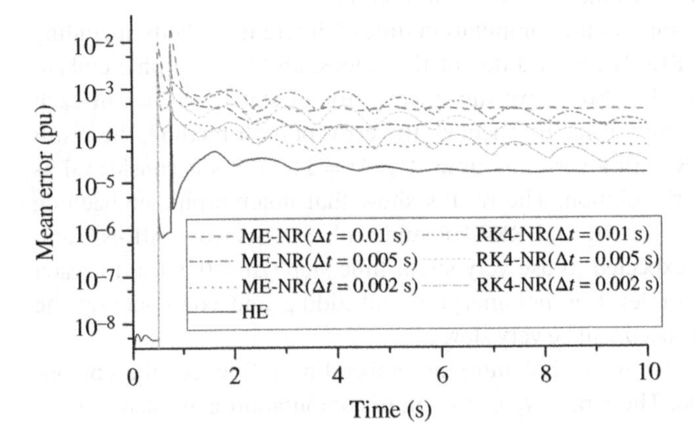

Figure 8.17 Comparison of errors in IEEE 39-bus system.

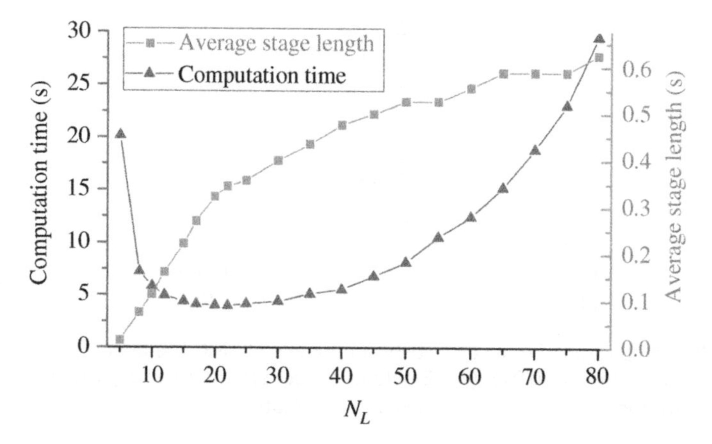

Figure 8.18 Computation time and average stage length under different N_L.

holomorphic embedding approximation. In the meantime, the effective range of holomorphic embedding approximation also enlarges with the increase of N_L, and thus fewer stages will be needed. Therefore, there might be a best N_L to achieve the least total computation time. In this simulation configuration, different N_L values are tested. Figure 8.18 shows the computation time as well as the average length of holomorphic embedding stages. The result indicates that the effective range of holomorphic embedding approximation grows as N_L increases; however, as the time consumption in each holomorphic embedding stage rises, the total computation time first drops and then increases. In this case, selecting an N_L value between 15 and 30 yields the best computation performance. Actually, per our experience on more dynamic simulation tests, the best N_L value usually falls between 15 and 30.

The previous test selects a relatively small disturbance to the system. Next, more severe disturbances are tested by decreasing the fault impedance Z_f. It is found that when Z_f is below $j0.0151$ (comparable to the impedance of adjacent lines), the ME-NR will fail to converge, while holomorphic embedding does not have such numerical issues. Moreover, the ME-NR may converge to some nonpractical solutions (as shown in Figure 8.19).

8.2.4.2 2383-Bus Polish System

The holomorphic embedding method is tested on a modified Polish test system with 2383 buses, 327 synchronous generators with turbine governors (TGs) and exciters, 1827 ZIP loads, and 1542 induction motors. The generators use sixth-order model. In the first case, we remove all the controllers on the generators, apply 3-phase fault on bus 1396 at 0.5 s, and clear the

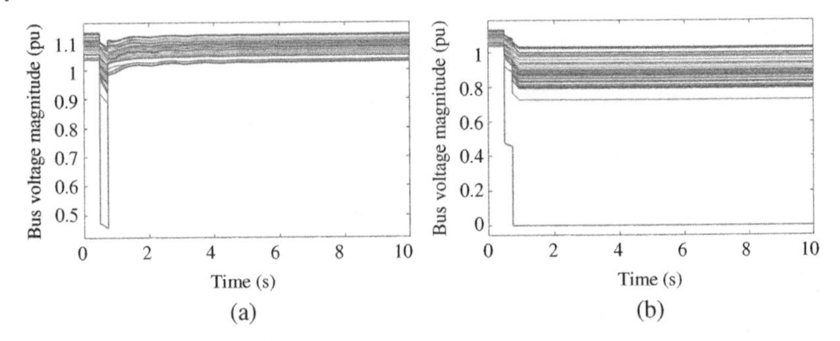

Figure 8.19 Comparison of robustness between HE and ME-NR. The fault is applied on bus 1 with $Z_f = j0.015$. (a) HE obtains correct system trajectory, while (b) ME-NR converges to a non-practical low-voltage solution.

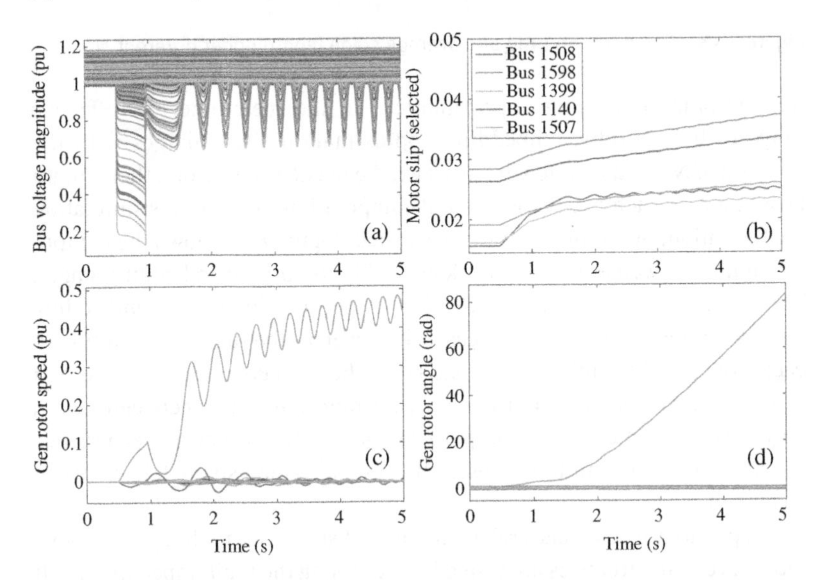

Figure 8.20 Dynamics of Polish system simulated by holomorphic embedding (unstable). (a) voltage magnitude; (b) motor slip; (c) generator rotor speed; (d) generator rotor angle.

fault at 0.95 s. The fault impedance $Z_f = j0.01$. In this case, the generator at bus 1140 accelerates significantly and finally runs out of step with the system. Figure 8.20 shows the system dynamics simulated by holomorphic embedding. In contrast, the traditional ME-NR method fails to converge in the simulation process. For the entire 5-s simulation, by setting $N_L = 15$, the holomorphic embedding takes 136.7 s to finish computation.

The system dynamics are also simulated with more complex scenarios. The automatic voltage regulators (AVRs) and TGs on the generators are

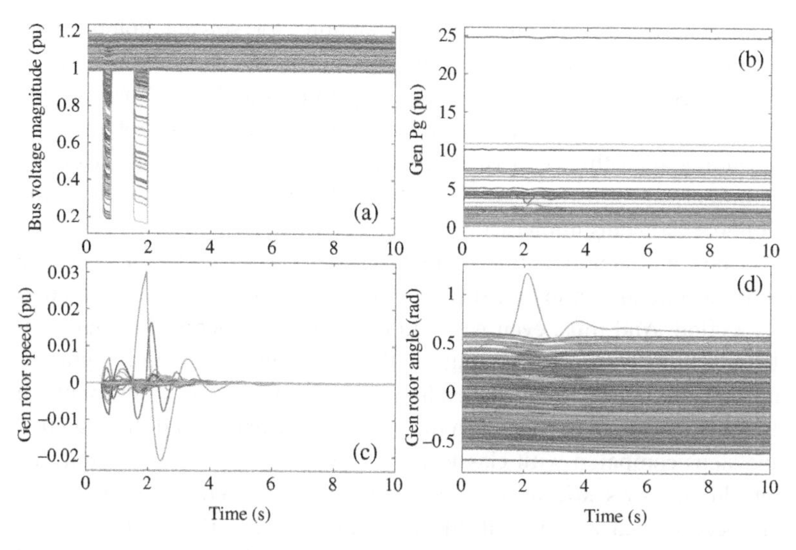

Figure 8.21 Dynamics of Polish system simulated by holomorphic embedding (stable). (a) voltage magnitude; (b) generator power; (c) rotor speed; (d) rotor angle.

activated, and the generators still use the sixth-order model. Double 3-phase faults are applied simultaneously on buses 42 and 540 at 0.5 s, and the faults are cleared at 0.75 s. Then another three-phase fault is applied on bus 1396 at 1.5 s, and the fault is cleared at 1.95 s. All the faults are assumed to have impedance of $Z_f = j0.01$. With $N_L = 15$, the simulation of 10 s dynamics takes 90.95 s to finish. Figure 8.21 shows the system dynamics simulated by holomorphic embedding. In comparison, the unstable case takes more computation time, because when system loses stability, the system states often change drastically (e.g. the rapid angle change shown in Figure 8.20(d)), which causes reduced effective range of holomorphic embedding approximation.

The above test cases in the modified Polish system demonstrate the robustness and efficiency of holomorphic embedding in computing dynamics of large-scale systems. The holomorphic embedding can well handle stable and unstable cases and avoid numerical instability or divergence, which shows advantages over classical numerical approaches in the analysis of systems with complex models and scenarios.

8.2.5 Summary of Section

This section explains more general holomorphic embedding algorithms for power system dynamic simulation. An important part of this section is the generic rule of deriving the equations for solving holomorphic embedding coefficients, with which a complex set of equations can be decomposed into atomic mathematical operations and thus holomorphic embedding coefficients can be solved using the rules. Because holomorphic embedding algorithms consists of linear algebraic computations, avoiding the numerical robustness issues of classical methods directly solving the nonlinear problems numerically.

For power system dynamic simulations that also have instant switching events, a general method for solving post-switch states is also explained in this section. And thus, even with multiple complex events, dynamic simulation can be performed by splitting the simulation period into segments and applying holomorphic embedding methods on each segments or at the instant of switching events. With the higher-order nature and the extended time steps compared with classical numerical methods, the holomorphic embedding shows advantages in both computation speed and accuracy.

In Sections 8.3–8.4, we will further enhance the performance of holomorphic embedding-powered simulation based on hybrid simulation and parallel or distributed computation techniques.

8.3 Extended-Term Hybrid Simulation

Power systems have various kinds of networked components as well as complex behaviors. The power system dynamics have multiple distinct time scales [41, 42]. For example, the timescales of the fast transients can be less than 0.01 s, and the actions of system-wide control (e.g. automatic generation control or AGC) [43] and some mechanical and thermal-driven processes [44] are usually in the timescales of seconds to several minutes. Changes in load levels and the economic dispatch actions take minutes to hours, and due to the interdependencies among the system components and various external impacts (e.g. weather, vegetation, natural disasters), complex event chains may occur. These complexities call for panoramic simulations of complex event processes involving various disturbances, system responses and control measures, and traditional security analysis methods and tools are insufficient for such computation tasks. For example, the transient stability analysis only assumes a single fault, and the duration is usually within one minute, which ignores the longer-timescale dynamics [45] such as AGC and dispatch. On the other hand, steady-state security analysis based on power flow models for longer-timescale analysis cannot capture system dynamics [23]. Because many security concerns in the power systems involve complex and extended-term processes [46], it is imperative to find new methods for robust and efficient extended-term simulation.

Traditional numerical computation methods have major limitations for extended-term simulations. The traditional numerical integration methods for solving differential equations are lower-order methods [40], and their efficiency is limited by tiny time steps. Such methods cannot flexibly adapt to the variations of dominant timescales. In contrast, holomorphic embedding (HE) [3, 4, 40] as a major semi-analytical simulation (SAS) method that features enhanced robustness and efficiency in simulations, has shown promising performance in the steady-state [3, 4] and dynamic analysis [40] tasks of power systems.

Particularly, holomorphic embedding shows natural advantages for handling events and multi-timescale simulation because of its analytical form in the time domain. This section will explain how holomorphic embedding enables hybrid extended-term simulation based on a simulation framework combining steady-state and dynamic simulation. The dynamic simulation can be performed during system transients, while QSS modeling can be adopted after the transients fade away. Switching from dynamic to QSS simulation can be efficiently performed with holomorphic embedding by using the holomorphic embedding solution parameters, which avoids the extra simulation burden in traditional approaches based on evaluating the trajectory variations. The extended-term simulation can be utilized for complex analysis tasks such as resilience analysis [47–49], cascading outages [42, 46, 50], restoration [51], and renewable energy control [52].

8.3.1 Steady-State and Dynamic Hybrid Simulation

8.3.1.1 Switching from Dynamic to Quasi-Steady-State (QSS) Models

In power system simulation, it is quite common to see that after disturbances, the transients will fade away after some time and the system enters steady state. If simulation can be switched from using dynamic models to a QSS simualtion, then we can reduce lots of computation burden and accelerate the computation progress. Here take synchronous generators as example. Usually, the generators are equipped with AVRs to maintain terminal voltage, so they can be converted to PV buses in the QSS model.

The QSS model also applies when the transient inside the generator fades away. The system-wide control, such as an AGC, has a much larger time constant than rotor transients, so after the generator transients fades away, the generator models can be converted to PV buses and the following QSS model [53], considering the AGC actions, can be used:

$$\begin{aligned} &(P_{Gi} - K_i\Delta f - jQ_{Gi})W_i^* - I_{Li} - \sum_l Y_{il}V_l = 0 \\ &\dot{P}_{Gi} = -\frac{\Delta f}{T_{gi}} \end{aligned} \tag{8.67}$$

where Δf is the difference between the frequency and the nominal frequency $\Delta f = f - f_s$, $K_i = D_i + 1/R_i$ is the coefficient representing the QSS frequency response [54]. T_{gi} is a control time constant of the AGC [53].

8.3.1.2 Switching from Steady-State to Dynamic Models

When there are no significant fast transients, the QSS simulation can provide satisfactory accuracy and significantly accelerate the computation. When the simulation comes across switch events, the sudden changes in algebraic variables triggers the transient process and the QSS model is converted back to full-dynamic model. For generator buses, the QSS PV buses model will be restored to dynamic models of synchronous generators with controllers.

8.3.1.3 Efficient Determination of Steady State Using Holomorphic Embedding Coefficients

Classical dynamic simulation usually uses the fluctuation of the trajectories to determine the steady state, but it requires an extra period of simulation and is time-consuming. In contrast, holomorphic embedding can enhance the switching between dynamic and steady state by making use of the analytical form of the solutions. Here we derive some criteria for determining steady state by using holomorphic embedding coefficients in power series (PS) and Padé appriximation (PA).

The determination of steady state through power series or PA will need efficient estimation of upper and lower bounds of polynomials within a given interval. So we first provide a general algorithm of estimating such bounds before introducing the steps for determining the steady state. Considering a polynomial $x(t) = \sum_{k=0}^{N} x_k t^k$ and an interval of t as $[0, T]$. First, the polynomial can be written as:

$$x(t) = (\cdots((x_N t + x_{N-1})t + x_{N-2})t + \cdots)t + x_0 \tag{8.68}$$

here t represents an interval $[0, T]$, and following the interval arithmetic, we can derive the interval of $x_N t + x_{N-1}$, and then the interval of $(x_N t + x_{N-1})t + x_{N-2}$, and all the way to the interval of the entire polynomial by unwrapping the parentheses. Then the terminal values of the polynomial are lower and upper bounds of the polynomial. The detailed computation procedures are in Algorithm 8.1. For a vector of polynomials $\mathbf{y}(t)$ with size N_y and order N, the Algorithm 8.1 has complexity of $O(N_y N)$, which is very efficient.

Next we explain the approach for determining steady state using holomorphic embedding coefficients. Assume the power-series approximate solution of a trajectory derived by holomorphic embedding is

$$x_{T,PS}(t) = \sum_{k=0}^{N} x[k]t^k \tag{8.69}$$

and the solution is effective within the interval $t \in [0, T_e]$, we aim at estimating the rate of changes of $x_{T,PS}(t)$ in $[0, T_e]$. The average rate of change of $x_{T,PS}(t)$ from 0 to t is

$$R_{T,PS}(t) \overset{\text{def}}{=} \frac{x_{T,PS}(t) - x_{T,PS}(0)}{t} = \sum_{k=1}^{N} x[k]t^{k-1} \tag{8.70}$$

So we can use the Algorithm 8.1 to estimate the bounds of $R_{T,PS}(t)$. Assume the upper and lower bounds of $R_{T,PS}(t)$ are $R_{T,PS,ub}$ and $R_{T,PS,lb}$, respectively. So the average rate of change of $x_{T,PS}(t)$ has a bound $\Delta_{T,PS} = \max\{|R_{T,PS,ub}|, |R_{T,PS,lb}|\}$. If $\Delta_{T,PS}$ is smaller than a preset threshold ε_T, then this variable can be considered as entered steady state.

Besides power series, holomorphic embedding-based simulation usually uses PA to obtain larger effective time steps than power series. Because power series has smaller effective range, the criteria based on (8.70) may be too conservative. So here we list another criterion for determining the

---

Algorithm 8.1 Calculate bounds of polynomial values in given interval.

Input: Polynomial $x(t) = \sum_{k=0}^{N} x_k t^k$, interval of t as $[0, T]$.

Output: Upper and lower bounds x_{ub}, x_{lb}, s.t. $x(t) \in [x_{lb}, x_{ub}]$ when $t \in [0, T]$.

1: $x_{ub} \leftarrow x_N$, $x_{lb} \leftarrow x_N$
2: **for** $k = N - 1 \rightarrow 0$ **do**
3: **if** $x_{ub} < 0$ //Interval arithmetic for x_{ub}
4: $x_{ub} \leftarrow x_k$
5: **else**
6: $x_{ub} \leftarrow x_{ub}T + x_k$
7: **endif**
8: **if** $x_{lb} > 0$ //Interval arithmetic for x_{lb}
9: $x_{lb} \leftarrow x_k$
10: **else**
11: $x_{lb} \leftarrow x_{lb}T + x_k$
12: **endif**
13: **end for**

steady state by using the coefficients in PA. Assume the trajectory of a variable approximated by the Padé approximation is

$$x_{T,PA}(t) = \frac{\sum_{k=0}^{N_A} x_A[k]t^k}{\sum_{k=0}^{N_B} x_B[k]t^k} \tag{8.71}$$

where $x_A[k]$ and $x_B[k]$ are the Padé coefficients on the numerator and denominator, respectively. To make the Padé approximation unique, it is usually set $x_B[0] = 1$. Assume the solution (8.71) is effective when $t \leq T_e$. The bounds of (8.71) when $0 \leq t \leq T_e$ by using the coefficients $x_A[k]$ and $x_B[k]$ will be derived next.

First we assume the denominator of (8.71) $\sum_{k=0}^{N_B} x_B[k]t^k$ does not change sign in interval $[0, T_e]$. This assumption usually holds because if the denominator changes sign, at least one time point in the interval will make the denominator zero, which makes (8.71) not well defined for the simulation. So without losing generality, we assume $\sum_{k=0}^{N_B} x_B[k]t^k$ be positive in the interval.

Second we assume $N_A = N_B$ for the following derivations. If $N_A \neq N_B$ in (8.71), e.g. $N_A > N_B$, simply replacing the denominator with $\sum_{k=0}^{N_B'} x_B'[k]t^k$, where $N_B' = N_A$, $x_B'[k] = x_B[k]$ if $k \leq N_B$ and otherwise $x_B'[k] = 0$. Denote $c = x_A[0]/x_B[0]$, then (8.71) can be written as:

$$x_{T,PA}(t) = c + \frac{\sum_{k=0}^{N_A}(x_A[k] - cx_B[k])t^k}{\sum_{k=0}^{N_B} x_B[k]t^k} = c + \frac{\sum_{k=1}^{N_A} \tilde{x}_A[k]t^{k-1}}{\sum_{k=0}^{N_B} x_B[k]t^k}t \tag{8.72}$$

where $\tilde{x}_A = x_A[k] - cx_B[k]$. Like (8.70), the average rate of change of $x_{T,PA}(t)$ from 0 to t is

$$R_{T,PA}(t) \overset{\text{def}}{=} \frac{x_{T,PA}(t) - x_{T,PA}(0)}{t} = \frac{\sum_{k=1}^{N_A} \tilde{x}_A[k]t^{k-1}}{\sum_{k=0}^{N_B} x_B[k]t^k} \tag{8.73}$$

By using the Algorithm 8.1, bounds of numerator and denominator in $R_{T,PA}(t)$ can be obtained. Assume the lower bound of $\sum_{k=0}^{N_B} x_B[k]t^k$ is x_{Blb}. If $x_{Blb} > 0$, i.e. a positive lower bound for the denominator of $R_{T,PA}(t)$ is gotten, then x_{Blb} can be used to estimate the bounds of $R_{T,PA}(t)$. Assume the upper and lower bounds of $\sum_{k=1}^{N_A} \tilde{x}_A[k]t^{k-1}$ by using Algorithm 8.1 are $\tilde{x}_{Aub}$ and $\tilde{x}_{Alb}$, respectively, then the upper and lower bounds of $R_{T,PA}(t)$ can be obtained:

$$\begin{aligned} R_{T,PA,lb} &= \frac{\tilde{x}_{Alb}}{x_{Blb}} \\ R_{T,PA,ub} &= \frac{\tilde{x}_{Aub}}{x_{Blb}} \end{aligned} \tag{8.74}$$

and the average rate of change of $x_{T,PA}(t)$ is bounded by $\Delta x_{T,PA} = \max\{|R_{T,PA,lb}|, |R_{T,PA,ub}|\}$. And once $\Delta x_{T,PA} < \varepsilon_T$, the studied variable is considered as entered steady state.

Here we can use both $\Delta x_{T,PA}$ and $\Delta x_{T,PS}$ for determining the steady state of variables. For a given variable, if $\Delta x_{T,PA} < \varepsilon_T$ or $\Delta x_{T,PS} < \varepsilon_T$, the variable is considered as entered steady state. Note that if $x_{Blb} \leq 0$, $\Delta x_{T,PA}$ is not well defined and thus the Padé approximation cannot be used to estimate the bounds for the studied variable.

In simulation, $\Delta x_{T,PA}$ and $\Delta x_{T,PS}$ of multiple variables need to be calculated and tracked, and only when all the variables satisfy the criteria above, the system can be considered to be in steady state. Most variables in the computation can directly leverage the above criteria for judging steady state, except the generator rotor angles. Because the center of inertia (COI) of the system may not be rotating at the nominal radius speed, the COI will not stop rotating in the nominal frequency coordinate even if system transients have well damped. Consequently, the rotor angles will keep changing even in the steady state. Therefore, the angle of one rotor can be selected as a reference, and the relative angles will be used in the above criteria to determine the steady-state.

8.3.2 Extended-Term Simulation Framework

8.3.2.1 Event-Driven Simulation Based on Holomorphic Embedding

Generally, there are two categories of events in the extended-term simulation:

(1) *System events* represent all the actual events in the system, such as switch actions and ramping start or stop.
(2) *Simulation events* correspond to changes in the simulation processes but are not actual events in the system, such as the switching to a new simulation stage and switching between dynamic and steady-state models.

Some system events are triggered by satisfying some condition $h(\mathbf{x}(t), \mathbf{y}(t), \mathbf{p}(t)) \geq 0, t \geq t_0$. Once the holomorphic embedding solution is obtained, the value of $h(\mathbf{x}(t), \mathbf{y}(t), \mathbf{p}(t))$ is tracked by substituting time t, and the time of the event can be approximately determined by binary searches. As Figure 8.22 shows, because holomorphic embedding provides a continuous trajectory of system states in the time domain, holomorphic embedding provides the instant of the event more accurately than the classical numerical integration methods that only provide values on discrete time steps. The simulator

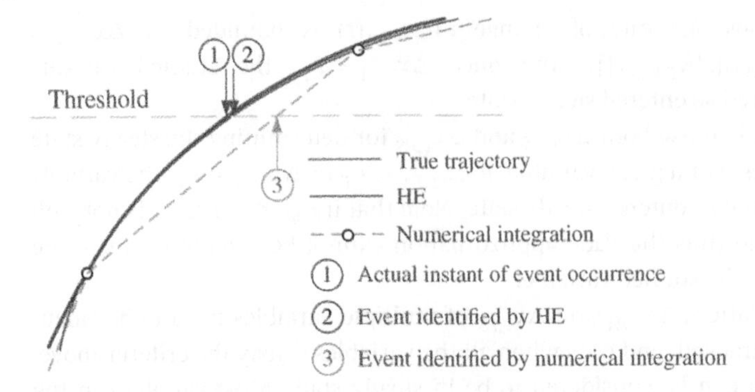

Figure 8.22 Illustration of event-tracking errors using holomorphic embedding and classical numerical integration simulation methods.

uses an event scheduler to manage the event-driven simulation. The event scheduler tracks all the events and determines the instant of the earliest event. At the instant of an event, the simulator simulates the event and the event scheduler updates the event list and tracks for the next event.

8.3.2.2 Overall Work Flow of Extended-Term Simulation

Figure 8.23 shows the overall work flow of the extended-term simulation. The simulation is driven by the event scheduler. The simulation is able to deal with multiple islands and the collapse of part or all of the system.

8.3.3 Experiments

8.3.3.1 2-Bus Test System

We used the 2-bus test system [40] to test the event-driven simulation, as shown in Figure 8.24. We use the same system parameters as [40], i.e. $E = 1.01$, $z = r + jx = 0.01 + j0.05$, and $P + jQ = 0.1 + j0.3$. Increase the load level λ at a constant rate $\frac{d\lambda}{dt} = 1$ from the initial value $\lambda(0) = 0$ until voltage collapse. The task is to determine the instant that the current of the line reaches a threshold I_{th}. The 2-bus system has a closed-form solution: The square of the line current is

$$I^2(t) = \frac{1}{r^2 + x^2}\left[\frac{E^2}{2} - (Pr + Qx)t - E\sqrt{\frac{E^2}{4} - (Pr + Qx)t - \frac{(Qr - Px)^2}{E^2}t^2}\right] \tag{8.75}$$

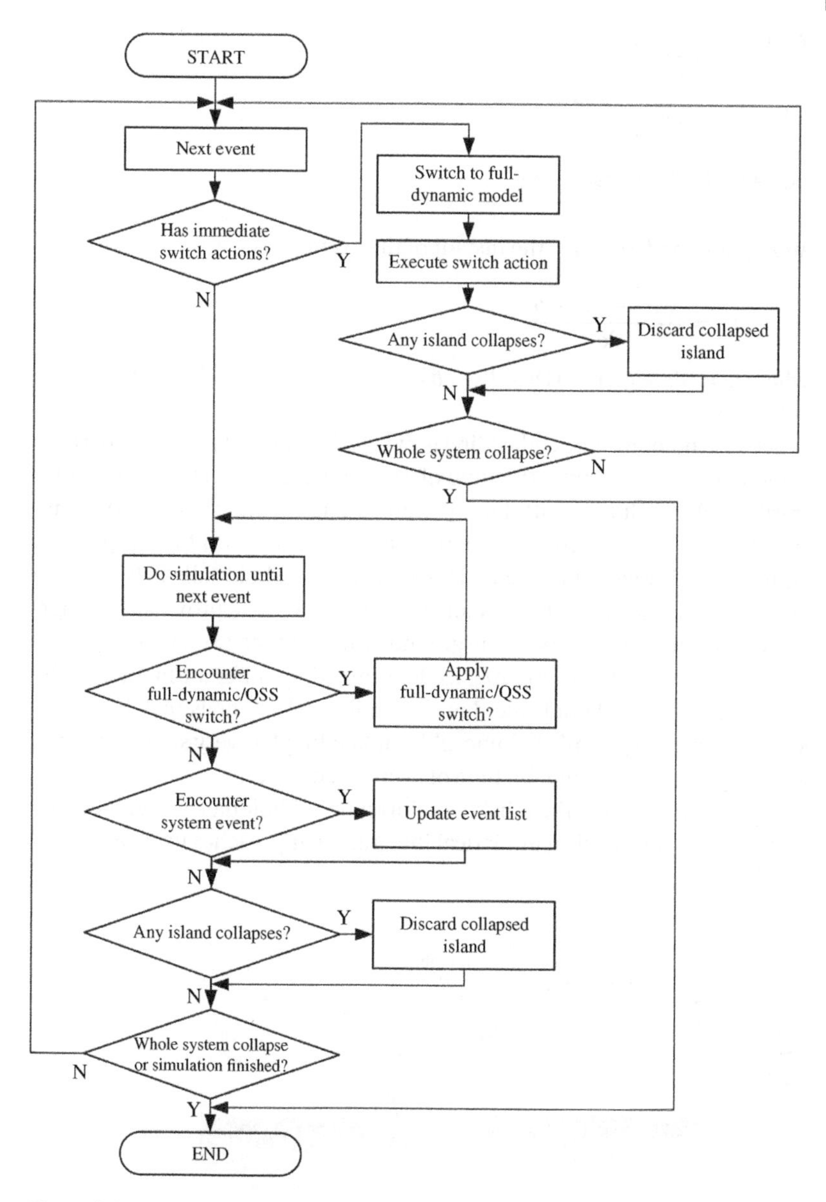

Figure 8.23 Flowchart of extended-term simulation.

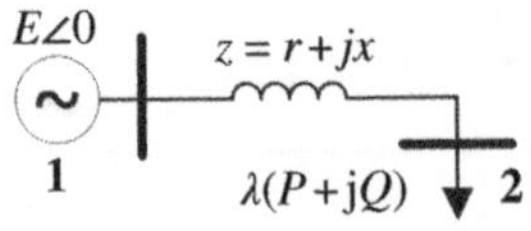

Figure 8.24 2-bus test system.

and by solving $I^2(t) = I_{th}^2$, the instant of the event t_{th} is

$$t_{th} = \frac{\sqrt{b^2 - 4ac} - b}{2a} \tag{8.76}$$

where $a = (Pr + Qx)^2 + (Qr - Px)^2$, $b = 2(Pr + Qx)(r^2 + x^2)I_{th}^2$, $c = (r^2 + x^2)^2 I_{th}^4 - E^2(r^2 + x^2)I_{th}^2$.

Because holomorphic embedding provides the trajectory of states as a continuous function of time, the time of an event can be determined at arbitrarily high resolution with binary search. The modified Euler (ME) and trapezoidal (TRAP) methods with time step 0.01 s are used for comparison. In numerical integration methods, when the threshold falls between two adjacent time steps, the time of an event needs to be approximately determined by interpolation. With the ground-truth solution (8.76), we can compare the error of event time Δt_{th} as determined by different methods. The results in Figure 8.25 indicate that the traditional methods may have substantial error of t_{th}, while holomorphic embedding has very stable and high accuracy for determining the time of event occurrence.

This test case shows the reliable performance of holomorphic embedding method over the classical numerical integration approaches in event-driven

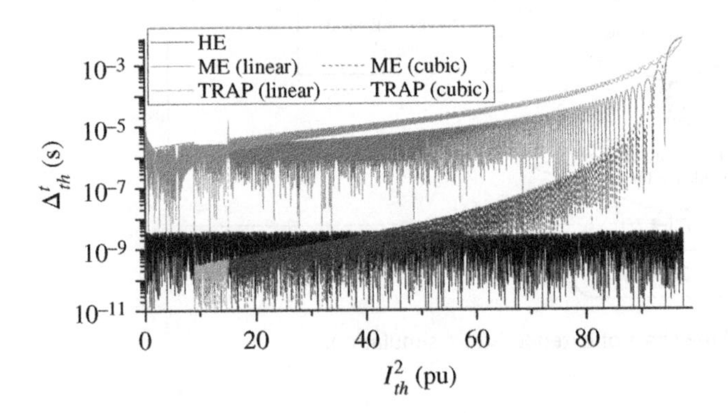

Figure 8.25 Event detection time error Δt_{th} with holomorphic embedding, modified Euler (ME) and trapezoidal (TRAP) methods (with linear and cubic interpolation).

simulation. In Sections 8.3.3.2–8.3.3.3, the extended-term hybrid steady-state and dynamic simulation approach will be demonstrated on other systems.

8.3.3.2 4-Bus Test System

In this section, we show the extended-term simulation method on a 4-bus test system. As Figure 8.26 shows, each bus has ZIP and induction motor loads, and buses 1 and 4 have synchronous generators with AVRs and TGs. The TGs have time constants $T_1 = 0.3$ s, $T_2 = 0.1$ s. The system is also equipped with AGC, and the time constant of the controller is $T_{gi} = 5$ s. In the beginning, each generator has an active power output of 1.1436 p.u.

During the 500 s simulation, the system periodically adds and cuts loads on buses 2 and 3 at time intervals of 30 s. The process involves multiple events, fast transients of generators and induction motors, and slower dynamics introduced by AGC, and the duration is much longer than the conventional dynamic security assessment. People can choose to perform the holomorphic embedding full-dynamic simulation with the conventional numerical simulation approaches or, alternatively, holomorphic embedding can be used to perform simulation for better accuracy and computation speed [40]. However, the full-dynamic simulation is still time-consuming for such a long process, and here we try the hybrid simulation approach.

Figure 8.27 shows the frequency and voltage of the system (for full dynamic simulation portions, the frequency is regarded as the mean value of the generator rotor speeds weighted by the inertia). The figure also shows the time intervals of the full-dynamic simulation and QSS simulation. In the whole 500 s process, only 66.32 s are in full-dynamic simulation, and about 87.7% of the process is simulated with QSS model. The holomorphic

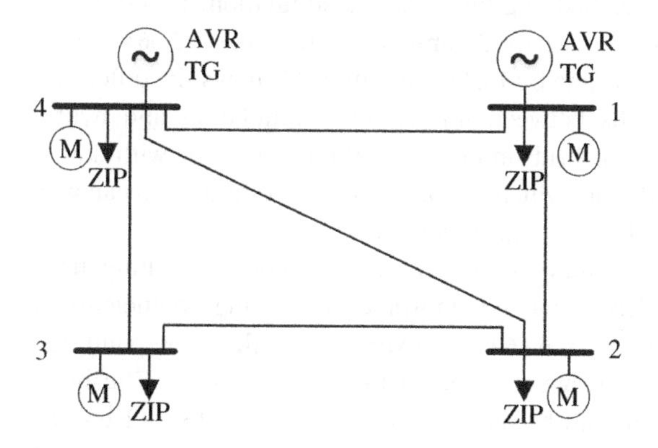

Figure 8.26 4-bus test system.

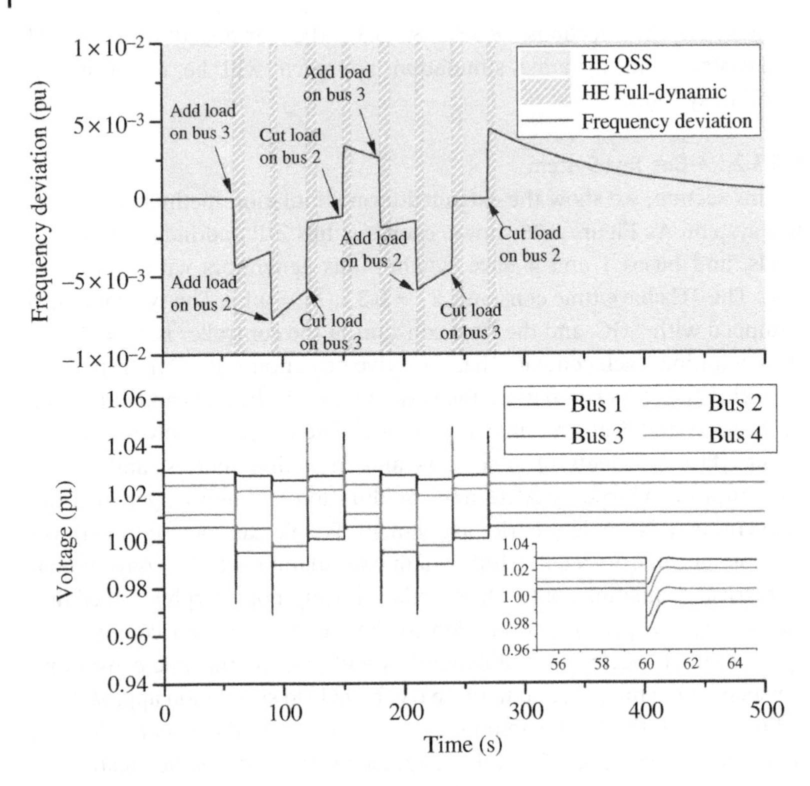

Figure 8.27 Frequency and voltage of 4-bus system.

embedding QSS and full-dynamic hybrid simulation is significantly faster than the holomorphic embedding full-dynamic simulation: Full-dynamic simulation takes 47.32 s, and hybrid simulation only takes 12.66 s, which reduces the computation time by 73.25%. Figure 8.28 shows the difference of the simulation results between full-dynamic simulation and hybrid simulation: The result of hybrid simulation is almost the same with that of the full-dynamic simulation, which verifies that hybrid simulation can well reproduce the result of full-dynamic simulation.

To better explain the holomorphic embedding-based switching from dynamic to QSS models, some holomorphic embedding coefficients at the model switching point of 69.69 s are extracted. At this time point, the simulation switches from dynamic model to the QSS model. The holomorphic embedding solution has an effective range $T_e = 0.137$ s. Here the difference of rotor speeds between the two generators ω', the square of voltage magnitude at bus 4 V_4^2 and AVR variable on generator 2 v_{m2} (on

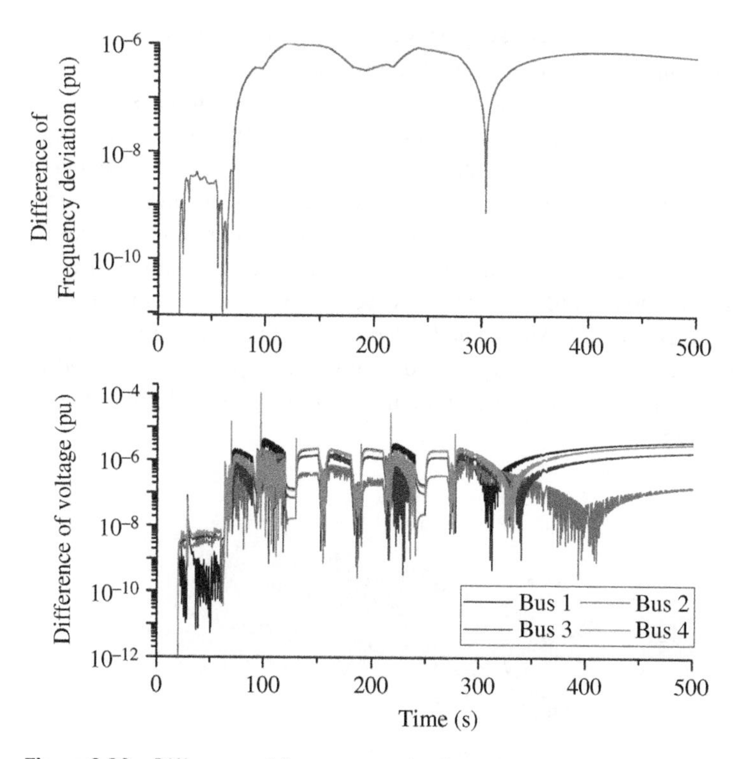

Figure 8.28 Difference of frequency and voltage between holomorphic embedding full dynamic and hybrid simulation on 4-bus system.

Table 8.7 Holomorphic embedding coefficients (power series) of rotor speed difference in 4-bus system.

k	ω'	V_4^2	v_{m2}
$\Delta x_{T,PS}$	6.11×10^{-4}	0.0279	3.76×10^{-4}
$\Delta x_{T,PA}$	1.26×10^{-4}	9.85×10^{-4}	0.0013

bus 4) are listed in Table 8.7. The threshold for determining steady state is selected as $\varepsilon_T = 10^{-3}$. The results show that for ω', both $\Delta x_{T,PS}$ and $\Delta x_{T,PA}$ satisfy the steady-state criteria, while only $\Delta x_{T,PA}$ for V_4^2 and $\Delta x_{T,PS}$ for v_{m2} satisfy the steady-state criteria. In any those cases, the variable are determined as entered steady state because at least one from $\Delta x_{T,PS}$ and $\Delta x_{T,PA}$ satisfy the criteria. This shows that both $\Delta x_{T,PS}$ and $\Delta x_{T,PA}$ are useful for the effective switching from dynamic to QSS simulation.

8.3.3.3 Simulation of Restoration on New England Test System

In this example, we demonstrate the process of system restoration on the IEEE 39-bus (New England) system. System restoration is a typical process involving complex dynamics in different timescales, with significant topology and system parameter changes, which is challenging to simulate [51, 55]. The generator on bus 39 acts as the black start generator for the restoration process. The buses, lines and generators are energized sequentially, and the loads and generation are picked up gradually. In this simulation task, the generators use 6th-order model, and the loads use ZIP+Motor model. The entire restoration process lasts 12 065 s, and full-dynamic and hybrid simulation approaches are used to simulate them respectively. The simulation was implemented and tested on Matlab 2017b.

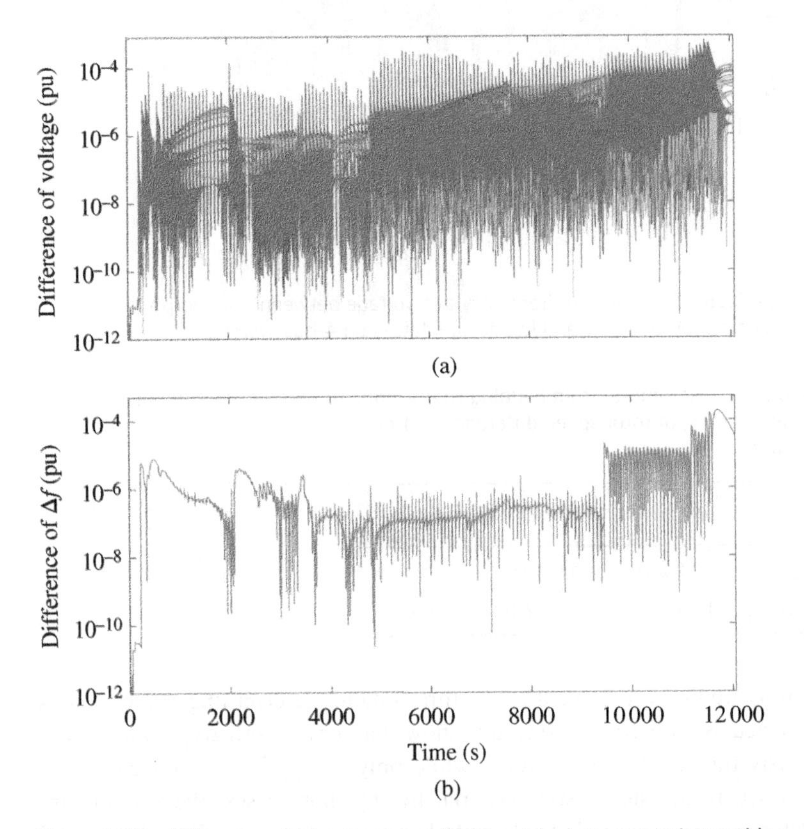

Figure 8.29 Difference of (a) voltage and (b) frequency between holomorphic embedding full-dynamic and holomorphic embedding hybrid simulation on 39-bus system.

During the whole 12 065 s restoration process there are 396 events, including adding lines, generators, static load, shunt capacitors, and induction motors, and ramping up generator power. In the hybrid simulation, 8057.3 s (i.e. 66.8%) of the restoration process is simulated in a QSS model. Figure 8.29 shows the difference in voltage and frequency between the full-dynamic simulation and the hybrid simulation, and the trajectories of some system states are shown in Figure 8.30. It can be seen that the results of hybrid simulation are very close to those of full-dynamic simulation. In terms of the computation speed, the full-dynamic simulation takes 5 909.3 s to finish computation. The hybrid simulation takes 2 779.2 s, which is a time savings of about 52.97%. Considering that QSS simulation covers 66.8% of the entire restoration process, the result also indicates that QSS simulation is significantly faster than full-dynamic simulation, and the hybrid simulation approach can significantly enhance the performance of extended-term simulation without losing accuracy.

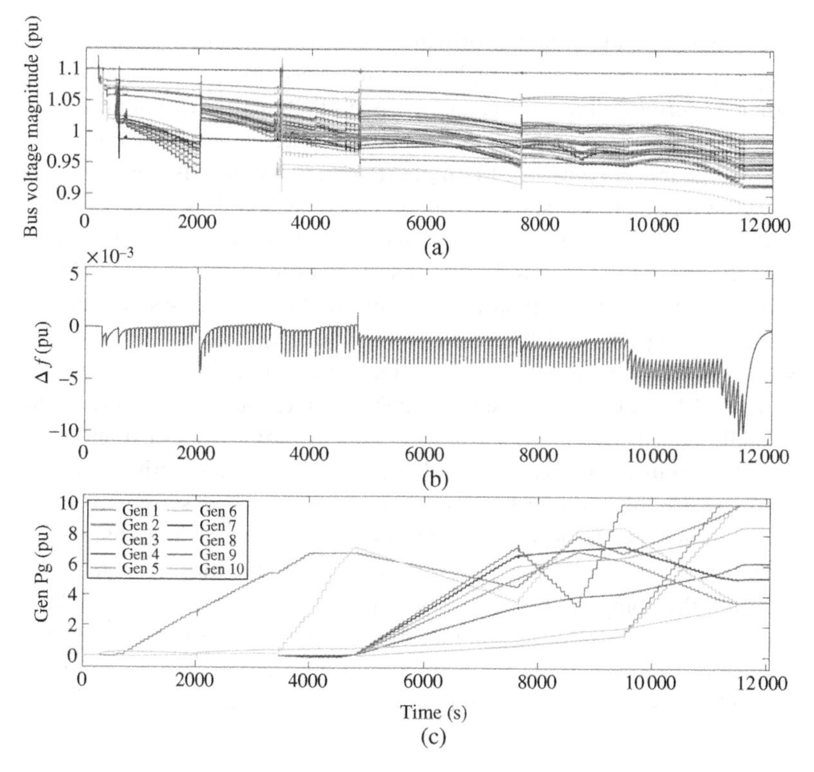

Figure 8.30 Selected system state trajectory of IEEE 39-bus system under restoration: (a) Voltage; (b) Frequency; (c) Reference generator active power.

8.3.4 Summary of Section

In this section, we explained how holomorphic embedding can accelerate extended-term simulation based on dynamic and steady-state hybrid simulation. The high accuracy and efficiency of semi-analytical simulation engine powered by holomorphic embedding (HE) method lays the foundation for the extended-term simulation. A key takeaway from this section is that holomorphic embedding solutions can provide very efficient determination of switching between dynamic and steady-state simulation thanks to its analytical nature. Another key feature is that because holomorphic embedding solution is a continuous function in time domain, it provides better accuracy for determining event instant than the classical numerical methods with discrete time steps.

8.4 Robust Parallel or Distributed Simulation

The major computational performance bottleneck of holomorphic embedding in simulation is on solving linear equations at the scale of system size. When performing analysis on very large-scale systems, the computation still could be intensive. A power system usually has hierarchical structures [56]: The buses and branches on the highest voltage levels constitute the bulk power system, and the sub-transmission systems and distribution systems with lower voltage levels are connected to the bulk system. With the increasing interdependency across the power system, events (such as outages) may propagate across the boundaries between the bulk power system and the lower-level systems [57]. Therefore, doing whole-system analysis is needed and consequently it is desirable to accelerate the simulation on large-scale systems. The bulk power system usually has meshed topology due to the reliability needs, while the lower-level systems are less meshed and have very few connection points to the bulk system. Therefore, other than the cumbersome method of directly simulating on the whole system, it would be more efficient to decompose the system and perform the analysis on each part concurrently with coordination on the boundaries. Some methods separate the computation on the bulk power system and the lower-level systems, and iterate on the boundaries [58] hoping that the process would converge. However, those methods do not guarantee convergence of the iterations. Also, the iterations may be very slow to reach convergence.

To overcome the convergence and efficiency issues, a partitioned holomorphic embedding (PHE) method that separates the holomorphic embedding models by the areas of systems and couples them with a generic voltage–current interface can be derived from the classic serial holomorphic embedding method. Such a PHE method preserves the numerical robustness of original holomorphic embedding and reduces the computational burden by splitting computation on smaller lower-level systems. Furthermore, the computation on lower-level systems is independent and can be parallelized, which forms the parallel partitioned holomorphic embedding (P^2HE) method. The PHE and P^2HE methods demonstrate promising numerical robustness and satisfactory efficiency on large-scale multi-area systems. To make the mathematical derivation and experiments more tangible, we formulate the parallel/distributed computation approach in the contingency analysis problem. In the meantime, it should be noted that the general idea of partitioning and the parallel/distributed computation algorithm also applies to other problems.

8.4.1 Steady-State Contingency Analysis: Problem Formulation and State of the Art

8.4.1.1 Problem Formulation

Again, classic power flow equations can be written as the following equations in the complex domain:

$$(P_i - jQ_i) - V_i^* \sum_j Y_{ij} V_j - V_i^* I_{Li} = 0 \tag{8.77}$$

where bus i can be a PQ or PV bus. V_i is the voltage of bus i, and Y_{ij} is the row-i, column-j element of the admittance matrix $\mathbf{Y}$. For a PV bus, the reactive power injection Q_i is unknown, but the voltage magnitude is given:

$$V_i V_i^* = |V_i^{sp}|^2, i \in S_{PV} \tag{8.78}$$

where S_{PV} is the set of PV buses.

Steady-state contingency analysis assumes the system loses one or several branches and tries to solve post-contingency power flow equations. Assume the contingency changes the admittance matrix from $\mathbf{Y}$ to $\mathbf{Y}' = \mathbf{Y} + \Delta\mathbf{Y}$, and then correspondingly, (8.77) is changed as:

$$(P_i - jQ_i) - V_i^* \sum_j (Y_{ij} + \Delta Y_{ij}) V_j - V_i^* I_{Li} = 0 \tag{8.79}$$

Assume that the solution of pre-contingency equations (8.77) and (8.78) are known, and contingency analysis tries to solve post-contingency equations (8.79) and (8.78).

8.4.1.2 Holomorphic Embedding-Based Contingency Analysis

The change of the admittance matrix causes the power flow solution to change. Assume that due to a contingency, the admittance matrix changes to $\mathbf{Y} + \Delta\mathbf{Y}$, and we can establish the holomorphic embedding (HE) formulation as follows:

$$\begin{aligned} &(P_i(\alpha) - jQ_i(\alpha))W_i^*(\alpha) \\ &\quad - \sum_l (Y_{ij} + \alpha \Delta Y_{ij}) V_j(\alpha) - I_{Li}(\alpha) = 0, i \notin S_{SL} \\ &\qquad V_i(\alpha) V_i^*(\alpha) = |V_i^{sp}|^2, i \in S_{PV} \end{aligned} \tag{8.80}$$

where S_{SL} is the set of slack bus, and a slack bus i is assumed to have a constant voltage phasor V_i. In (8.80), $\alpha = 0$ corresponds to the pre-contingency state, and $\alpha = 1$ corresponds to the post-contingency state. In the steady-state contingency analysis, $P_i(\alpha)$ is given as a constant P_i. For a PV bus, $Q_i(\alpha)$ is to be solved, while for other buses, $Q_i(\alpha)$ is given as a constant Q_i. holomorphic embedding aims to derive the solution of (8.80) as power series of α, e.g.

$$\mathbf{V}(\alpha) = \mathbf{V}[0] + \mathbf{V}[1]\alpha + \mathbf{V}[2]\alpha^2 + \cdots \tag{8.81}$$

or the corresponding PA [59].

According to the rules for deriving holomorphic embedding coefficients, we can obtain the linear equations in (8.82), where we reorder the buses so that the PV buses follow the PQ buses. **G** and **B** are real and imaginary parts of admittance matrix **Y**. **Y**, **V**, and **W** only include PQ and PV buses, and $\mathbf{V}_{SL}$ is the voltage of the Vθ bus. $\mathbf{Y}_{SL}$ has rows corresponding to PQ and PV buses and a column corresponding to the Vθ bus. $\mathcal{D}(\cdot)$ stands for a diagonal matrix. **C** and **D** are the real and imaginary parts of bus voltage **V**, and **E**, and **F** are the real and imaginary parts of **W**. $(\cdot)_{PQ}$, $(\cdot)_{PV}$ and $(\cdot)_{SL}$ stand for the variables corresponding to PQ, PV, and Vθ buses, respectively.

$$\left[\begin{array}{c:c:c:c:c} -\mathbf{G} & \mathbf{B} & \mathcal{D}(\mathbf{P}_0) & -\mathcal{D}(\mathbf{Q}_0) & \begin{matrix}\mathbf{0}\\ -\mathcal{D}(\mathbf{F}_{PV}[0])\end{matrix} \\ \hdashline -\mathbf{B} & -\mathbf{G} & -\mathcal{D}(\mathbf{Q}_0) & -\mathcal{D}(\mathbf{P}_0) & \begin{matrix}\mathbf{0}\\ -\mathcal{D}(\mathbf{E}_{PV}[0])\end{matrix} \\ \hdashline \mathcal{D}(\mathbf{E}[0]) & -\mathcal{D}(\mathbf{F}[0]) & \mathcal{D}(\mathbf{C}[0]) & -\mathcal{D}(\mathbf{D}[0]) & \mathbf{0} \\ \hdashline \mathcal{D}(\mathbf{F}[0]) & \mathcal{D}(\mathbf{E}[0]) & \mathcal{D}(\mathbf{D}[0]) & \mathcal{D}(\mathbf{C}[0]) & \mathbf{0} \\ \hdashline \mathbf{0}\ \mathcal{D}(\mathbf{C}_{PV}[0]) & \mathbf{0}\ \mathcal{D}(\mathbf{D}_{PV}[0]) & \mathbf{0} & \mathbf{0} & \mathbf{0} \end{array}\right] \begin{bmatrix} \mathbf{C}[n] \\ \mathbf{D}[n] \\ \mathbf{E}[n] \\ \mathbf{F}[n] \\ \mathbf{Q}_{PV}[n] \end{bmatrix}$$

$$
= \begin{bmatrix} \Re\left(\mathbf{I}_{LPQ}[n]\right) \\ \Re\left(j\sum_{k=1}^{n-1}\mathbf{Q}_{PV}[k]\circ\mathbf{W}^*_{PV}[n-k]+\mathbf{I}_{LPV}[n]\right) \\ \hdashline \Im\left(\mathbf{I}_{LPQ}[n]\right) \\ \Im\left(j\sum_{k=1}^{n-1}\mathbf{Q}_{PV}[k]\circ\mathbf{W}^*_{PV}[n-k]+\mathbf{I}_{LPV}[n]\right) \\ \hdashline \Re\left(-\sum_{k=1}^{n-1}\mathbf{W}[k]\circ\mathbf{V}[n-k]\right) \\ \hdashline \Im\left(-\sum_{k=1}^{n-1}\mathbf{W}[k]\circ\mathbf{V}[n-k]\right) \\ \hdashline -\frac{1}{2}\sum_{k=1}^{n-1}\mathbf{V}_{PV}[k]\circ\mathbf{V}^*_{PV}[n-k] \end{bmatrix} + \begin{bmatrix} \Re\left(\Delta\mathbf{YV}[n-1]+\Delta\mathbf{Y}_{SL}\mathbf{V}_{SL}[n-1]\right) \\ \hdashline \Im\left(\Delta\mathbf{YV}[n-1]+\Delta\mathbf{Y}_{SL}\mathbf{V}_{SL}[n-1]\right) \\ \hdashline \mathbf{0} \\ \hdashline \mathbf{0} \\ \hdashline \mathbf{0} \end{bmatrix} \tag{8.82}
$$

The 0th-order holomorphic embedding coefficients are known, i.e. the pre-contingency system states. Then the arbitrarily higher-order holomorphic embedding coefficients can be calculated recursively by solving (8.82) [60]. If the computation successfully reaches $\alpha = 1$, then the solution at $\alpha = 1$ is the post-contingency solution. Otherwise, the system is considered as collapsed after the contingency.

8.4.2 Partitioned Holomorphic Embedding (PHE)

8.4.2.1 Interface-Based Partitioning

A power system consists of the network and the components connected to the network. The I_{Li} terms in holomorphic embedding formulation (8.80) for contingency analysis can be generalized as a flexible voltage-current interface compatible with various loads, generation models or lower-level systems. Figure 8.31 illustrates the structure of a main system with

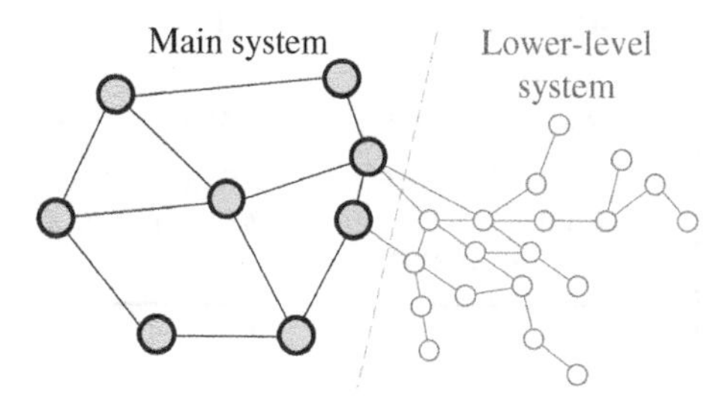

Figure 8.31 Illustration of main system and lower-level systems.

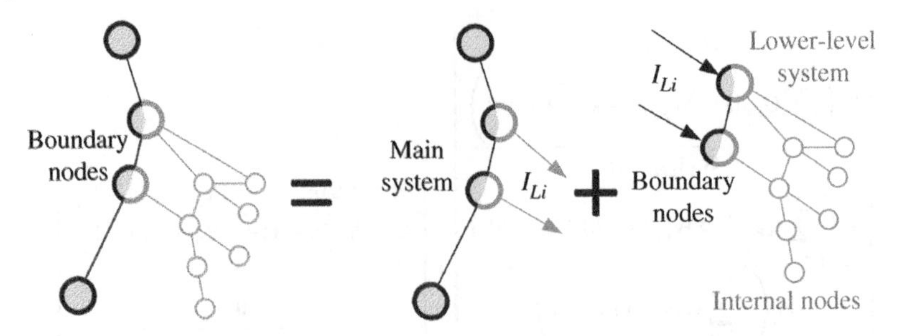

Figure 8.32 Separate modeling of main system and lower-level system(s).

subsystems connected to it. Each lower-level system is connected to a single node or a limited number of nodes in the main system.

In this section, we will show that the computation of holomorphic embedding can also be adapted to the inter-area interfaces corresponding to the hierarchical structure of power systems. The nodes on the main system that lower-level systems connect to are called boundary nodes. From the perspective of the main system, the current from the boundary nodes to the lower-level systems is equivalent of the lower-level system, and vice versa: the influence of the main system on the lower-level system can be represented by the voltage and current injection. As Figure 8.32 shows, the coupled system with the main system and the subsystem can be viewed as the superposition of the main system and the lower-level system with injection currents, and then they can be modeled separately with holomorphic embedding. When modeling the lower-level system, the boundary node only acts as a bridge to the main system; all the other components (loads, generators, or other shunt components) on the boundary nodes will be modeled with the main system. All other nodes belonging to the lower-level system are called internal nodes.

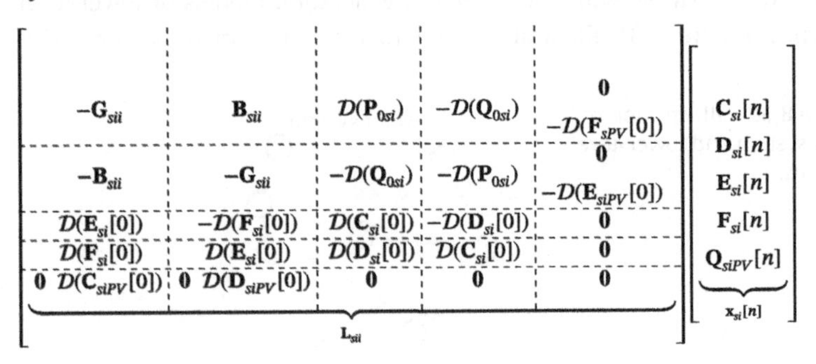

$$\left[\underbrace{\begin{array}{c:c:c:c:c} -\mathbf{G}_{sii} & \mathbf{B}_{sii} & \mathcal{D}(\mathbf{P}_{0si}) & -\mathcal{D}(\mathbf{Q}_{0si}) & \begin{matrix}\mathbf{0}\\ -\mathcal{D}(\mathbf{F}_{sPV}[0])\end{matrix} \\ \hdashline -\mathbf{B}_{sii} & -\mathbf{G}_{sii} & -\mathcal{D}(\mathbf{Q}_{0si}) & -\mathcal{D}(\mathbf{P}_{0si}) & \begin{matrix}\mathbf{0}\\ -\mathcal{D}(\mathbf{E}_{siPV}[0])\end{matrix} \\ \hdashline \mathcal{D}(\mathbf{E}_{si}[0]) & -\mathcal{D}(\mathbf{F}_{si}[0]) & \mathcal{D}(\mathbf{C}_{si}[0]) & -\mathcal{D}(\mathbf{D}_{si}[0]) & \mathbf{0} \\ \hdashline \mathcal{D}(\mathbf{F}_{si}[0]) & \mathcal{D}(\mathbf{E}_{si}[0]) & \mathcal{D}(\mathbf{D}_{si}[0]) & \mathcal{D}(\mathbf{C}_{si}[0]) & \mathbf{0} \\ \hdashline \mathbf{0}\ \mathcal{D}(\mathbf{C}_{siPV}[0]) & \mathbf{0}\ \mathcal{D}(\mathbf{D}_{siPV}[0]) & \mathbf{0} & \mathbf{0} & \mathbf{0} \end{array}}_{\mathbf{L}_{sii}}\right] \left[\underbrace{\begin{array}{c} \mathbf{C}_{si}[n] \\ \mathbf{D}_{si}[n] \\ \mathbf{E}_{si}[n] \\ \mathbf{F}_{si}[n] \\ \mathbf{Q}_{siPV}[n] \end{array}}_{\mathbf{x}_{si}[n]}\right]$$

$$
= -\underbrace{\begin{bmatrix} -\mathbf{G}_{sib} & \mathbf{B}_{sib} \\ -\mathbf{B}_{sib} & -\mathbf{G}_{sib} \\ \mathbf{0} & \mathbf{0} \\ \mathbf{0} & \mathbf{0} \\ \mathbf{0} & \mathbf{0} \end{bmatrix}}_{\mathbf{L}_{sib}} \begin{bmatrix} \underbrace{\begin{matrix} \mathbf{C}_{sb}[n] \\ \mathbf{D}_{sb}[n] \end{matrix}}_{\mathbf{x}_{sb}[n]} \end{bmatrix}
$$

$$
+ \underbrace{\begin{bmatrix} \Re\left(\mathbf{I}_{siLPQ}[n]\right) \\ \Re\left(j\sum_{k=1}^{n-1}\mathbf{Q}_{siPV}[k] \circ \mathbf{W}^*_{siPV}[n-k] + \mathbf{I}_{siLPV}[n]\right) + \Re\left(\Delta\mathbf{Y}_{si\cdot}\mathbf{V}_s[n-1]\right) \\ \Im\left(\mathbf{I}_{siLPQ}[n]\right) \\ \Im\left(j\sum_{k=1}^{n-1}\mathbf{Q}_{siPV}[k] \circ \mathbf{W}^*_{siPV}[n-k] + \mathbf{I}_{siLPV}[n]\right) + \Im\left(\Delta\mathbf{Y}_{si\cdot}\mathbf{V}_s[n-1]\right) \\ \Re\left(-\sum_{k=1}^{n-1}\mathbf{W}_{si}[k] \circ \mathbf{V}_{si}[n-k]\right) \\ \Im\left(-\sum_{k=1}^{n-1}\mathbf{W}_{si}[k] \circ \mathbf{V}_{si}[n-k]\right) \\ -\frac{1}{2}\sum_{k=1}^{n-1}\mathbf{V}_{siPV}[k] \circ \mathbf{V}^*_{siPV}[n-k] \end{bmatrix}}_{\mathbf{R}_{si}[n]} \tag{8.83}
$$

Assume that the slack bus of the whole system is on the main system. For the main system, the holomorphic embedding formulation is the same as (8.80) where I_{Li} terms include the current injections to the lower-level systems. The equations of holomorphic embedding coefficients are the same as (8.24). The $I_{Li}[n]$ terms depend on the voltage of the boundary nodes as well as the states of the lower-level systems. For the lower-level system, the equations of the holomorphic embedding coefficients have some differences, as (8.83) shows. $\mathbf{V}_s$ is the vector containing the voltages of the internal nodes and the boundary nodes of a lower-level system. $\mathbf{V}_{sb}$ is the voltage of boundary nodes, whose real and imaginary parts are $\mathbf{C}_{sb}$ and $\mathbf{D}_{sb}$. $\mathbf{V}_{si}$ is the voltage of internal nodes, whose real and imaginary parts are $\mathbf{C}_{si}$ and $\mathbf{D}_{si}$. $\mathbf{E}_{si}$ and $\mathbf{F}_{si}$ are the real and imaginary parts of $\mathbf{W}_{si}$, i.e. the reciprocal of $\mathbf{V}_{si}$. $\mathbf{Y}_s$ is the admittance matrix with the following blocks, whose rows and columns correspond to the internal and boundary nodes as shown on the subscripts:

$$
\mathbf{Y}_s = \begin{bmatrix} \mathbf{Y}_{sbb} & \mathbf{Y}_{sbi} \\ \mathbf{Y}_{sib} & \mathbf{Y}_{sii} \end{bmatrix} \tag{8.84}
$$

here $\mathbf{Y}_{si\cdot} = [\mathbf{Y}_{sib}, \mathbf{Y}_{sii}]$. $\mathbf{G}_{s\cdot}$ and $\mathbf{B}_{s\cdot}$ are the real and imaginary parts of the admittance matrix or its sub-matrices. $\mathbf{I}_{siL}$ is the current to other components on the internal nodes. On the boundary nodes, all the other components have been modeled with the main system, and here on the lower-level system, the boundary nodes act only as bridges absorbing injection currents from the main system and distributing them to the lower-level system. The boundary node does not have any other injection currents. Therefore, for the boundary nodes, the equations of holomorphic embedding coefficients are:

$$\underbrace{\left[\begin{array}{c:c:c:c:c} -\mathbf{G}_{sii} & \mathbf{B}_{sii} & \mathbf{0} & \mathbf{0} & \mathbf{0} \\ \hdashline -\mathbf{B}_{sii} & -\mathbf{G}_{sii} & \mathbf{0} & \mathbf{0} & \mathbf{0} \end{array}\right]}_{\mathbf{L}_{sbi}} \begin{bmatrix} \mathbf{C}_{si}[n] \\ \mathbf{D}_{si}[n] \\ \mathbf{E}_{si}[n] \\ \mathbf{F}_{si}[n] \\ \mathbf{Q}_{siPV}[n] \end{bmatrix} + \underbrace{\left[\begin{array}{c:c} -\mathbf{G}_{sib} & \mathbf{B}_{sib} \\ \hdashline -\mathbf{B}_{sib} & -\mathbf{G}_{sib} \end{array}\right]}_{\mathbf{L}_{sbb}} \begin{bmatrix} \mathbf{C}_{sb}[n] \\ \mathbf{D}_{sb}[n] \end{bmatrix} = -\left[\begin{array}{c} \Re\left(\mathbf{I}_L[n]\right) \\ \hdashline \Im\left(\mathbf{I}_L[n]\right) \end{array}\right] + \underbrace{\left[\begin{array}{c} \Re\left(\Delta\mathbf{Y}_{sb\cdot}\mathbf{V}_s[n-1]\right) \\ \hdashline \Im\left(\Delta\mathbf{Y}_{sb\cdot}\mathbf{V}_s[n-1]\right) \end{array}\right]}_{\mathbf{R}_{sb}[n]} \tag{8.85}$$

where on the RHS, the $\mathbf{I}_L$ term is the current injection from the main system. From (8.83) the internal variables can be expressed as:

$$\mathbf{x}_{si}[n] = -\mathbf{L}_{sii}^{-1}\mathbf{L}_{sib}\mathbf{x}_{sb}[n] + \mathbf{L}_{sii}^{-1}\mathbf{R}_{si}[n], \tag{8.86}$$

and combining with (8.85), the internal variables are eliminated:

$$\underbrace{(\mathbf{L}_{sbb} - \mathbf{L}_{sbi}\mathbf{L}_{sii}^{-1}\mathbf{L}_{sib})}_{\mathbf{L}_s} \begin{bmatrix} \mathbf{C}_{sb}[n] \\ \mathbf{D}_{sb}[n] \end{bmatrix} = -\left[\begin{array}{c} \Re\left(\mathbf{I}_L[n]\right) \\ \hdashline \Im\left(\mathbf{I}_L[n]\right) \end{array}\right] \underbrace{+\mathbf{R}_{sb}[n] - \mathbf{L}_{sbi}\mathbf{L}_{sii}^{-1}\mathbf{R}_{si}[n]}_{\mathbf{R}_s[n]}. \tag{8.87}$$

Equation (8.87) is substituted in the *main-system problem* to eliminate $\mathbf{I}_L$. The matrix $\mathbf{L}_s$ is merged into the left-hand side matrix $\mathbf{L}$, and $\mathbf{R}_s[n]$ is merged into the RHS $\mathbf{R}[n]$. Then the main system variables can be solved, which also means that the boundary variables of each lower-level system problem $\mathbf{x}_{sb}[n]$ are obtained. Finally, the holomorphic embedding coefficients of the lower-level system $\mathbf{x}_{si}[n]$ are solved from (8.86). Note that the PHE solution is the same with holomorphic embedding. And PHE does not compromise the numerical robustness of the original holomorphic embedding method.

8.4.2.2 Comparative Complexity Analysis

A computational complexity analysis can be done to compare the theoretical speed of PHE and holomorphic embedding. To simplify the analysis, assume the entire system consists of a main system with N_m buses and K lower-level systems each with N_s buses. Each lower-level system is connected to the main system through n_b boundary nodes. For the ordinary holomorphic embedding approach (8.24) that directly solves the entire system, first we factorize the LHS matrix of (8.24), which requires $c_0(N_m + KN_s)^3$ operations, here c_0 as well as c_1, c_2, c_{Rn} below are constants depending on the system structure and bus types. Here we ignore some minor computation costs, e.g. generating matrices. Then for each order n of holomorphic embedding coefficients ($1 \leq n \leq N$), assume generating the RHS vector costs $c_{Rn}(N_m + KN_s)$ operations, and using forward and backward substitution to solve the equation costs $c_1(N_m + KN_s)^2$ operations. So the total number of operations of holomorphic embedding approach is:

$$
\begin{aligned}
m_1 = {} & c_0(N_m + KN_s)^3 + Nc_1(N_m + KN_s)^2 \\
& + \sum_{n=1}^{N} c_{Rn}(N_m + KN_s)
\end{aligned}
\tag{8.88}
$$

Next we analyze the complexity of the PHE approach. At the beginning, for each system $\mathbf{L}_{sii}$ is factorized, and then $\mathbf{L}_{sbi}\mathbf{L}_{sii}^{-1}$ and $\mathbf{L}_{sii}^{-1}\mathbf{L}_{sib}$ are computed, which constitutes $Kc_0N_s^3 + 2Kn_bc_1N_s^2$ operations for all the lower-level systems. For the main system, factorizing $\mathbf{L}$ costs $c_0N_m^3$ operations. And then for each order n of holomorphic embedding coefficient, generating $\mathbf{R}_{si}[n]$ costs $Kc_{Rn}N_s$ operations, calculating $\mathbf{R}_s[n]$ in (8.87) costs $Kc_1N_s^2 + Kc_2n_bN_s$ operations. On the main system, generating $\mathbf{R}[n]$ costs $c_{Rn}N_m$ operations, and solving the coefficients on the main system costs $c_1N_m^2$. Finally $Kc_2n_bN_s$ operations are needed to calculate holomorphic embedding coefficients on all the lower-level systems based on (8.86).

$$
\begin{aligned}
m_2 = {} & c_0(N_m^3 + KN_s^3) + 2Kn_bc_1N_s^2 + Nc_1(N_m^2 + KN_s^2) \\
& + \sum_{n=1}^{N} c_{Rn}(N_m + KN_s) + 2NKc_2n_bN_s
\end{aligned}
\tag{8.89}
$$

Some reference values for the constants are c_0=60, c_1=40, and c_2=9. Considering that $K \geq 1$ and normally $n_b \ll N_m$, $n_b \ll N_s$, we can get

$m_2 < m_1$, which means that PHE costs fewer operations than holomorphic embedding, and thus should be faster than holomorphic embedding. Also we can conjecture that a larger N_m, N_s, K or a smaller n_b would help PHE gain a larger advantage over holomorphic embedding.

However, some factors that add to computational burden of PHE are not directly reflected in (8.89) but should be considered in practice. First, the $\mathbf{L}_s$ may add to nonzero elements of $\mathbf{L}$, which leads to a larger effective c_0 and c_1 in (8.89). Second, the computational procedures of PHE is more complicated than holomorphic embedding, which means higher overhead for program execution (e.g. memory access and management). Third, the complexity of matrix factorization being $O(N_{(\cdot)}^3)$ ($N_{(\cdot)}$ is the size of a matrix) is based on dense matrix. On sparse matrices, the effective exponent will be lower than three, which also diminishes the advantage of PHE. Generally, PHE should have more significant advantages over holomorphic embedding on large systems (larger N_m, N_s) that have multiple lower-level systems (larger K) with clear boundaries (smaller n_b), so that m_2 is significantly lower than m_1 and the overheads can be ignored. Fortunately, power systems usually have such traits and thus we can expect the advantage of PHE over holomorphic embedding on computational efficiency.

8.4.3 Parallel and Distributed Computation

The parallelism has two aspects. First, the parallel or distributed computation based on network partition. It should be noted that the generalized inter-area interface enables very compact data transfer, which favors distributed computation. Second, different outages under the same base state can be computed in parallel due to shared data.

8.4.3.1 Parallel Partitioned Holomorphic Embedding (P^{2}HE)

Section 8.3 has demonstrated the computation of PHE based on system partition and voltage–current interfaces among partitions. Such a partitioned scheme enables parallel or distributed computation. The computation is separated into sub-tasks on the main system and lower-level systems, and the data are transferred among the sub-tasks. When there are multiple lower-level systems, the sub-tasks for lower-level systems are independent of each other and can be parallelized. Algorithm 8.2 shows the procedures of calculating holomorphic embedding solutions with parallel or distributed computation. We assume that each partition of the system corresponds to a process in computer, and the processes either use shared memory or local memory and can communicate with each other.

The Algorithm 8.2 is presented in the form compatible with the commonly used message passing interface (MPI) [61] for parallel or distributed computation. The *send* and *receive* operations directly correspond to the standard calls for data exchange in MPI (e.g. `MPI_Send` and `MPI_Recv`) among processes. In Algorithm 8.2, steps 1–9 prepare the matrices for holomorphic embedding computation and only needs to be done once. Steps 10–21 are the procedures of calculating the holomorphic embedding coefficients from order 1 to N. The steps of the main process (MP) and the lower-level system processes (SP) are marked in two different color (light or dark gray) blocks. The codes in the same color block can be parallelized among different processes, but there are barriers [61] between adjacent blocks with different colors i.e., the next block can start only after the previous code block is finished on all relevant processes. Algorithm 8.2 shows that the computation steps on the SPs are independent of each other and thus can be parallelized (in the ***parallel* foreach** blocks). For the systems that have many lower-level systems, the P^{2}HE method can substantially enhance efficiency compared with the holomorphic embedding or PHE methods that directly solve the entire coupled system. Note that unlike some parallel numerical methods based on iterations on the boundaries that usually undermine numerical robustness (i.e. are more likely to be slow or even diverge), the P^{2}HE method does not affect the numerical stability: It can be verified that the P^{2}HE method produces the same result as the holomorphic embedding method that directly solves the entire coupled system.

The P^{2}HE method also favors distributed computation. Algorithm 8.2 shows that during the computation, the SPs send $\mathbf{L}_s$ (only once, with size of $2n_b \times 2n_b$) and $\mathbf{R}_s[n]$ (with size of $2n_b \times 1$) to the MP, and MP sends back $\mathbf{C}_{sb}[n]$ (with size of $n_b \times 1$) and $\mathbf{D}_{sb}[n]$ (with size of $n_b \times 1$) to each SP. Since n_b is usually quite small, the exchanged data are highly compact and will not cause a high communication burden. Also, such a message-passing scheme does not exchange lots of internal information about each system partition, so the privacy on each partition is well preserved. This is desirable for coordinated analysis among different system owners or operators.

8.4.3.2 Parallelism Among Contingency Analysis Tasks

In practice, contingency screening often involves assessing different contingencies from the same initial state [62]. Note that steps 1–5 of Algorithm 8.2 are dependent only on the initial state and are independent of the contingencies. Therefore, when assessing different contingencies or outages based on the same initial stage, steps 1–5 only needs to be performed once. For the following steps, the procedures of different contingencies are independent of each other and thus can be parallelized.

Algorithm 8.2 Parallel computation of holomorphic embedding on partitioned system.

System model:

Admittance matrices of main system $\mathbf{Y}$ and all lower-level systems $\mathbf{Y}_s$.
Power injections of the main system $\mathbf{P}_0$, $\mathbf{Q}_0$.
Power injections of the lower-level systems $\mathbf{P}_{0s}$, $\mathbf{Q}_{0s}$.
Other component models on the main system and lower-level systems.

Inputs:

Initial states of the main system: $\mathbf{C}[0]$, $\mathbf{D}[0]$, $\mathbf{E}[0]$, $\mathbf{F}[0]$, $\mathbf{Q}_{PV}[0]$.
Initial states of lower-level systems: $\mathbf{C}_{si}[0]$, $\mathbf{D}_{si}[0]$, $\mathbf{E}_{si}[0]$, $\mathbf{F}_{si}[0]$, $\mathbf{Q}_{siPV}[0]$.
Initial states of boundary nodes $\mathbf{C}_{sb}[0]$, $\mathbf{D}_{sb}[0]$ (from $\mathbf{C}[0]$, $\mathbf{D}[0]$).

Outputs: HE coefficients for $n = 1 \cdots N$

Main system: $\mathbf{C}[n]$, $\mathbf{D}[n]$, $\mathbf{E}[n]$, $\mathbf{F}[n]$, $\mathbf{Q}_{PV}[n]$.
Lower-level systems: $\mathbf{C}_{si}[n]$, $\mathbf{D}_{si}[n]$, $\mathbf{E}_{si}[n]$, $\mathbf{F}_{si}[n]$, $\mathbf{Q}_{siPV}[n]$.

Processes:

Main process (MP): computation process of the main system.
Sub-process (SP(s)): computation process of lower-level system s.

1: ***parallel* foreach** SP(s) **do**
2: Calculate $\mathbf{L}_{sbb}$, $\mathbf{L}_{sbi}$, $\mathbf{L}_{sib}$, $\mathbf{L}_{sii}$. Factorize $\mathbf{L}_{sii}$.
3: Calculate $\mathbf{L}_s = \mathbf{L}_{sbb} - \mathbf{L}_{sbi}\mathbf{L}_{sii}^{-1}\mathbf{L}_{sib}$ and *send* to the MP.
4: **end foreach**
5: MP *receives* all $\mathbf{L}_s$ and calculate LHS matrix $\mathbf{L}$ in (8.80). Factorize $\mathbf{L}$.
6: ***parallel* foreach** SP(s) **do**
7: Prepare $\Delta\mathbf{Y}_s$ based on contingency information.
8: **end foreach**
9: MP prepares $\Delta\mathbf{Y}$ based on contingency information.
10: **for** $n = 1 \to N$ **do**
11: ***parallel* foreach** SP(s) **do**
12: Calculate $\mathbf{R}_s[n]$ based on (8.87) and *send* to MP.
13: **end foreach**
14: MP *receives* $\mathbf{R}_s[n]$ and calculate RHS term $\mathbf{R}[n]$ in (8.80).
15: MP solves (8.80) and obtains $\mathbf{C}[n]$, $\mathbf{D}[n]$, $\mathbf{E}[n]$, $\mathbf{F}[n]$, $\mathbf{Q}_{PV}[n]$.
16: MP *sends* $\mathbf{C}_{sb}[n]$, $\mathbf{D}_{sb}[n]$ to each SP(s).
17: ***parallel* foreach** SP(s) **do**
18: SP(s) *receives* $\mathbf{C}_{sb}[n]$, $\mathbf{D}_{sb}[n]$ from MP.
19: Solve $\mathbf{C}_{si}[n]$, $\mathbf{D}_{si}[n]$, $\mathbf{E}_{si}[n]$, $\mathbf{F}_{si}[n]$, $\mathbf{Q}_{siPV}[n]$ from (8.86).
20: **end foreach**
21: **end for**

8.4.4 Experiment on Large-Scale System

Here we test several related methods and compare the performance. A large-scale synthetic system is created by connecting 9×2383-bus Polish systems. The whole system has 21 447 buses, 26 078 branches, 2934 PV buses, 18 514 PQ buses, and 756 ZIP loads. The connecting branches are listed in Table 8.8.

100 randomly selected $N-1$ contingencies are tested with holomorphic embedding, PHE and P^2HE methods on the synthetic system, and the computation time of each method is listed in Table 8.9. Results show that the PHE method is significantly faster than holomorphic embedding, which verifies its advantage when analyzing large-scale systems. And P^2HE can further accelerate the computation by making use of parallelism.

8.4.5 Summary of Section

This section demonstrates a partitioned and parallel holomorphic embedding approach that can further enhance the simulation performance

Table 8.8 Connections between main system and lower-level systems.

Sub-system	Branches (Main bus#-Sub bus#)
2	18–18
3	448–445, 474–475
4	2254–2248, 2247–2250
5	1089–1092, 1100–1095
6	673–665
7	1354–1356, 1544–1547, 738–739
8	1100–1092, 1089–1095
9	146–148

Table 8.9 Computation time of 100 $N-1$ analysis on 21 447-bus system.

Method	Computation time (s)
HE	414.65
PHE	118.84
P^2HE	48.64

and maintaining the numerical robustness of the serial holomorphic embedding. Considering the ubiquitous structure of power systems with main system and lower-level systems, the PHE method is extended from the basic holomorphic embedding formulation based on the generic voltage-w-current interface on the boundaries. By partitioning the system and computation in PHE, the computation burden can be substantially reduced. Moreover, the PHE formulation has lower-level system computation tasks independent of each other, which is parallelized as a P^2HE method. The data exchange is quite compact and thus also favors distributed computation. Note that unlike the traditional participation methods requiring numerical iterations on the boundaries, the PHE and P^2HE methods are completely equivalent to the basic holomorphic embedding method and the computational robustness is not compromised. The partitioned and parallel algorithm also has a good fit with popular high-performance computing architecture and developer tools (such as MPI), which is convenient to implement in high-performance computation platforms.

References

1 Trias, A. (2012). The holomorphic embedding load flow method. *Power and Energy Society General Meeting*, 1–8. IEEE.

2 Chiang, H.-D., Wang, T., and Sheng, H. (2017). A novel fast and flexible holomorphic embedding method for power flow problems. *IEEE Transactions on Power Systems* 33 (3): 2551–2562.

3 Rao, S., Feng, Y., Tylavsky, D.J., and Subramanian, M.K. (2016). The holomorphic embedding method applied to the power-flow problem. *IEEE Transactions on Power Systems* 31 (5): 3816–3828.

4 Liu, C., Wang, B., Hu, F. et al. (2017). Online voltage stability assessment for load areas based on the holomorphic embedding method. *IEEE Transactions on Power Systems* 33 (4): 3720–3734.

5 Trias, A. and Marín, J.L. (2016). The holomorphic embedding load-flow method for DC power systems and nonlinear DC circuits. *IEEE Transactions on Circuits and Systems I: Regular Papers* 63 (2): 322–333.

6 Basiri-Kejani, M. and Gholipour, E. (2017). Holomorphic embedding load-flow modeling of thyristor-based facts controllers. *IEEE Transactions on Power Systems* 32 (6): 4871–4879.

7 Rao, S.D., Tylavsky, D.J., and Feng, Y. (2017). Estimating the saddle-node bifurcation point of static power systems using the holomorphic embedding method. *International Journal of Electrical Power & Energy Systems* 84: 1–12.

8 Liu, C., Wang, B., Xu, X. et al. (2017). A multi-dimensional holomorphic embedding method to solve ac power flows. *IEEE Access* 5: 25270–25285.

9 Wang, B., Liu, C., and Sun, K. (2018). Multi-stage holomorphic embedding method for calculating the power-voltage curve. *IEEE Transactions on Power Systems* 33 (1): 1127–1129.

10 Pereira, L., Kosterev, D., Mackin, P. et al. (2002). An interim dynamic induction motor model for stability studies in the WSCC. *IEEE Transactions on Power Systems* 17 (4): 1108–1115.

11 Kosterev, D., Meklin, A., Undrill, J. et al. (2008). Load modeling in power system studies: WECC progress update. *IEEE PES General Meeting*, 1–8.

12 Price, W.W., Wirgau, K.A., Murdoch, A. et al. (1988). Load modeling for power flow and transient stability computer studies. *IEEE Transactions on Power Systems* 3 (1): 180–187.

13 Sekine, Y. and Ohtsuki, H. (1990). Cascaded voltage collapse. *IEEE Transactions on Power Systems* 5 (1): 250–256.

14 Balanathan, R., Pahalawaththa, N.C., and Annakkage, U.D. (2002). Modelling induction motor loads for voltage stability analysis. *International Journal of Electrical Power & Energy Systems* 24 (6): 469–480.

15 Van Cutsem, T. and Vournas, C.D. (1996). Voltage stability analysis in transient and mid-term time scales. *IEEE Transactions on Power Systems* 11 (1): 146–154.

16 Hill, D.J. (1993). Nonlinear dynamic load models with recovery for voltage stability studies. *IEEE Transactions on Power Systems* 8 (1): 166–176.

17 Wang, B., Fang, B., Wang, Y. et al. (2016). Power system transient stability assessment based on big data and the core vector machine. *IEEE Transactions on Smart Grid* 7 (5): 2561–2570.

18 Pulgar-Painemal, H., Wang, Y., and Silva-Saravia, H. (2018). On inertia distribution, inter-area oscillations and location of electronically-interfaced resources. *IEEE Transactions on Power Systems* 33 (1): 995–1003.

19 Milano, F. (2010). *Power System Modelling and Scripting*. Springer Science & Business Media.

20 Yao, R., Liu, F., He, G. et al. (2012). Static security region calculation with improved CPF considering generation regulation. *IEEE International Conference on Power System Technology (POWERCON)*, 1–6.

21 Yao, R., Huang, S., Sun, K. et al. (2016). A multi-timescale quasi-dynamic model for simulation of cascading outages. *IEEE Transactions on Power Systems* 31 (4): 3189–3201.

22 Yao, R., Sun, K., Liu, F., and Mei, S. (2018). Management of cascading outage risk based on risk gradient and Markovian tree search. *IEEE Transactions on Power Systems* 33 (4): 4050–4060.

23 Vournas, C.D. and Manos, G.A. (1998). Modelling of stalling motors during voltage stability studies. *IEEE Transactions on Power Systems* 13 (3): 775–781.

24 Chow, J.H. and Cheung, K.W. (1992). A toolbox for power system dynamics and control engineering education and research. *IEEE Transactions on Power Systems* 7 (4): 1559–1564.

25 Zimmerman, R.D., Murillo-Sánchez, C.E., and Thomas, R.J. (2011). MATPOWER: steady-state operations, planning, and analysis tools for power systems research and education. *IEEE Transactions on Power Systems* 26 (1): 12–19.

26 Chiang, H.-D. (2011). *Direct Methods for Stability Analysis of Electric Power Systems: Theoretical Foundation, BCU Methodologies, and Applications*. Wiley.

27 Anghel, M., Milano, F., and Papachristodoulou, A. (2013). Algorithmic construction of Lyapunov functions for power system stability analysis. *IEEE Transactions on Circuits and Systems I: Regular Papers* 60 (9): 2533–2546.

28 Lof, P.-A., Smed, T., Andersson, G., and Hill, D.J. (1992). Fast calculation of a voltage stability index. *IEEE Transactions on Power Systems* 7 (1): 54–64.

29 Stott, B. (1979). Power system dynamic response calculations. *Proceedings of the IEEE* 67 (2): 219–241.

30 Maffezzoni, P., Codecasa, L., and D'Amore, D. (2007). Time-domain simulation of nonlinear circuits through implicit Runge–Kutta methods. *IEEE Transactions on Circuits and Systems I: Regular Papers* 54 (2): 391–400.

31 Khaitan, S.K., McCalley, J.D., and Chen, Q. (2008). Multifrontal solver for online power system time-domain simulation. *IEEE Transactions on Power Systems* 23 (4): 1727–1737.

32 Jin, S., Huang, Z., Diao, R. et al. (2017). Comparative implementation of high performance computing for power system dynamic simulations. *IEEE Transactions on Smart Grid* 8 (3): 1387–1395.

33 Joo, S.-K., Liu, C.-C., Jones, L.E., and Choe, J.-W. (2004). Coherency and aggregation techniques incorporating rotor and voltage dynamics. *IEEE Transactions on Power Systems* 19 (2): 1068–1075.

34 Kim, J.-Y., Jeon, J.-H., Kim, S.-K. et al. (2010). Cooperative control strategy of energy storage system and microsources for stabilizing the microgrid during islanded operation. *IEEE Transactions on Power Electronics* 25 (12): 3037–3048.

35 Diao, R., Vittal, V., and Logic, N. (2010). Design of a real-time security assessment tool for situational awareness enhancement in modern power systems. *IEEE Transactions on Power Systems* 25 (2): 957–965.

36 Liu, C.-W. and Thorp, J.S. (2000). New methods for computing power system dynamic response for real-time transient stability prediction. *IEEE Transactions on Circuits and Systems I: Fundamental Theory and Applications* 47 (3): 324–337.

37 Stahl, H. (1997). The convergence of Padé approximants to functions with branch points. *Journal of Approximation Theory* 91 (2): 139–204.

38 Liu, Y. and Sun, K. (2019). Solving power system differential algebraic equations using differential transformation. *IEEE Transactions on Power Systems* 35 (3): 2289–2299.

39 Liu, Y., Sun, K., and Dong, J. (2020). A dynamized power flow method based on differential transformation. *IEEE Access* 8: 182441–182450.

40 Yao, R., Liu, Y., Sun, K. et al. (2019). Efficient and robust dynamic simulation of power systems with holomorphic embedding. *IEEE Transactions on Power Systems* 35 (2): 938–949.

41 Chen, J. and Crow, M.L. (2008). A variable partitioning strategy for the multirate method in power systems. *IEEE Transactions on Power Systems* 23 (2): 259–266.

42 Yao, R., Huang, S., Sun, K. et al. (2016). A multi-timescale quasi-dynamic model for simulation of cascading outages. *IEEE Transactions on Power Systems* 31 (4): 3189–3201. Doi https://doi.org/10.1109/TPWRS.2015.2466116.

43 Ma, F. and Luo, X. (2013). SimAGC-an open-source power system dynamic simulator for AGC study. *2013 IEEE Power & Energy Society General Meeting*, 1–5. IEEE.

44 Chen, L., Zhang, H., Wu, Q., and Terzija, V. (2017). A numerical approach for hybrid simulation of power system dynamics considering extreme icing events. *IEEE Transactions on Smart Grid* 9 (5): 5038–5046.

45 Korkali, M. and Min, L. (2017). GMLC Extreme Event Modeling–Slow-Dynamics Models for Renewable Energy Resources. *Technical Report*. Livermore, CA (United States): Lawrence Livermore National Lab.(LLNL).

46 Fu, C. (2011). High-speed extended-term time-domain simulation for online cascading analysis of power systemalysis of power system. PhD dissertation. Iowa State University.

47 Wang, C., Hou, Y., Qiu, F. et al. (2016). Resilience enhancement with sequentially proactive operation strategies. *IEEE Transactions on Power Systems* 32 (4): 2847–2857.

48 Panteli, M. and Mancarella, P. (2015). The grid: Stronger, bigger, smarter?: Presenting a conceptual framework of power system resilience. *IEEE Power and Energy Magazine* 13 (3): 58–66.

49 Huang, G., Wang, J., Chen, C. et al. (2017). Integration of preventive and emergency responses for power grid resilience enhancement. *IEEE Transactions on Power Systems* 32 (6): 4451–4463.

50 Song, J., Cotilla-Sanchez, E., Ghanavati, G., and Hines, P.D.H. (2015). Dynamic modeling of cascading failure in power systems. *IEEE Transactions on Power Systems* 31 (3): 2085–2095.

51 Qiu, F. and Li, P. (2017). An integrated approach for power system restoration planning. *Proceedings of the IEEE* 105 (7): 1234–1252.

52 Qureshi, M. (2019). A fast quasi-static time series simulation method using sensitivity analysis to evaluate distributed PV impacts. PhD thesis. Georgia Institute of Technology.

53 Dobson, I. and McCalley, J. (2008). PSERC Public Report S-26: Risk of Cascading Outages. *Technical Report*. Power Systems Engineering Research Center (United States).

54 Ju, W., Sun, K., and Yao, R. (2018). Simulation of cascading outages using a power-flow model considering frequency. *IEEE Access* 6: 37784–37795.

55 Hou, Y., Liu, C.-C., Sun, K. et al. (2011). Computation of milestones for decision support during system restoration. *2011 IEEE Power and Energy Society General Meeting*, 1–10. IEEE.

56 Cao, X., Wang, H., Liu, Y. et al. (2017). Coordinating self-healing control of bulk power transmission system based on a hierarchical top-down strategy. *International Journal of Electrical Power & Energy Systems* 90: 147–157.

57 Li, Z., Wang, J., Sun, H., and Guo, Q. (2015). Transmission contingency analysis based on integrated transmission and distribution power flow in smart grid. *IEEE Transactions on Power Systems* 30 (6): 3356–3367.

58 Huang, Q. and Vittal, V. (2016). Integrated transmission and distribution system power flow and dynamic simulation using mixed three-sequence/three-phase modeling. *IEEE Transactions on Power Systems* 32 (5): 3704–3714.

59 Yao, R., Sun, K., and Qiu, F. (2019). Vectorized efficient computation of Padé approximation for semi-analytical simulation of large-scale power systems. *IEEE Transactions on Power Systems* 34 (5): 3957–3959.

60 Yao, R. and Qiu, F. (2020). Novel AC distribution factor for efficient outage analysis. *IEEE Transactions on Power Systems* 35 (6): 4960–4963.

61 Gropp, W., Lusk, E., Doss, N., and Skjellum, A. (1996). A high-performance, portable implementation of the MPI message passing interface standard. *Parallel Computing* 22 (6): 789–828.

62 Yao, R., Huang, S., Sun, K. et al. (2017). Risk assessment of multi-timescale cascading outages based on Markovian tree search. *IEEE Transactions on Power Systems* 32: 2887–2900.

Index

Note: *Italicized* and **bold** page numbers refer to figures and tables, respectively.

a

Power System Simulation Using Semi-Analytical Methods, First Edition. Edited by Kai Sun.

Published 2024 by John Wiley & Sons, Inc.

Printed and bound by CPI Group (UK) Ltd, Croydon, CR0 4YY
16/04/2025
14658366-0001